6

Laboratory Experiments for General, Organic and Biochemistry

Frederick A. Bettelheim
Late of Adelphi University

Joseph M. Landesberg
Adelphi University

THOMSON
BROOKS/COLE

Australia • Canada • Mexico • Singapore • Spain
United Kingdom • United States

Laboratory Experiments for General, Organic and Biochemistry, sixth edition
Frederick Bettelheim and Joseph Landesberg

Publisher: David Harris
Acquisitions Editor: Lisa Lockwood
Development Editor: Ellen Bitter
Editorial Assistant: Lauren Oliviera
Technology Project Manager: Ericka Yeoman
Marketing Manager: Amee Mosley
Marketing Assistant: Michele Colella
Project Manager, Editorial Production: Belinda Krohmer
Creative Director: Rob Hugel
Print Buyer: Barbara Britton
Permissions Editor: Kiely Sisk

Production Service: Interactive Composition Corporation
Text Designer: Carolyn Deacy
Cover Image: Thomas G. Barnes @ USDA-NRCS PLANTS Database/Barnes, T. G. & S. W. Francis, 2004. *Wildflowers and Ferns of Kentucky*. University Press of Kentucky
Copy Editor: Anna Trabucco
Cover Printer: West Group
Compositor: Interactive Composition Corporation
Printer: West Group

Printed in the United States of America
2 3 4 5 6 7 09 08 07

Library of Congress Control Number: 2005933648

ISBN-13: 978-0-495-01504-8
ISBN-10: 0-495-01504-0

Thomson Higher Education
10 Davis Drive
Belmont, CA 94002-3098
USA

Contents

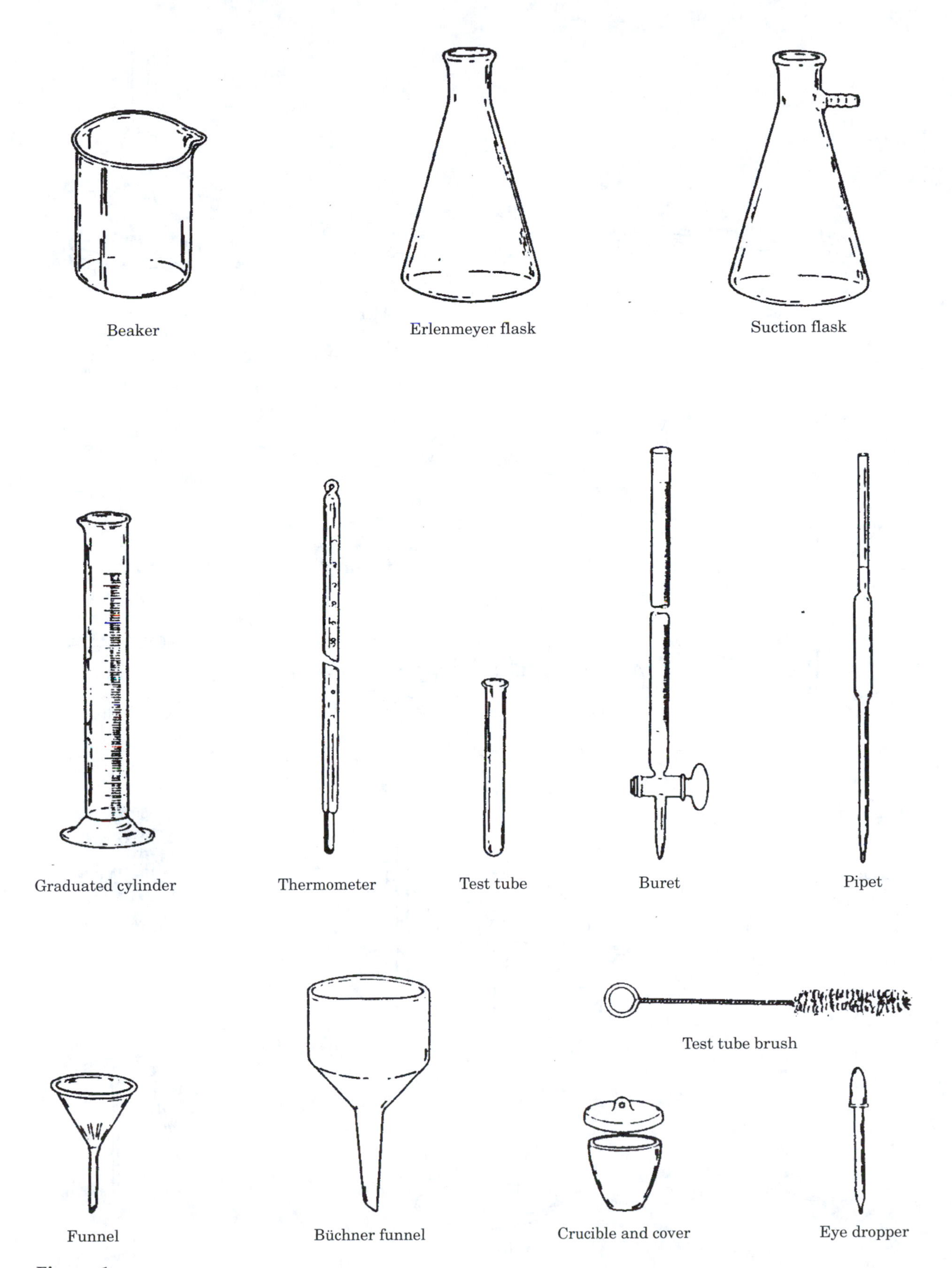

Figure 1
Common laboratory equipment.

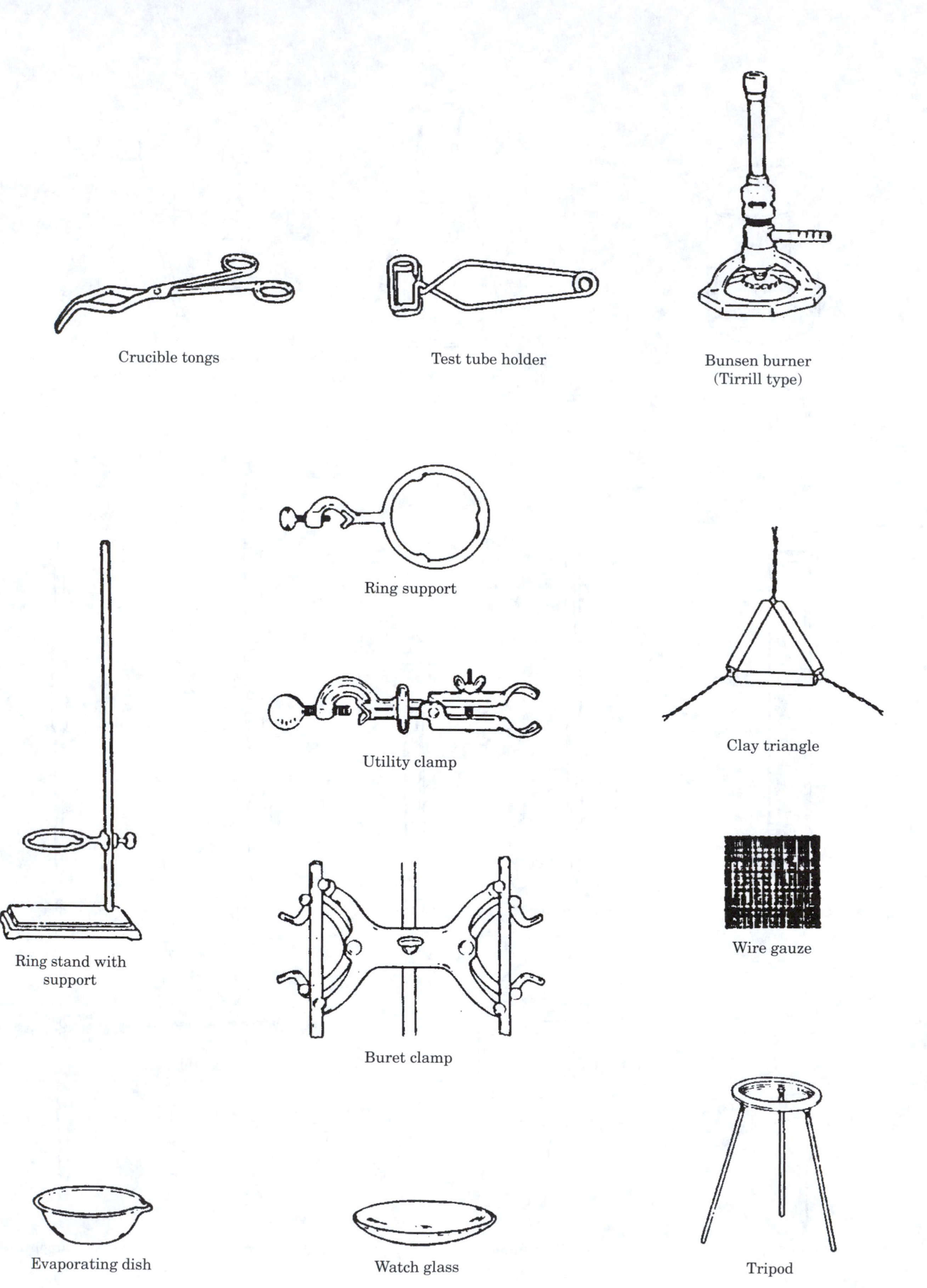
Crucible tongs
Test tube holder
Bunsen burner
(Tirrill type)
Ring support
Ring stand with
support
Utility clamp
Clay triangle
Wire gauze
Buret clamp
Evaporating dish
Watch glass
Tripod

Figure 1
Continued

Preface

In preparing the sixth edition of this Laboratory Manual, we wish to thank our colleagues who made this new edition possible by adopting our manual for their courses. This sixth edition coincides with the publication of the eighth edition of the textbook: *Introduction to General, Organic, and Biochemistry* by F. A. Bettelheim, W. H. Brown, M. K. Campbell, and S. O. Farrell. This laboratory manual shares the outline and pedagogical philosophy of the textbook. As in previous editions, we have strived for the clearest possible writing in the procedures. The experiments give the student a meaningful, reliable laboratory experience that consistently works, while covering the basic principles of general, organic, and biochemistry. Throughout the years, feedback from colleges and universities has confirmed that this manual not only eases the student's task in performing experiments, but also is student friendly. Our new edition maintains this standard and improves upon it.

The major changes in this new edition are as follows: (1) We added one new experiment, combined the first two, and eliminated four old experiments. This gives a total of 48 experiments and still provides a wide selection of experiments from which the instructor can choose. The new experiment demonstrates the chemistry and action of enzymes. (2) We improved the procedures of a number of experiments as a result of our observations of how our students carried out these experiments in our laboratories at Adelphi. (3) Safety issues and waste disposal are emphasized throughout this edition. (4) We further improved our aim to minimize the use of hazardous chemicals where possible and to design experiments that work on a small scale or on a semi-micro scale. (5) Most Pre-Lab and Post-Lab Questions have been changed or modified.

As in previous editions, three basic goals were followed in all the experiments: (1) the experiments should illustrate the concepts learned in the classroom; (2) the experiments should be clearly and concisely written so that students will easily understand the task at hand, will work with minimal supervision because the manual provides enough information on experimental procedures, and will be able to perform the experiments in a two-and-a-half-hour laboratory period; (3) the experiments should not only be simple demonstrations, but also should contain a sense of discovery.

It did not escape our attention that in adopting this manual of Laboratory Experiments, the instructor must pay attention to budgetary constraints. All experiments in this manual generally use only inexpensive pieces of equipment and glassware. A few spectrophotometers and pH meters are necessary in a number of experiments. A few experiments may require more specialized, albeit inexpensive, equipment—for example, a few viscometers. The instructor might wish to do these experiments as a demonstration.

The 48 experiments in this book will provide suitable choices for the instructor to select about 24 experiments for a two-semester or three-quarter course. The following are the principal features of this book:

1. The **Table of Contents** is organized so that the first 20 experiments illustrate the principles of general chemistry, the next 11 those of organic chemistry, and the remaining 17 those of biochemistry.
2. Each experiment begins with a **Background** section that contains all the information necessary to understand the experiment. All the relevant principles and their applications are reviewed in this section.
3. The **Procedure** section of the experiment provides a step-by-step description of the experiment. Clarity of writing in this section is of the utmost importance for successful execution of the experiments. **Caution!** signs alert students when dealing with dangerous chemicals, such as strong acids or bases.
4. **Pre-Lab Questions** are provided to familiarize students with the concepts and procedures before they begin the experiments. By requiring students to answer these questions and then grading their answers, we accomplish the task of preparing the students for the experiments and getting them to read the experiments before coming to the laboratory.
5. In the **Report Sheet** we not only ask for the recording of raw data, but we also require some calculations to yield secondary data.
6. The **Post-Lab Questions** are designed to encourage students to reflect upon the results, interpret them, and relate their significance.
7. At the end of the book in **Appendix 3**, we provide stockroom personnel with detailed instructions on preparation of solutions and other chemicals for each experiment. We also give detailed amounts of materials needed for a class of 25 students.

An **Instructor's Manual** that accompanies this book is **solely for the use of the instructor.** It helps in the grading process by providing ranges of the experimental results we obtained from class use. It alerts the instructor to some of the difficulties that may be encountered in certain experiments. These experiments have been "field tested" in our laboratories so that we believe that the "bugs" have been worked out.

The disposal of waste material is discussed for each experiment. For further information, we recommend *Prudent Practices in the Laboratory*, National Academy Press, Washington, DC (1995). A sample chemical MSDS (Material Safety Data Sheet) is included to alert students to information available regarding chemical safety. Laboratories should have these sheets on file for all chemicals that are used in the experiments, and these sheets should be made available to students on demand.

We hope that you will find our book of Laboratory Experiments helpful in instructing your students. We anticipate that students will find it inspiring in studying different aspects of chemistry.

Garden City, NY
September 2005

Frederick A. Bettelheim (deceased)
Joseph M. Landesberg

Acknowledgments

I wish to acknowledge my colleague Frederick A. Bettelheim, who has died since the last edition. Without his encouragement and guidance, I would have bypassed the opportunity to participate in this manual. Thus, with the publication of this manual, we were able to bring the Adelphi experience to others.

These experiments have been used by our colleagues over many years and their criticism and expertise were instrumental in refinement of the experiments. Early on we had the advice of Robert Halliday (dec.), Cathy Ireland, Mahadevappa Kumbar, Jerry March (dec.), Sung Moon, Donald Opalecky, Charles Shopsis, Kevin Terrance, and Stanley Windwer. As later editions appeared, Susan Charnick, Stephen Goldberg, Robert Lippman, Simeon Moshchitsky, Reuben Rudman, and Suzanne Sitkowski provided important contributions. We acknowledge the early contributions of Dr. Jessie Lee, Community College of Philadelphia.

We also thank the following reviewers for their thoughtful comments and suggestions in early editions: R. E. Bozak, California State University, Hayward; Robert Bruner, Contra Costa College; Bridget Dube, San Antonio College; Katherine Jimison, Cuesta College; Jo Kohn, Olympic College; Eli C. Minkoff, Bates College; Margaret Sequin, San Francisco State University; Conrad Stanitski, University of Central Arkansas; and Mona Washby, Macomb Community College.

We extend our appreciation to the entire staff at Thomson Learning-Brooks/Cole, especially to David Harris, Publisher, Physical Sciences, and Ellen Bitter, Assistant Editor, Physical Sciences, for their encouragement and excellent efforts in producing this book.

Practice Safe Laboratory

A few precautions can make the laboratory experience relatively hazard free and safe. These experiments are on a small scale and thus many of the dangers found in the chemistry laboratory have been minimized. In addition to specific regulations that you may have for your laboratory, the following **DO** and **DON'T RULES** should be observed at all times.

DO RULES

- ❑ **Do wear approved safety glasses or goggles at all times.**
The first thing you should do after you enter the laboratory is to put on your safety eyewear. The last thing you should do before you leave the laboratory is to remove them. Contact lens wearers must wear additional safety goggles. Prescription glasses are not suitable safety glasses; you must wear safety goggles over them.
- ❑ **Do wear protective clothing.**
Wear sensible clothing in the laboratory (i.e., no shorts, no tank tops, no sandals). Be covered from the neck to the feet. Laboratory coats or aprons are recommended. Tie back long hair, out of the way of flames.
- ❑ **Do know the location and use of all safety equipment.**
This includes eyewash facilities, fire extinguishers, fire showers, and fire blankets. In case of fire, do not panic, clear out of the immediate area, and call your instructor for help.
- ❑ **Do use proper techniques and procedures.**
Closely follow the instructions given in this laboratory manual. These experiments have been student tested; however, accidents do occur but can be avoided if the steps for an experiment are followed. Pay heed to the **Caution!** signs in a procedure.
- ❑ **Do discard waste material properly.**
Organic chemical waste should be collected in appropriate waste containers and *not flushed down sink drains*. Dilute, nontoxic solutions may be washed down the sink with plenty of water. Insoluble and toxic waste chemicals should be collected in properly labeled waste containers. Follow the directions of your instructor for alternative or special procedures.
- ❑ **Do be alert, serious, and responsible.**
The best way you can prepare for an experiment is to read the procedure carefully and be aware of the hazards before stepping foot into the laboratory.

DON'T RULES

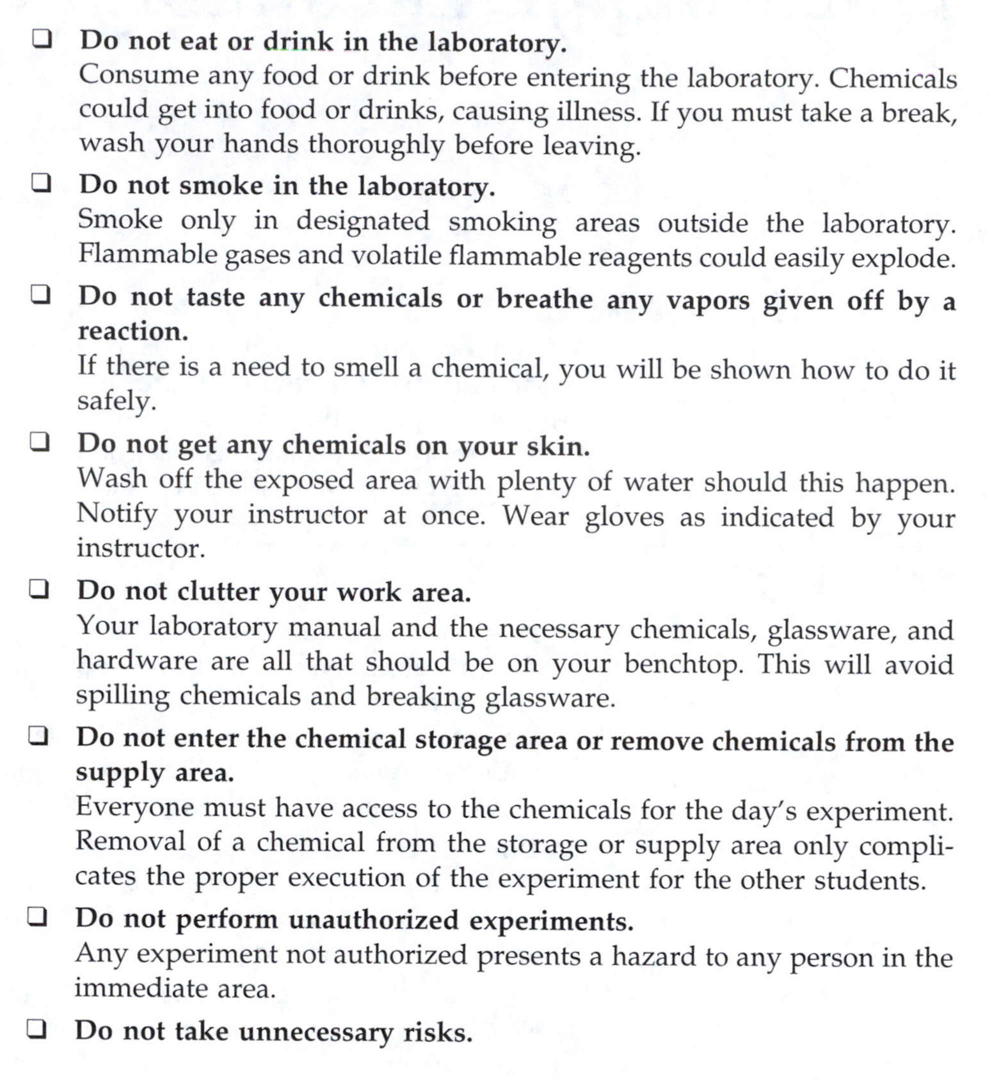

- ❑ **Do not eat or drink in the laboratory.**
  Consume any food or drink before entering the laboratory. Chemicals could get into food or drinks, causing illness. If you must take a break, wash your hands thoroughly before leaving.
- ❑ **Do not smoke in the laboratory.**
  Smoke only in designated smoking areas outside the laboratory. Flammable gases and volatile flammable reagents could easily explode.
- ❑ **Do not taste any chemicals or breathe any vapors given off by a reaction.**
  If there is a need to smell a chemical, you will be shown how to do it safely.
- ❑ **Do not get any chemicals on your skin.**
  Wash off the exposed area with plenty of water should this happen. Notify your instructor at once. Wear gloves as indicated by your instructor.
- ❑ **Do not clutter your work area.**
  Your laboratory manual and the necessary chemicals, glassware, and hardware are all that should be on your benchtop. This will avoid spilling chemicals and breaking glassware.
- ❑ **Do not enter the chemical storage area or remove chemicals from the supply area.**
  Everyone must have access to the chemicals for the day's experiment. Removal of a chemical from the storage or supply area only complicates the proper execution of the experiment for the other students.
- ❑ **Do not perform unauthorized experiments.**
  Any experiment not authorized presents a hazard to any person in the immediate area.
- ❑ **Do not take unnecessary risks.**

These **DO** and **DON'T RULES** for a safe laboratory are not an exhaustive list, but are a minimum list of precautions that will make the laboratory a safe and fun activity. Should you have any questions about a hazard, ask your instructor *first*—not your laboratory partner. Finally, if you wish to know about the dangers of any chemical you work with, read the Material Safety Data Sheet (MSDS). These sheets should be on file in the chemistry department office. A sample sheet is included here so you know what one looks like. This is the MSDS for glucose. Read it and see the kind of data included in there. Imagine all the additional cautions and precautions that the sheets would contain were you dealing with a chemical that is toxic or carcinogenic.

SIGMA-ALDRICH

MATERIAL SAFETY DATA SHEET

Date Printed: 10/04/2004
Date Updated: 03/11/2004
Version 1.6

Section 1 - Product and Company Information

Product Name	D- (+) -GLUCOSE
Product Number	G8270
Brand	SIGMA
Company	Sigma-Aldrich
Street Address	3050 Spruce Street
City, State, Zip, Country	SAINT LOUIS MO 63103 US
Technical Phone:	314 771 5765
Emergency Phone:	414 273 3850 Ext. 5996
Fax:	800 325 5052

Section 2 - Composition/Information on Ingredient

Substance Name	CAS #	SARA 313
GLUCOSE	50-99-7	No

Formula	$C_6H_{12}O_6$
Synonyms	Anhydrous dextrose * Cartose * Cerelose * Corn sugar * Dextropur * Dextrose * Dextrose, anhydrous * Dextrosol * Glucolin * Glucose * Glucose, anhydrous * D-Glucose, anhydrous * Glucose liquid * Grape sugar * Sirup * Sugar, grape
RTECS Number:	LZ6600000

Section 3 - Hazards Identification

HMIS RATING

HEALTH: 0
FLAMMABILITY: 0
REACTIVITY: 0

NFPA RATING

HEALTH: 0
FLAMMABILITY: 0
REACTIVITY: 0

For additional information on toxicity, please refer to Section 11.

Section 4 - First Aid Measures

ORAL EXPOSURE

If swallowed, wash out mouth with water provided person is conscious. Call a physician.

INHALATION EXPOSURE

If inhaled, remove to fresh air. If breathing becomes difficult, call a physician.

DERMAL EXPOSURE

In case of contact, immediately wash skin with soap and copious amounts of water.

EYE EXPOSURE
In case of contact with eyes, flush with copious amounts of water for at least 15 minutes. Assure adequate flushing by separating the eyelids with fingers. Call a physician.

Section 5 - Fire Fighting Measures

FLASH POINT
N/A

AUTOIGNITION TEMP
N/A

FLAMMABILITY
N/A

EXTINGUISHING MEDIA
Suitable: Water spray. Carbon dioxide, dry chemical powder, or appropriate foam.

FIREFIGHTING
Protective Equipment: Wear self-contained breathing apparatus and protective clothing to prevent contact with skin and eyes.
Specific Hazard(s): Emits toxic fumes under fire conditions.

Section 6 - Accidental Release Measures

PROCEDURE(S) OF PERSONAL PRECAUTION(S)
Exercise appropriate precautions to minimize direct contact with skin or eyes and prevent inhalation of dust.

METHODS FOR CLEANING UP
Sweep up, place in a bag and hold for waste disposal. Avoid raising dust. Ventilate area and wash spill site after material pickup is complete.

Section 7 - Handling and Storage

HANDLING
User Exposure: Avoid inhalation. Avoid contact with eyes, skin, and clothing. Avoid prolonged or repeated exposure.

STORAGE
Suitable: Keep tightly closed.

Section 8 - Exposure Controls / PPE

ENGINEERING CONTROLS
Safety shower and eye bath. Mechanical exhaust required.

PERSONAL PROTECTIVE EQUIPMENT
Respiratory: Wear dust mask.
Hand: Protective gloves.
Eye: Chemical safety goggles.

GENERAL HYGIENE MEASURES
Wash thoroughly after handling.

Section 9 – Physical/Chemical Properties

Appearance — Physical State: Solid

Property	Value	At Temperature or Pressure
Molecular Weight	180.16 AMU	
pH	N/A	
BP/BP Range	N/A	
MP/MP Range	153 – 156 °C	
Freezing Point	N/A	
Vapor Pressure	N/A	
Vapor Density	N/A	
Saturated Vapor Conc.	N/A	
SG/Density	N/A	
Bulk Density	N/A	
Odor Threshold	N/A	
Volatile%	N/A	
VOC Content	N/A	
Water Content	N/A	
Solvent Content	N/A	
Evaporation Rate	N/A	
Viscosity	N/A	
Surface Tension	N/A	
Partition Coefficient	N/A	
Decomposition Temp.	N/A	
Flash Point	N/A	
Explosion Limits	N/A	
Flammability	N/A	
Autoignition Temp	N/A	
Refractive Index	N/A	
Optical Rotation	N/A	
Miscellaneous Data	N/A	
Solubility	N/A	

N/A = not available

Section 10 – Stability and Reactivity

STABILITY

Stable: Stable.

Materials to Avoid: Strong oxidizing agents.

HAZARDOUS DECOMPOSITION PRODUCTS

Hazardous Decomposition Products: Carbon monoxide, Carbon dioxide.

HAZARDOUS POLYMERIZATION

Hazardous Polymerization: Will not occur

Section 11 – Toxicological Information

ROUTE OF EXPOSURE

Skin Contact: May cause skin irritation.
Skin Absorption: May be harmful if absorbed through the skin.
Eye Contact: May cause eye irritation.
Inhalation: May be harmful if inhaled. Material may be irritating to mucous membranes and upper respiratory tract.
Ingestion: May be harmful if swallowed.

SIGNS AND SYMPTOMS OF EXPOSURE

To the best of our knowledge, the chemical, physical, and toxicological properties have not been thoroughly investigated.

TOXICITY DATA

Oral
Rat
25800 mg/kg
LD50
Remarks: Behavioral:Coma. Lungs, Thorax, or Respiration:Cyanosis. Gastrointestinal:Hypermotility, diarrhea.

Intraperitoneal
Mouse
18 GM/KG
LD50

Intravenous
Mouse
9 GM/KG
LD50

CHRONIC EXPOSURE - CARCINOGEN

Species: Rat
Route of Application: Subcutaneous
Dose: 15400 GM/KG
Exposure Time: 22W
Frequency: C
Result: Tumorigenic:Equivocal tumorigenic agent by RTECS criteria. Tumorigenic:Tumors at site or application.

CHRONIC EXPOSURE - TERATOGEN

Species: Woman
Dose: 2 GM/KG
Route of Application: Oral
Exposure Time: (28W PREG)
Result: Specific Developmental Abnormalities: Craniofacial (including nose and tongue). Specific Developmental Abnormalities: Other developmental abnormalities.

Species: Woman
Dose: 1057 UG/KG
Route of Application: Intravenous
Exposure Time: (39W PREG)
Result: Specific Developmental Abnormalities: Hepatobiliary system.

Species: Hamster
Dose: 20 GM/KG
Route of Application: Intraperitoneal
Exposure Time: (6-8D PREG)
Result: Specific Developmental Abnormalities: Eye, ear.

Species: Hamster
Dose: 20 GM/KG
Route of Application: Subcutaneous
Exposure Time: (6-8D PREG)
Result: Specific Developmental Abnormalities: Eye, ear.

Species: Hamster
Dose: 20 GM/KG
Route of Application: Multiple
Exposure Time: (6-8D PREG)
Result: Specific Developmental Abnormalities: Eye, ear. Specific Developmental Abnormalities: Musculoskeletal system.

CHRONIC EXPOSURE - MUTAGEN

Species: Human
Dose: 30 MMOL/L
Cell Type: Other cell types
Mutation test: DNA damage

Species: Mouse
Dose: 179 MMOL/L
Cell Type: lymphocyte
Mutation test: Mutation in mammalian somatic cells.

CHRONIC EXPOSURE - REPRODUCTIVE HAZARD

Species: Woman
Dose: 2 GM/KG
Route of Application: Intravenous
Exposure Time: (39W PREG)
Result: Maternal Effects: Other effects. Effects on Embryo or Fetus: Other effects to embryo.

Species: Woman
Dose: 1300 MG/KG
Route of Application: Intravenous
Exposure Time: (39W PREG)
Result: Effects on Newborn: Biochemical and metabolic. Effects on Newborn: Behavioral.

Species: Rat
Dose: 300 GM/KG
Route of Application: Intraperitoneal
Exposure Time: (30D PRE)
Result: Maternal Effects: Ovaries, fallopian tubes. Maternal Effects: Uterus, cervix, vagina.

Species: Hamster
Dose: 20 MG/KG
Route of Application: Multiple
Exposure Time: (6-8D PREG)
Result: Effects on Fertility: Post-implantation mortality (e.g., dead and/or resorbed implants per total number of implants). Specific Developmental Abnormalities: Urogenital system.

Section 12 - Ecological Information

No data available.

Section 13 - Disposal Considerations

APPROPRIATE METHOD OF DISPOSAL OF SUBSTANCE OR PREPARATION

Contact a licensed professional waste disposal service to dispose of this material. Dissolve or mix the material with a combustible solvent and burn in a chemical incinerator equipped with an afterburner and scrubber. Observe all federal, state, and local environmental regulations.

Section 14 - Transport Information

DOT

Proper Shipping Name: None
Non-Hazardous for Transport: This substance is considered to be non-hazardous for transport.

IATA

Non-Hazardous for Air Transport: Non-hazardous for air transport.

Section 15 - Regulatory Information

UNITED STATES REGULATORY INFORMATION

SARA LISTED: No
TSCA INVENTORY ITEM: Yes

CANADA REGULATORY INFORMATION

WHMIS Classification: This product has been classified in accordance with the hazard criteria of the CPR, and the MSDS contains all the information required by the CPR.
DSL: Yes
NDSL: No

Section 16 - Other Information

DISCLAIMER

For R&D use only. Not for drug, household or other uses.

WARRANTY

The above information is believed to be correct but does not purport to be all inclusive and shall be used only as a guide. The information in this document is based on the present state of our knowledge and is applicable to the product with regard to appropriate safety precautions. It does not represent any guarantee of the properties of the product. Sigma-Aldrich Inc., shall not be held liable for any damage resulting from handling or from contact with the above product. See reverse side of invoice or packing slip for additional terms and conditions of sale.

Laboratory Techniques: Using the Laboratory Gas Burner; Making Laboratory Measurements

BACKGROUND

Using the Laboratory Gas Burner

Tirrill or Bunsen burners provide a ready source of heat in the chemistry laboratory. In general, because chemical reactions proceed faster at elevated temperatures, the use of heat enables the experimenter to accomplish many experiments more quickly than would be possible at room temperature. The burner illustrated in Figure 1.1 is typical of the burners used in most general chemistry laboratories.

A burner is designed to allow gas and air to mix in a controlled manner. The gas often used is "natural gas," mostly the highly flammable and odorless hydrocarbon methane, CH_4. When ignited, the flame's temperature can be adjusted by altering the various proportions of gas and air. The gas flow can be controlled either at the main gas valve or at the gas control valve at the base of the burner. Manipulation of the air vents at the bottom of the barrel allows air to enter and mix with the gas. The hottest flame has a nonluminous violet outer cone, a pale-blue middle cone, and a dark-blue inner cone; the air vents, in this case, are opened sufficiently to assure complete combustion of the gas. Combustion of the gas yields carbon dioxide and water.

$$CH_4(g) + 2O_2(g) \rightarrow CO_2(g) + 2H_2O(g)$$

Lack of air produces a cooler, luminous yellow flame. This flame lacks the inner cone, most likely is smokey, and often deposits soot on objects it contacts. The soot is due to small particles of carbon from unburned fuel, which luminesce and give the yellow flame. Too much air blows out the flame.

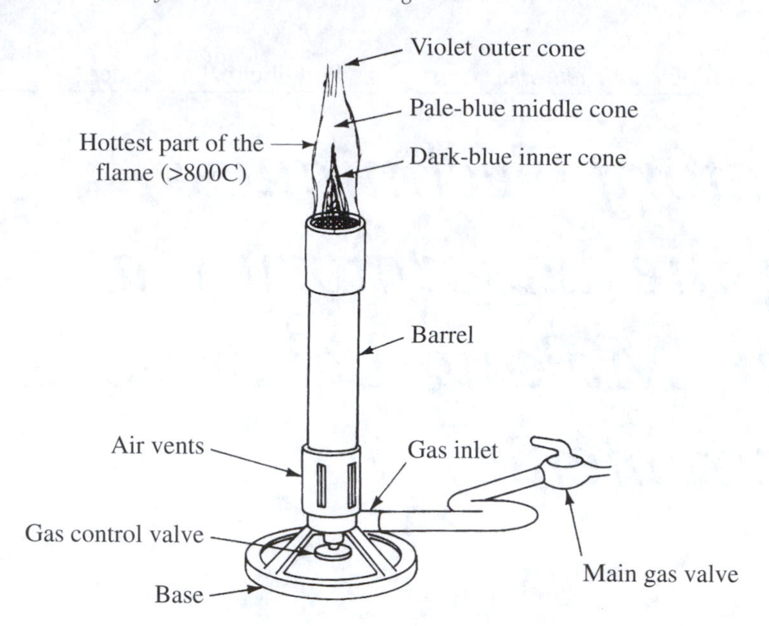

Figure 1.1
The Tirrill burner.

In the chemistry laboratory, much of the work that is carried out is with specialized glassware. It is important to understand the properties of the glass used in the make-up of this glassware and the limitations that these properties impose.

Glass is a super-cooled liquid. Unlike crystalline solids, which have sharp melting points, glass softens when heated and flows. However, not all glass is the same; there are different grades and compositions. Most laboratory glassware is made from borosilicate glass (containing silica and borax compounds). Commercially, this type of glass is known as *Pyrex*® (made by Corning Glass) or *Kimax*® (made by Kimble Glass). This glass does not soften very much below 800°C and is able to withstand most heating operations done in the laboratory. In addition, borosilicate glass has a low thermal coefficient of expansion. This refers to the material's change in volume per degree change in temperature. Borosilicate glass expands or contracts slowly when heated or cooled. Thus, glassware composed of this material can withstand rapid changes in temperature and can resist cracking.

Soft glass, on the other hand, is a poor material for laboratory glassware. Glass of this type is composed primarily of silica sand, SiO_2, and softens in the region of 300–400°C; this low softening temperature is not suitable for most laboratory work. Besides, soft glass has a high thermal coefficient of expansion. This means that soft glass expands or contracts very rapidly when heated or cooled; sudden, rapid changes in temperature introduce too much stress into the material, and the glass cracks.

The beakers, Erlenmeyer flasks, and test tubes used in our laboratory experiments are composed of borosilicate glass and will withstand the heating and cooling required.

LABORATORY MEASUREMENTS

Units of Measurement

The metric system of weights and measures is used by scientists of all fields, including chemists. This system uses the base 10 for measurements; for conversions, measurements may be multiplied or divided by 10.

Table 1.1 *Frequently Used Factors*

Prefix	*Power of 10*	*Decimal Equivalent*	*Abbreviation*
Micro	10^{-6}	0.000001	μ
Milli	10^{-3}	0.001	m
Centi	10^{-2}	0.01	c
Kilo	10^{3}	1000	k

Table 1.2 *Units and Equipment*

Measure	*SI Unit*	*Metric Unit*	*Equipment*
Length	Meter (m)	Meter (m)	Meterstick
Volume	Cubic meter (m^3)	Liter (L)	Pipet, graduated cylinder, Erlenmeyer flask, beaker
Mass	Kilogram (kg)	Gram (g)	Balance
Energy	Joule (J)	Calorie (cal)	Calorimeter
Temperature	Kelvin (K)	Degree Celsius (°C)	Thermometer

Table 1.1 lists the most frequently used factors in the laboratory, which are based on powers of 10.

The measures of length, volume, mass, energy, and temperature are used to evaluate our physical and chemical environment. Table 1.2 compares the metric system with the more recently accepted SI system (International System of Units). The laboratory equipment associated with obtaining these measures is also listed.

Accuracy, precision, and significant figures

Chemistry is a science that depends on experience and observation for data. It is an empirical science. An experiment that yields data requires the appropriate measuring devices in order to get accurate measurements. Once the data is in hand, calculations are done with the numbers obtained. How good the calculations are depends on a number of factors: (1) how careful you are in taking the measurements (laboratory techniques), (2) how good your measuring device is in getting a true measure (accuracy), and (3) how reproducible the measurement is (precision).

The measuring device usually contains a scale. The scale, with its subdivisions or graduations, tells the limits of the device's accuracy. You cannot expect to obtain a measurement better than your instrument is capable of reading. Consider the portion of the ruler shown in Figure 1.2.

There are major divisions labeled at intervals of 1 cm and subdivisions of 0.1 cm or 1 mm. The accuracy of the ruler is to 0.1 cm (or 1 mm); that is the measurement that is known for certain. However, it is possible to estimate to 0.01 cm (or 0.1 mm) by reading in between the subdivisions; this number is less accurate and, of course, is less certain. In general, you should be able to record the measured value to one more place than the

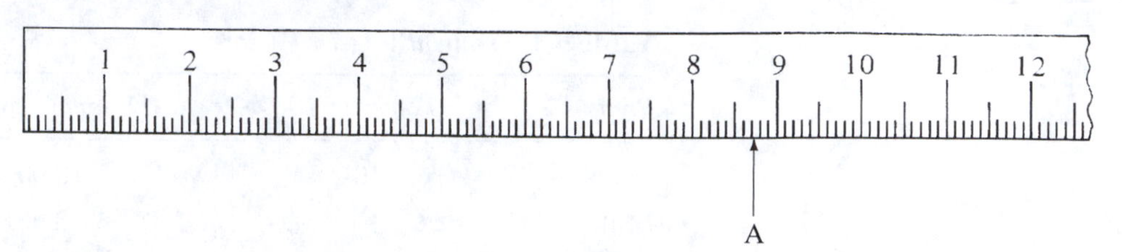

Figure 1.2
Reading a metric ruler.

scale is marked. For example, in Figure 1.2 there is a reading marked on the ruler. This value is 8.75 cm: two numbers are known with certainty, *8.7*, and one number, 0.05, is uncertain since it is the *best estimate* of the fractional part of the subdivision. The number recorded, 8.75, contains 3 significant figures, 2 certain plus 1 uncertain. When dealing with *significant figures*, remember: (1) the uncertainty is in the last recorded digit, and (2) the number of significant figures contains the number of digits definitely known, plus one more that is estimated.

The manipulation of significant figures in multiplication, division, addition, and subtraction is important. It is particularly important when using electronic calculators, which give many more digits than are useful or significant. If you keep in mind the principle that the final answer can be no more accurate than the least accurate measurement, you should not go wrong. A few examples will demonstrate this.

Example 1

Divide 9.3 by 4.05. If this calculation is done by a calculator, the answer found is 2.296296296. However, *a division should have as an answer the same number of significant figures as the least accurately known (fewest significant figures) of the numbers being divided.* One of the numbers, 9.3, contains only 2 significant figures. Therefore, the answer can only have 2 significant figures, i.e., 2.3 (rounded off).

Example 2

Multiply 0.31 by 2.563. Using a calculator, the answer is 0.79453. *As in division, a multiplication can have as an answer the same number of significant figures as the least accurately known (fewest significant figures) of the numbers being multiplied.* The number 0.31 has 2 significant figures (the zero fixes the decimal point). Therefore, the answer can only have 2 significant figures, i.e., 0.79 (rounded off).

Example 3

Add 3.56 + 4.321 + 5.9436. A calculator gives 13.8246. *With addition (or subtraction), the answer is significant to the least number of decimal places of the numbers added (or subtracted).* The least accurate number is 3.56, measured only to the hundredth's place. The answer should be to this accuracy, i.e., 13.82 (rounded off to the hundredth's place).

Example 4

Do the subtraction 6.532 – 1.3. A calculator gives 5.232 as the answer. However, since the least accurate number is 1.3, measured to the tenth's place, the answer should be to this accuracy, i.e., 5.2 (rounded off to the tenth's place).

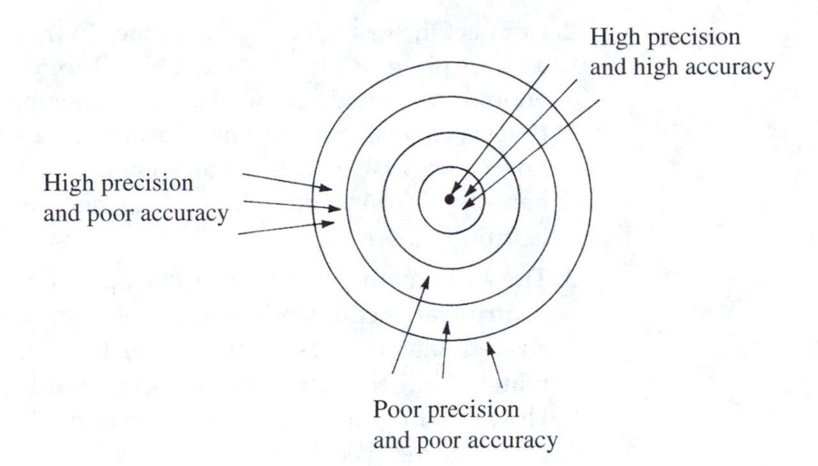

Figure 1.3
Precision and accuracy illustrated by a target.

Finally, how do precision and accuracy compare? *Precision* is a determination of the reproducibility of a measurement. It tells you how closely several measurements agree with one another. Several measurements of the same quantity showing high precision will cluster together with little or no variation in value; however, if the measurements show a wide variation, the precision is low. *Random errors* are errors that lead to differences in successive values of a measurement and affect precision; some values will be off in one direction or another. One can estimate the precision for a set of values for a given quantity as follows: estimate = $\pm\Delta/2$, where Δ is the difference between the highest and lowest values.

Accuracy is a measure of how closely the value determined agrees with a known or accepted value. Accuracy is subject to *systematic errors*. These errors cause measurements to vary from the known value and will be off in the same direction, either too high or too low. A consistent error in a measuring device will affect the accuracy, but always in the same direction. It is important to use properly calibrated measuring devices. If a measuring device is not properly calibrated, it may give high precision, but none of the measurements will be accurate. However, a properly calibrated measuring device will be both precise and accurate. (See Figure 1.3.) A systematic error is expressed as the difference between the known value and the average of the values obtained by measurement in a number of trials.

OBJECTIVES

1. To learn how to use a Bunsen or Tirrill burner.
2. To learn how to use simple, common equipment found in the laboratory.
3. To learn to take measurements.
4. To be able to record these measurements with precision and accuracy using the proper number of significant figures.

PROCEDURE

The Laboratory Gas Burner; Use of the Bunsen Burner

1. Before connecting the Bunsen burner to the gas source, examine the burner and compare it to Figure 1.1. Be sure to locate the gas control valve and the air vents and see how they work.

2. Connect the gas inlet of your burner to the main gas valve by means of a short piece of thin-walled rubber tubing. Be sure the tubing is long enough to provide some slack for movement on the bench top. Close the gas control valve. If your burner has a screw-needle valve, turn the knob clockwise. Close the air vents. This can be done by rotating the barrel of the burner (or sliding the ring over the air vents if your burner is built this way).
3. Turn the main gas valve to the open position. Slowly open the gas control valve counterclockwise until you hear the hiss of gas. Quickly strike a match or use a gas striker to light the burner. With a lighted match, hold the flame to the top of the barrel. The gas should light. How would you describe the color of the flame? Hold a Pyrex test tube in this flame. What do you observe?
4. Carefully turn the gas control valve, first clockwise and then counterclockwise. What happens to the flame size? (If the flame should go out, or if the flame did not light initially, shut off the main gas valve and start over, as described above.)
5. With the flame on, adjust the air vents by rotating the barrel (or sliding the ring). What happens to the flame as the air vents open? Adjust the gas control valve and the air vents until you obtain a flame about 3 or 4 in. high, with an inner cone of blue (Figure 1.1). The tip of the pale blue inner cone is the hottest part of the flame.
6. Too much air will blow out the flame. Should this occur, close the main gas valve immediately. Relight following the procedure in step 3.
7. Too much gas pressure will cause the flame to rise away from the burner and "roar" (Figure 1.4). If this happens, reduce the gas flow by closing the gas control valve until a proper flame results.
8. "Flashback" sometimes occurs. If so, the burner will have a flame at the bottom of the barrel. Quickly close the main gas valve. Allow the barrel to cool. Relight following the procedures in step 3.

Figure 1.4
The flame rises away from the burner.

Laboratory Measurements

Length: use of the meterstick (or metric ruler)

1. The meterstick is used to measure length. Examine the meterstick in your kit. You will notice that one side has its divisions in inches (in.) with subdivisions in sixteenths of an inch; the other side is in centimeters (cm) with subdivisions in millimeters (mm). Some useful conversion factors are listed below.

1 km = 1000 m	1 in. = 2.54 cm
1 m = 100 cm	1 ft. = 30.48 cm
1 cm = 10 mm	1 yd. = 91.44 cm
1 m = 1000 mm	1 mi. = 1.6 km

A meterstick that is calibrated to 0.1 cm can be read to the hundredth's place; however, only a 0 (0.00) or a 5 (0.05) may appear. A measurement falling directly on a subdivision is read as a 0 in the hundredth's place. A measurement falling anywhere between adjacent subdivisions is read as a 5 in the hundredth's place.

2. With your meterstick (or metric ruler), measure the length and width of this laboratory manual. Take the measurements in inches (to the nearest sixteenth of an inch) and in centimeters (to the nearest 0.5 cm). Record your response on the Report Sheet (1).

3. Convert the readings in cm to mm and m (2).
4. Calculate the area of the manual in in.^{2}, cm^2, and mm^2 (3). Be sure to express your answers to the proper number of significant figures.

Example 5

A student measured a piece of paper and found it to be 20.30 cm by 29.25 cm. The area was found to be

$$\mathbf{20.30\ cm \times 29.25\ cm = 593.8\ cm^2}$$

Volume: use of a graduated cylinder, an Erlenmeyer flask, and a beaker

1. Volume in the metric system is expressed in liters (L) and milliliters (mL). Another way of expressing milliliters is in cubic centimeters (cm^3 or cc). Several conversion factors for volume measurements are listed below.

1 L = 1000 mL	1 qt. = 0.96 L
1 mL = 1 cm^3 = 1 cc	1 gal. = 3.79 L
1 L = 0.26 gal.	1 fl. oz. = 29.6 mL

2. The graduated cylinder is a piece of glassware used for measuring the volume of a liquid. Graduated cylinders come in various sizes with different degrees of accuracy. A convenient size for this experiment is the 100-mL graduated cylinder. Note that this cylinder is marked in units of 1 mL; major divisions are of 10 mL and subdivisions are of 1 mL. Estimates can be made to the nearest 0.1 mL. When a liquid is in the graduated cylinder, you will see that the level in the cylinder is curved with the lowest point at the center. This is the *meniscus*, or the dividing line between liquid and air. When reading the meniscus for the volume, be sure to read the *lowest* point on the curve and not the upper edge. To avoid errors in reading the meniscus, the eye's line of sight must be perpendicular to the scale (Figure 1.5). In steps 3 and 4, use the graduated cylinder to see how well the marks on an Erlenmeyer flask and a beaker measure the indicated volume.
3. Take a 50-mL graduated Erlenmeyer flask (Figure 1.6) and fill with water to the 50-mL mark. Transfer the water, completely and without spilling, to a 100-mL graduated cylinder. Record the volume on the Report Sheet (4) to the nearest 0.1 mL; convert to L.

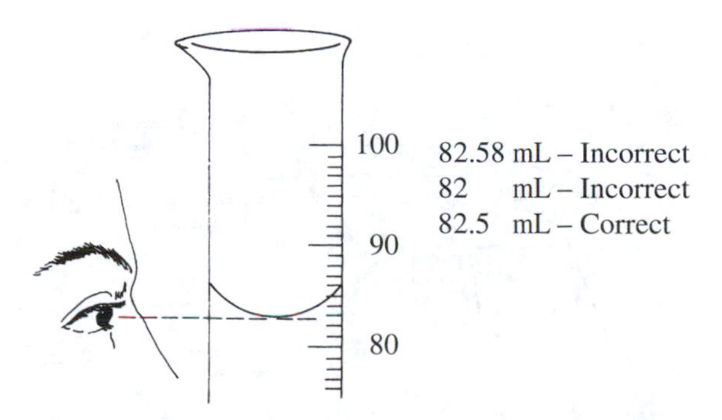

Figure 1.5
Reading the meniscus on a graduated cylinder.

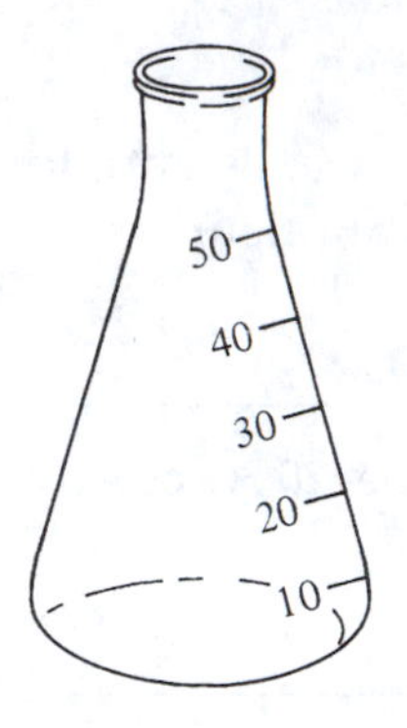

Figure 1.6
A 50-mL graduated Erlenmeyer flask.

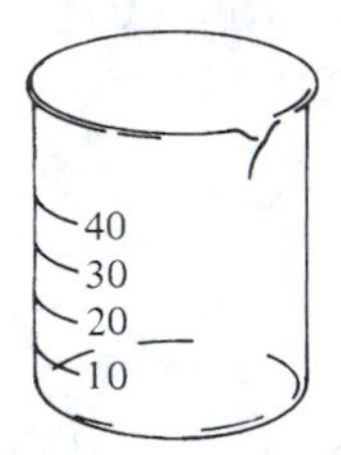

Figure 1.7
A 50-mL graduated beaker.

4. Take a 50-mL graduated beaker (Figure 1.7), and fill with water to the 40-mL mark. Transfer the water, completely and without spilling, to a dry 100-mL graduated cylinder. Record the volume on the Report Sheet (5) to the nearest 0.1 mL; convert to L.
5. What is the error in mL and in percent for obtaining 50.0 mL for the Erlenmeyer flask and 40.0 mL for the beaker (6)? Calculate the % error in the following way: [(volume by graduated cylinder) minus (volume using the mark of the Erlenmeyer or the beaker) divided by (volume by graduated cylinder)] times 100.
6. Which piece of glassware will give you a more accurate measure of liquid: the graduated cylinder, the Erlenmeyer flask, or the beaker (7)?

Mass: use of the laboratory balance

1. Mass measurements of objects are carried out with the laboratory balance. Many types of balances are available for laboratory use. The proper choice of a balance depends upon what degree of accuracy is needed for a measurement. The standard units of mass are the kilogram (kg) in the SI system and the gram (g) in the metric system. Some conversion factors are listed below.

   1 kg = 1000 g   1 lb. = 454 g
   1 g = 1000 mg   1 oz. = 28.35 g

   Three types of balances are illustrated in Figures 1.8, 1.10, and 1.12. A platform triple beam balance is shown in Figure 1.8. This balance can weigh objects up to 2610 g. Because the scale is marked in 0.1-g divisions, it is mostly used for rough weighing; weights to 0.01 g can be estimated. Figure 1.9 illustrates how to take a reading on this balance.

   The single pan, triple beam (Centogram) balance is shown in Figure 1.10. This Centogram balance has a higher degree of accuracy because the divisions are marked in 0.01-g (estimates can be made to 0.001 g) increments.

   Smaller quantities of material can be weighed on this balance (to a maximum of 311 g). Figure 1.11 illustrates how a reading on this balance would be taken.

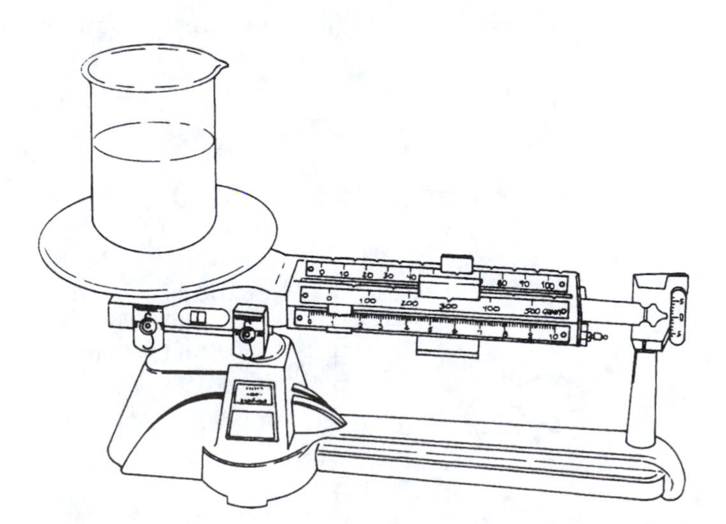

Figure 1.8
A platform triple beam balance.

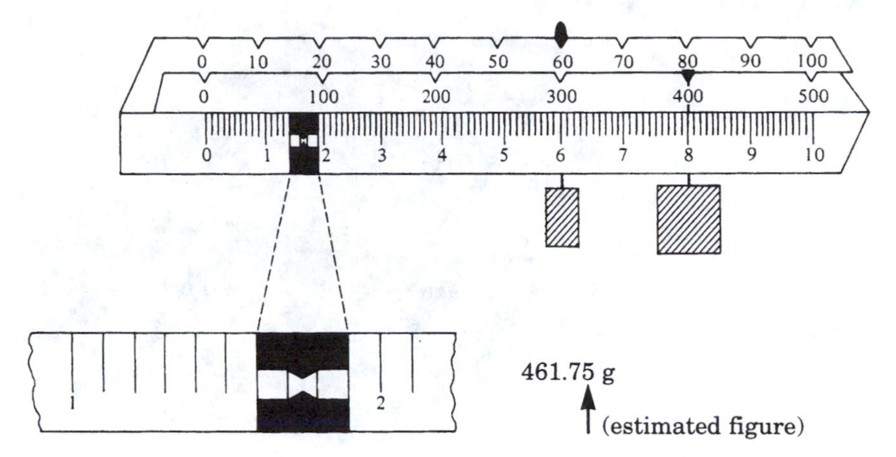

Figure 1.9
Reading on a platform triple beam balance.

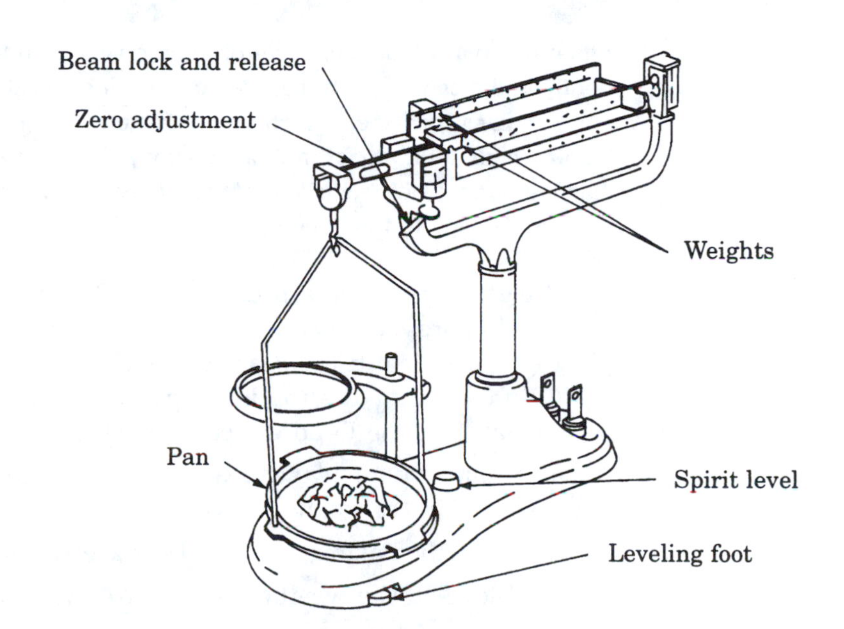

Figure 1.10
A single pan, triple beam balance (Centogram).

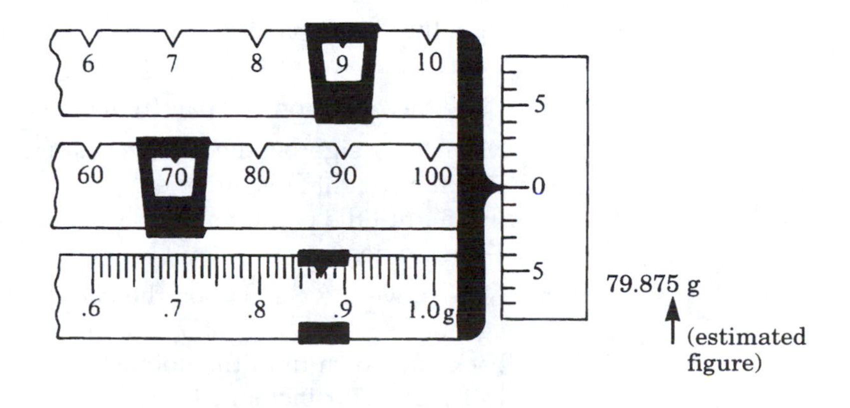

Figure 1.11
Reading on a single pan, triple beam balance.

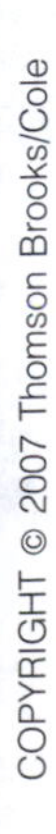

A top loading balance shows the highest accuracy (Figure 1.12). Objects can be weighed very rapidly with this balance because the total weight, to the nearest 0.001 g, can be read directly from a digital readout (Figure 1.12). A balance of this type is very expensive and should be used only after the instructor has demonstrated its use.

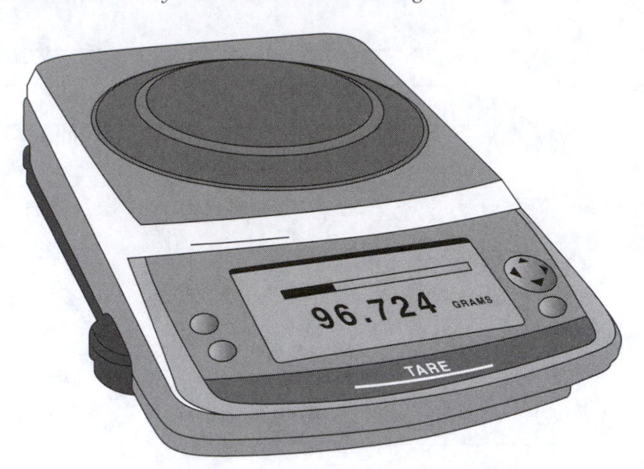

Figure 1.12
A top loading balance with a digital readout.

CAUTION

When using any balance, never drop an object onto the pan; place it gently in the center of the pan. Never place chemicals directly on the pan; use either a glass container (watch glass, beaker, weighing bottle) or weighing paper. Never weigh a hot object; hot objects may mar the pan. Buoyancy effects will cause incorrect weights. Clean up any chemical spills in the balance area to prevent damage to the balance.

2. Weigh a quarter, a test tube (100 × 13 mm), and a 125-mL Erlenmeyer flask. Express each weight to the proper number of significant figures. Use a platform triple beam balance, a single pan, triple beam balance (Centogram), and a top loading balance for these measurements. Use the table on the Report Sheet to record each weight.
3. The single pan, triple beam balance (Centogram) (Figure 1.10) is operated in the following way.
   - **a.** Place the balance on a level surface; use the leveling foot to level it.
   - **b.** Move all the weights to the zero position at left.
   - **c.** Release the beam lock.
   - **d.** The pointer should swing freely in an equal distance up and down from the zero or center mark on the scale. Use the zero adjustment to make any correction to the swing.
   - **e.** Place the object on the pan (remember the caution).
   - **f.** Move the weight on the middle beam until the pointer drops; make sure the weight falls into the "V" notch. Move the weight back one notch until the pointer swings up. This beam weighs up to 10 g, in 1-g increments.
   - **g.** Now move the weights on the back beam until the pointer drops; again be sure the weight falls into the "V" notch. Move the weight back one notch until the pointer swings up. This beam weighs up to 1 g, in 0.1-g increments.
   - **h.** Lastly, move the smallest weight (the cursor) on the front beam until the pointer balances, that is, swings up and down an equal distance from the zero or center mark on the scale. This last beam weighs up to 0.1 g, in 0.01-g increments.
   - **i.** The weight of the object on the pan is equal to the weights shown on each of the three beams (Figure 1.11). Weights to 0.005 g may be estimated.

j. Repeat the movement of the cursor to check your precision.

k. When finished, move the weights to the left, back to zero, and arrest the balance with the beam lock.

Temperature: use of the thermometer

1. Routine measurements of temperature are done with a thermometer. Thermometers found in chemistry laboratories may use either mercury or a colored fluid as the liquid, and degrees Celsius (°C) as the units of measurement. The fixed reference points on this scale are the freezing point of water, 0°C, and the boiling point of water, 100°C. Between these two reference points, the scale is divided into 100 units, with each unit equal to 1°C. Temperature can be estimated to 0.1°C. Other thermometers use either the Fahrenheit (°F) or the Kelvin (K) temperature scale and use the same reference points, that is, the freezing and boiling points of water. Conversion between the scales can be accomplished using the formulas below.

$$^\circ\mathrm{F} = \frac{9}{5}{}^\circ\mathrm{C} + 32.0 \qquad ^\circ\mathrm{C} = \frac{5}{9}(^\circ\mathrm{F} - 32.0) \qquad \mathrm{K} = {}^\circ\mathrm{C} + 273.15$$

Example 6

Convert 37.0°C to °F and K.

$$^\circ\mathrm{F} = \frac{9}{5}(37.0^\circ\mathrm{C}) + 32.0 = 98.6^\circ\mathrm{F}$$

$$\mathrm{K} = 37.0^\circ\mathrm{C} + 273.15 = 310.2\ \mathrm{K}$$

---

2. Use the thermometer in your kit and record to the nearest 0.1°C the temperature of the laboratory at *room temperature*. Use the Report Sheet to record your results.
3. Record the temperature of boiling water. Set up a 250-mL beaker containing 100 mL water, and heat on a hot plate until boiling. Hold the thermometer in the boiling water for at least 1 min. before reading the temperature (*be sure not to touch the sides of the beaker*). Using the Report Sheet, record your results to the nearest 0.1°C.
4. Record the temperature of ice water. Into a 250-mL beaker, add enough crushed ice to fill halfway. Add distilled water to the level of the ice. Stir the ice water gently with a glass rod for 1 min. before reading the thermometer. Hold the thermometer in the ice water for at least 1 min. before reading the temperature. *Use caution; be careful not to touch the walls of the beaker with the thermometer or to hit the thermometer with the glass rod.* Read the thermometer to the nearest 0.1°C. Record your results on the Report Sheet.

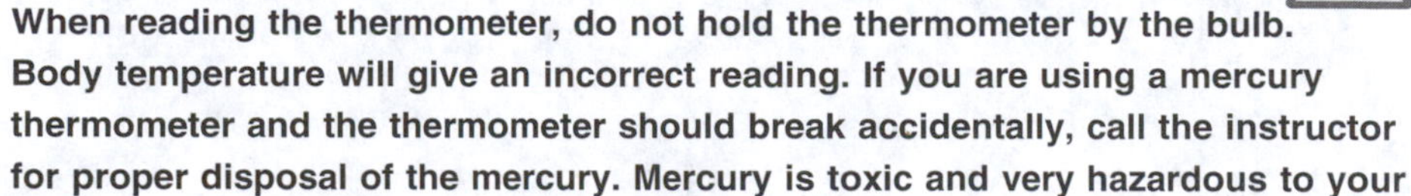

CAUTION

When reading the thermometer, do not hold the thermometer by the bulb. Body temperature will give an incorrect reading. If you are using a mercury thermometer and the thermometer should break accidentally, call the instructor for proper disposal of the mercury. Mercury is toxic and very hazardous to your health. Do not handle the liquid or breathe its vapor.

5. Convert your answers to questions 2, 3, and 4 into °F and K.

CHEMICALS AND EQUIPMENT

1. Bunsen or Tirrill burner
2. 50-mL graduated beaker
3. 50-mL graduated Erlenmeyer flask
4. 100-mL graduated cylinder
5. Meterstick or ruler
6. Quarter
7. Balances
8. Hot plates

Density Determination

BACKGROUND

Samples of matter can be identified by using characteristic physical properties. A substance may have a unique color, odor, melting point, or boiling point. These properties do not depend on the quantity of the substance and are called *intensive properties*. Density also is an intensive property and may serve as a means of identification.

The *density* of a substance is the *ratio of its mass per unit volume*. Density can be found mathematically by dividing the mass of a substance by its volume. The formula is $\mathbf{d} = \frac{\mathbf{m}}{\mathbf{V}}$, where **d** is density, **m** is mass, and **V** is volume. Whereas mass and volume do depend on the quantity of a substance (these are *extensive properties*), the ratio is constant at a given temperature. The units of density, reported in standard references, is in terms of g/mL (or g/cc or g/cm^3) at 20°C. The temperature is reported because the volume of a sample will change with temperature and, thus, so does the density.

Example

A bank received a yellow bar, marked gold, of mass 453.6 g, and volume 23.5 cm^3. Is it gold? (Density of gold = 19.3 g/cm^3 at 20°C.)

$$\mathbf{d = \frac{m}{V} = \frac{453.6\ g}{23.5\ cm^3} = 19.3\ g/cm^3}$$

Yes, it is gold.

OBJECTIVES

1. To determine the densities of regular- and irregular-shaped objects and use them as a means of identification.
2. To determine the density of water.
3. To determine the density of a liquid and use this as a means of identification.

PROCEDURE

Density of a Regular-Shaped Object

1. Obtain a solid block from the instructor. Record the code number.
2. Using your metric ruler, determine the dimensions of the block (length, width, height) and record the values to the nearest 0.05 cm (1). Calculate the volume of the block (2). Repeat the measurements for a second trial.
3. Using a single pan, triple beam balance (Centogram) or a top loading balance (if available), determine the mass of the block (3). Record the mass to the nearest 0.001 g. Calculate the density of the block (4). Repeat the measurements for a second trial.

Density of an Irregular-Shaped Object

1. Obtain a sample of unknown metal from your instructor. Record the code number.
2. Obtain a mass of the sample of approximately 5 g. Be sure to record the exact quantity to the nearest 0.001 g (5).
3. Choose either a 10-mL or a 25-mL graduated cylinder. For small metal pieces, you should use the 10-mL graduated cylinder. For large metal pieces, you should use the 25-mL graduated cylinder. Fill the graduated cylinder approximately halfway with water. Record the exact volume to the nearest 0.05 mL (6). [Depending on the number of subdivisions between milliliter divisions, you can read to the hundredth's place. For example, you can read the 10-mL graduated cylinder to the hundredth's place where only a 0 (the reading is directly on a subdivision, e.g., 0.00) or a 5 (anywhere between two adjacent subdivisions, e.g., 0.05) can be used.]
4. Place the metal sample into the graduated cylinder. (If the pieces of metal are too large for the opening of the 10-mL graduated cylinder, use the larger graduated cylinder.) Be sure all of the metal is below the water line. Gently tap the sides of the cylinder with your fingers to ensure that no air bubbles are trapped in the metal. Read the new level of the water in the graduated cylinder to the nearest 0.05 mL (7). Assuming that the metal does not dissolve or react with the water, the difference between the two levels represents the volume of the metal sample (8) (Figure 2.1).

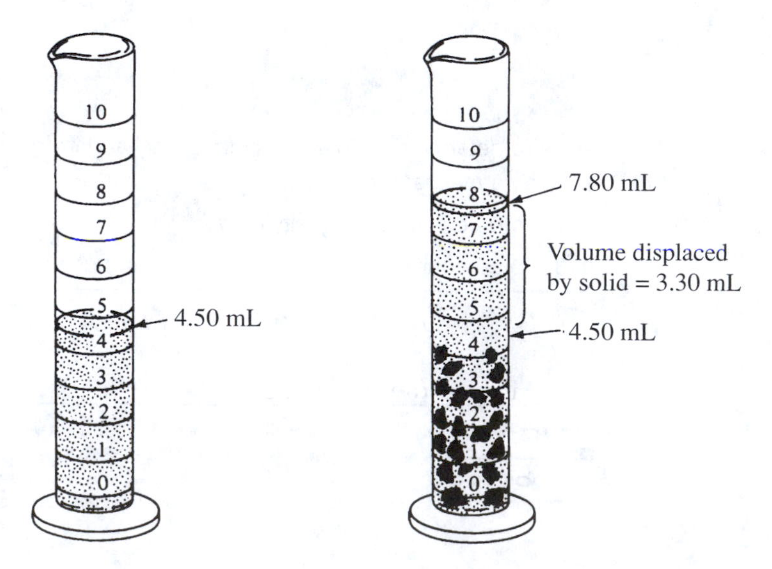

Figure 2.1
Measurement of volume of an irregular-shaped object.

Table 2.1 *Densities of Selected Metals*

Sample	*Formula*	*Density* (g/cm^3)
Aluminum	Al	2.70
Iron	Fe	7.86
Tin (white)	Sn	7.29
Zinc	Zn	7.13
Lead	Pb	11.30

5. Carefully recover the metal sample and dry it with a paper towel. Repeat the experiment.
6. Calculate the density of the metal sample from your data (9). Determine the average density from your trials, reporting to the proper number of significant figures.
7. Determine the identity of your metal sample by comparing its density to the densities listed in Table 2.1 (10).
8. Recover your metal sample and return it as directed by your instructor.

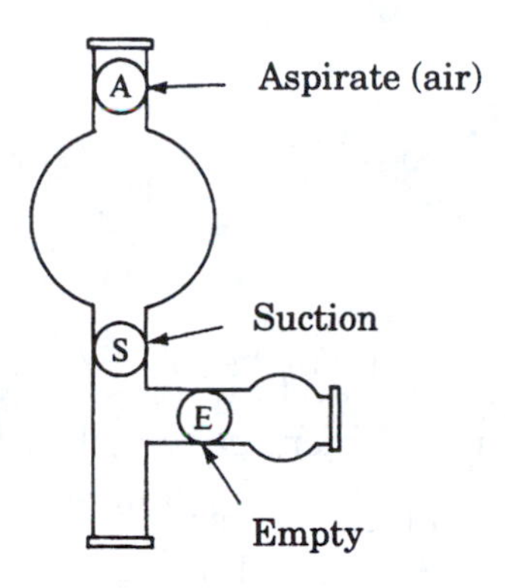

Figure 2.2
The Spectroline pipet filler.

CAUTION 

Do not discard the samples in waste containers or in the sink. Use a labeled collection container that is specific for each sample.

Use of the Spectroline Pipet Filler

1. Examine the Spectroline pipet filler and locate the valves marked "A," "S," and "E" (Figure 2.2). These operate by pressing the flat surfaces between the thumb and forefinger.
2. Squeeze the bulb with one hand while you press valve "A" with two fingers of the other hand. The bulb flattens as air is expelled. If you release your fingers when the bulb is flattened, the bulb remains collapsed.
3. Carefully insert the pipet end into the Spectroline pipet filler (Figure 2.3). The end should insert easily and not be forced.

CAUTION 

Before inserting the pipet end into the Spectroline pipet filler, lubricate the glass by rubbing the opening with a drop of water or glycerin.

4. Place the tip of the pipet into the liquid to be pipetted. Make sure that the tip is below the surface of the liquid at all times.
5. With your thumb and forefinger, press valve "S." Liquid will be drawn up into the pipet. By varying the pressure applied by your fingers, the rise of the liquid into the pipet can be controlled. Allow the liquid to fill the pipet to a level slightly above the etched mark on the stem. Release the valve; the liquid should remain in the pipet.
6. Withdraw the pipet from the liquid. Draw the tip of the pipet lightly along the wall of the beaker to remove excess liquid.

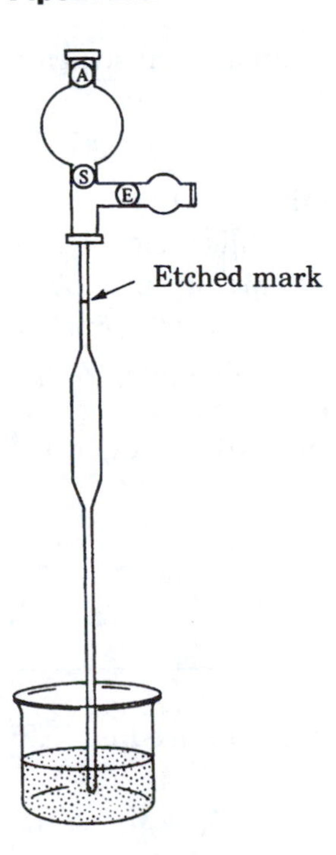

Figure 2.3
Using the Spectroline pipet filler to pipet.

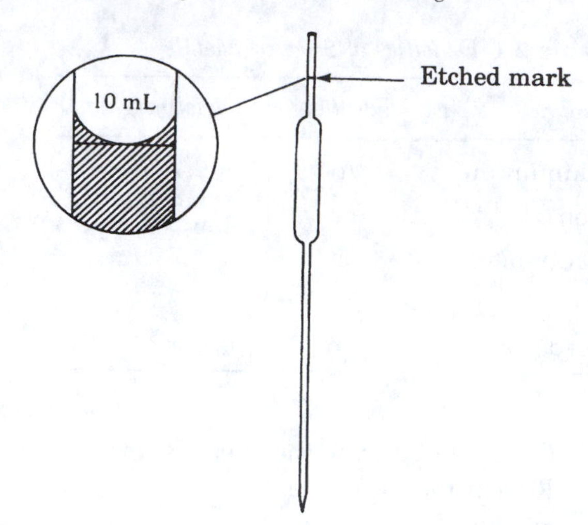

Figure 2.4
Adjusting the curved meniscus of the liquid to the etched mark.

7. Adjust the level of the meniscus of the liquid by carefully pressing valve "E." The level should lower until the curved meniscus touches the etched mark (Figure 2.4). Carefully draw the tip of the pipet lightly along the wall of the beaker to remove excess liquid.
8. Drain the liquid from the pipet into a collection flask by pressing valve "E." Remove any drops on the tip by touching the tip of the pipet against the inside walls of the collection flask. Liquid should remain inside the tip; the pipet is calibrated with this liquid in the tip.

Density of Water

1. Obtain approximately 50 mL of distilled water from your instructor. Record the temperature of the water (11).
2. Take a clean, dry 50-mL beaker; weigh to the nearest 0.001 g (12).
3. With a 10-mL volumetric pipet, transfer 10.00 mL of distilled water into the preweighed beaker using a Spectroline pipet filler (Figure 2.3). (Before transferring the distilled water, be sure there are no air bubbles trapped in the volumetric pipet. If there are, gently tap the pipet to dislodge the air bubbles, and then refill to the line.) Immediately weigh the beaker and water and record the weight to the nearest 0.001 g (13). Calculate the weight of the water by subtraction (14). Calculate the density of the water at the temperature recorded (15).

CAUTION

Never use your mouth when pipetting. Always use a pipet filler.

4. Repeat step 3 for a second trial. Be sure all the glassware used is clean and dry.
5. Calculate the average density (16). Compare your average value at the recorded temperature to the value reported for that temperature in a standard reference. (The instructor should have available a CRC *Handbook of Chemistry and Physics*.)

Table 2.2 *Densities of Selected Liquids*

Sample	*Density (g/mL)*	*T°C*
Hexane	0.659	20
Ethanol	0.791	20
Olive oil	0.918	15
Sea water	1.025	20 (3.15% NaCl, g/100 g sol)
Milk	1.028–1.035	20
Ethylene glycol	1.109	20

Density of an Unknown Liquid

1. Obtain approximately 25 mL of an unknown liquid from your instructor. Record the code number. Determine the temperature of the liquid (17).
2. Weigh a clean, dry 50-mL beaker to the nearest 0.001 g (18).
3. Transfer 10.00 mL of the liquid with a 10-mL volumetric pipet into the pre-weighed beaker using the Spectroline pipet filler (Figure 2.2). Immediately reweigh the beaker to the nearest 0.001 g (19).
4. Calculate the weight of the unknown liquid by subtraction (20).
5. Calculate the density of the unknown liquid at the temperature recorded (21).
6. Repeat the procedure following steps 1–5 for a second trial. When you repeat the steps, be sure all the glassware is clean and dry.
7. Calculate the average density. Determine the identity of your unknown liquid by comparing its density to the densities listed in Table 2.2 (22).
8. Discard your used liquid samples into containers provided by your instructor. Do not pour them into the sink.

CHEMICALS AND EQUIPMENT

1. Spectroline pipet filler
2. 10-mL volumetric pipet
3. Solid wood block
4. Aluminum
5. Iron
6. Lead
7. Tin
8. Zinc
9. Hexane
10. Ethanol
11. Ethylene glycol
12. Milk
13. Olive oil
14. Sea water

Separation of the Components of a Mixture

BACKGROUND

Mixtures are not unique to chemistry; we use and consume them on a daily basis. The beverages we drink each morning, the fuel we use in our automobiles, and the ground we walk on are mixtures. Very few materials we encounter are pure. Any material made up of two or more substances that are not chemically combined is a mixture.

The isolation of pure components of a mixture requires the separation of one component from another. Chemists have developed techniques for doing this. These methods take advantage of the differences in physical properties of the components. The techniques to be demonstrated in this laboratory are the following:

1. *Sublimation*. This involves heating a solid until it passes directly from the solid phase into the gaseous phase. The reverse process, when the vapor goes back to the solid phase without a liquid state in between, is called condensation or deposition. Some solids that sublime are iodine, caffeine, and *para*-dichlorobenzene (mothballs).
2. *Extraction*. This uses a solvent to selectively dissolve one or more components from a solid mixture. With this technique, a soluble solid can be separated from an insoluble solid.
3. *Decantation*. This separates a liquid from an insoluble solid sediment by carefully pouring the liquid from the solid without disturbing the solid (Figure 3.1).
4. *Filtration*. This separates a solid from a liquid through the use of a porous material as a filter. Paper, charcoal, or sand can serve as a filter. These materials allow the liquid to pass through but not the solid (see Figure 3.4 in the **Procedure** section).
5. *Evaporation*. This is the process of heating a mixture in order to drive off, in the form of vapor, a volatile liquid, so as to make the remaining component dry.

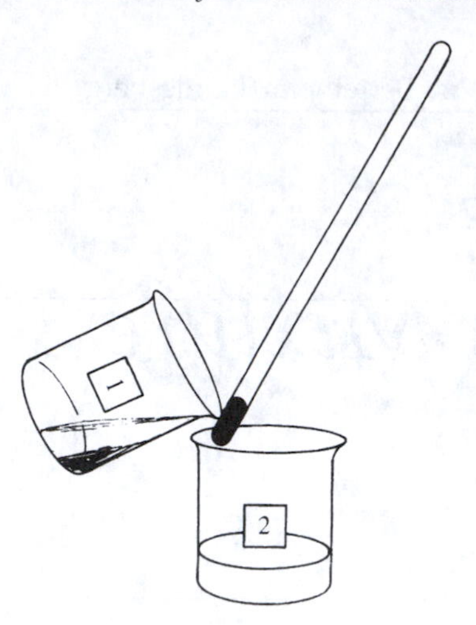

Figure 3.1
Decantation.

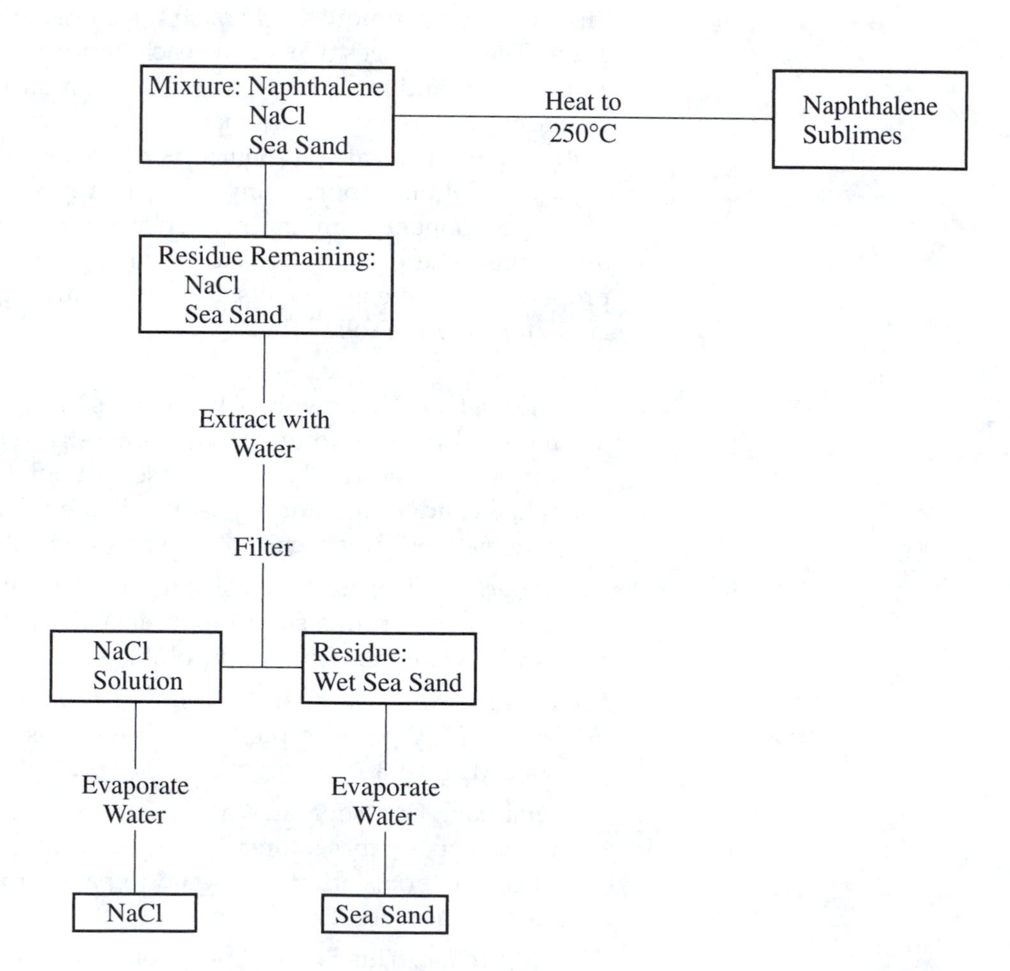

Figure 3.2
Separation scheme.

The mixture that will be separated in this experiment contains three components: naphthalene, $C_{10}H_8$, common table salt, NaCl, and sea sand, SiO_2. The separation will be done according to the scheme in Figure 3.2 by

1. heating the mixture to sublime the naphthalene,
2. dissolving the table salt with water to extract, and
3. evaporating water to recover dry NaCl and sand.

OBJECTIVES

1. To demonstrate the separation of a mixture.
2. To examine some techniques for separation using physical methods.

PROCEDURE

1. Obtain a clean, dry 150-mL beaker and carefully weigh it to the nearest 0.001 g. Record this weight for beaker 1 on the Report Sheet (1). Obtain a sample of the unknown mixture from your instructor; use a mortar and pestle to grind the mixture into a fine powder. With the beaker still on the balance, carefully transfer approximately 2 g of the unknown mixture into the beaker. Record the weight of the beaker with the contents to the nearest 0.001 g (2). Calculate the exact sample weight by subtraction (3).
2. Set up and do the sublimation in the hood. Place an evaporating dish on top of the beaker containing the mixture. Place the beaker and evaporating dish on a wire gauze with an iron ring and ring stand assembly as shown in Figure 3.3. Place ice in the evaporating dish, being careful not to get any water on the underside of the evaporating dish or inside the beaker.
3. Carefully heat the beaker with a Bunsen burner, increasing the intensity of the flame until vapors appear in the beaker. A solid should collect on the underside of the evaporating dish. After 10 min. of heating, remove the Bunsen burner from under the beaker. Carefully remove the evaporating dish from the beaker and collect the solid by scraping it off the dish with a spatula onto a weighing paper. Drain away any water from the evaporating dish and add ice to it, if necessary. Stir the contents of the beaker with a glass rod. Return the evaporating dish to the beaker and apply the heat again. Continue heating and scraping off solid until no more solid collects. Weigh all

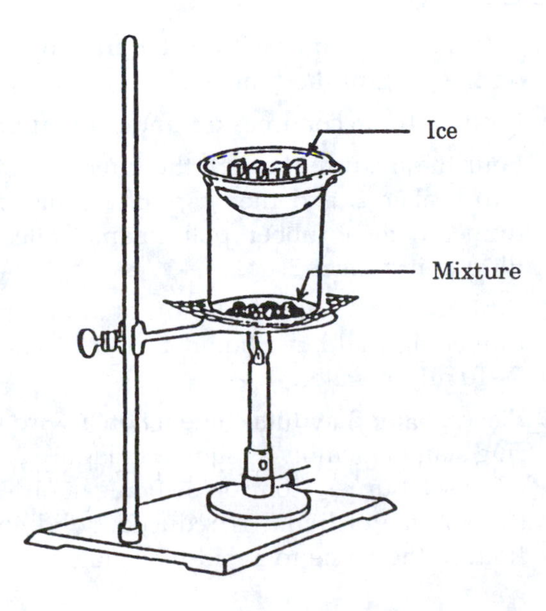

Figure 3.3
Assembly for sublimation.

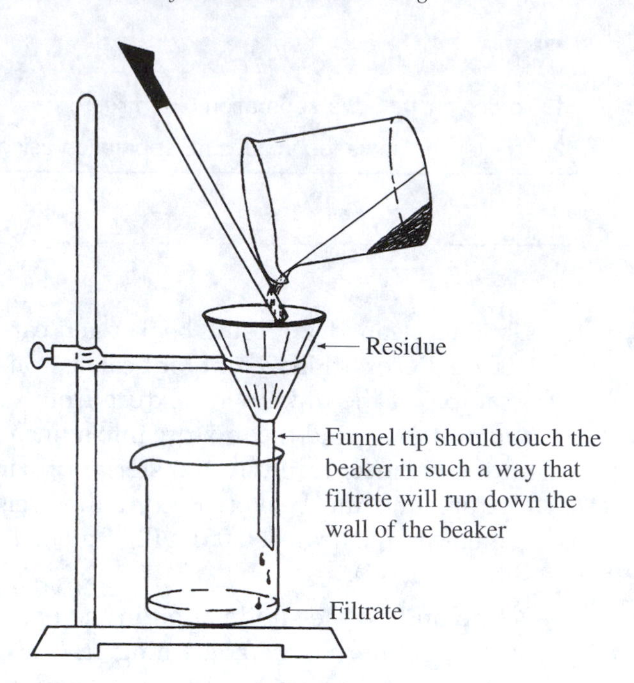
Residue
Funnel tip should touch the beaker in such a way that filtrate will run down the wall of the beaker
Filtrate

Figure 3.4
Gravity filtration.

the naphthalene collected and record on the Report Sheet to the nearest 0.001 g (4). Discard the naphthalene into a special container provided by your instructor.

4. Allow the beaker to cool until it reaches room temperature. Weigh the beaker with the contained solid (5a). Calculate the weight of the naphthalene that sublimed (5b).
5. Add 25 mL of distilled water to the solid in the beaker. Heat gently and stir continuously for 5 min. Do not let the mixture boil because it will "bump" and spatter the solid.
6. Weigh a second clean, dry 150-mL beaker with 2 or 3 boiling chips, to the nearest 0.001 g (6).
7. Assemble the apparatus for gravity filtration as shown in Figure 3.4.
8. Fold a piece of filter paper following the technique shown in Figure 3.5.
9. Wet the filter paper with water and adjust the paper so that it lies flat on the glass of the funnel.
10. Position the second beaker under the funnel.
11. Pour the mixture through the filter, first decanting most of the liquid into beaker 2, and then carefully transferring the wet solid into the funnel with a rubber policeman. Collect all the liquid (called the filtrate) in beaker 2.
12. Rinse beaker 1 with 5–10 mL of water, pour over the residue in the funnel, and add the liquid to the filtrate; repeat with an additional 5–10 mL of water.
13. Place beaker 2 and its contents on a wire gauze with an iron ring and ring stand assembly as shown in Figure 3.6a. Begin to heat gently with a Bunsen burner. Control the flame in order to prevent boiling over. As the volume of liquid is reduced, solid sodium chloride will appear. Reduce the flame to avoid bumping of the solution and spattering of

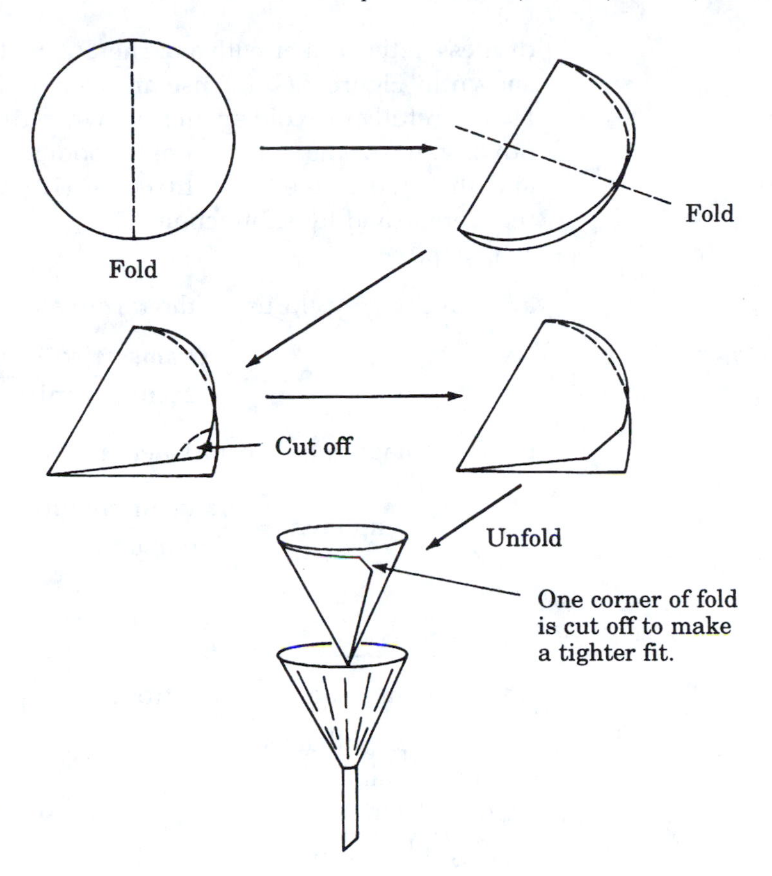

Figure 3.5
Steps for folding a filter paper for gravity filtration.

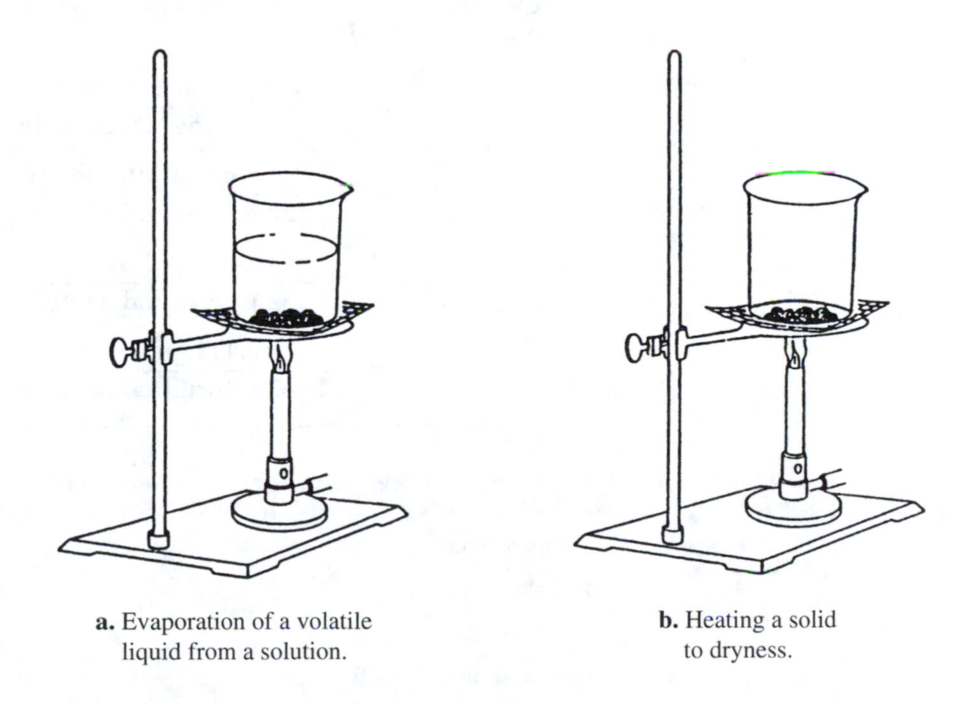

Figure 3.6
Assembly for evaporation.

a. Evaporation of a volatile liquid from a solution.

b. Heating a solid to dryness.

the solid. When all of the liquid is gone, cool the beaker to room temperature. Weigh the beaker, chips, and the solid residue to the nearest 0.001 g (7). Calculate the weight of the recovered NaCl by subtraction (8).

14. Carefully weigh a third clean, dry 150-mL beaker to the nearest 0.001 g (9). Transfer the sand from the filter paper to beaker 3. Heat the sand to dryness in the beaker with a burner, using the ring stand and assembly shown in Figure 3.6b (or use an oven at T = 90–100°C, if available). Heat carefully to avoid spattering; when dry, the sand should be freely flowing. Allow the sand to cool to room temperature. Weigh the beaker and the sand to the nearest 0.001 g (10). Calculate the weight of the recovered sand by subtraction (11).

15. Calculate

a. Percentage yield using the formula:

$$\%\ \textbf{yield} = \frac{\textbf{grams of solid recovered}}{\textbf{grams of initial sample}} \times 100$$

b. Percentage of each component in the mixture by using the formula:

$$\%\ \textbf{component} = \frac{\textbf{grams of component isolated}}{\textbf{grams of initial sample}} \times 100$$

Example

A student isolated the following from a sample of 1.132 g:

0.170 g of naphthalene
0.443 g of NaCl
0.499 g of sand
1.112 g solid recovered

The student calculated the percentage yield and percentage of each component as follows:

$$\%\ \textbf{yield} = \frac{\textbf{1.112 g (solid recovered)}}{\textbf{1.132 g (original sample)}} \times 100 = 98.2\%$$

$$\%\ C_{10}H_8 = \frac{\textbf{0.170 g (naphthalene)}}{\textbf{1.132 g (original sample)}} \times 100 = 15.0\%$$

$$\%\ \textbf{NaCl} = \frac{\textbf{0.443 g (NaCl)}}{\textbf{1.132 g (original sample)}} \times 100 = 39.1\%$$

$$\%\ \textbf{sand} = \frac{\textbf{0.499 g (sand)}}{\textbf{1.132 g (original sample)}} \times 100 = 44.1\%$$

CHEMICALS AND EQUIPMENT

1. Unknown mixture
2. Balances
3. Boiling chips
4. Evaporating dish, 6 cm
5. Filter paper, 15 cm
6. Mortar and pestle
7. Oven (if available)
8. Ring stands (3)
9. Rubber policeman

Resolution of a Mixture by Distillation

BACKGROUND

Distillation is one of the most common methods of purifying a liquid. It is a very simple method: a liquid is brought to a boil, the liquid becomes a gas, the gas condenses and returns to the liquid state, and the liquid is collected.

Everyone has had an opportunity to heat water to a boil. As heat is applied, water molecules increase their kinetic energy. Some molecules acquire sufficient energy to escape from the liquid phase and enter into the vapor phase. Evaporation occurs this way. The vapor above the liquid exerts a pressure, called the vapor pressure. As more and more molecules obtain enough energy to escape into the vapor phase, the vapor pressure of these molecules increases. Eventually the vapor pressure equals the pressure exerted externally on the liquid (this external pressure usually is caused by the atmosphere). Boiling occurs when this condition is met, and the temperature where this occurs is called the boiling point.

In distillation, the process described is carried out in an enclosed system, such as is illustrated in Figure 4.2. The liquid in the boiling flask is heated to a boil, and the vapor rises through tubing. The vapor then travels into a tube cooled by water, which serves as a condenser, where the vapor returns to the liquid state. If the mixture has a low-boiling component (a volatile substance with a high vapor pressure), it will distill first and can be collected. Higher-boiling and nonvolatile components (substances with low vapor pressure) remain in the boiling flask. Only by applying more heat will the higher-boiling component be distilled. Nonvolatile substances will not distill. For example, pure or "distilled" water for steam irons or car batteries is prepared this way.

Normal distillations, procedures carried out at atmospheric pressure, require "normal" boiling points. However, when boiling takes place in a closed system, it is possible to change the boiling point of the liquid by changing the pressure in the closed system. If the external pressure is reduced, usually by using a vacuum pump or a water aspirator, the boiling point of the liquid is reduced. Thus, heat-sensitive liquids, some of which

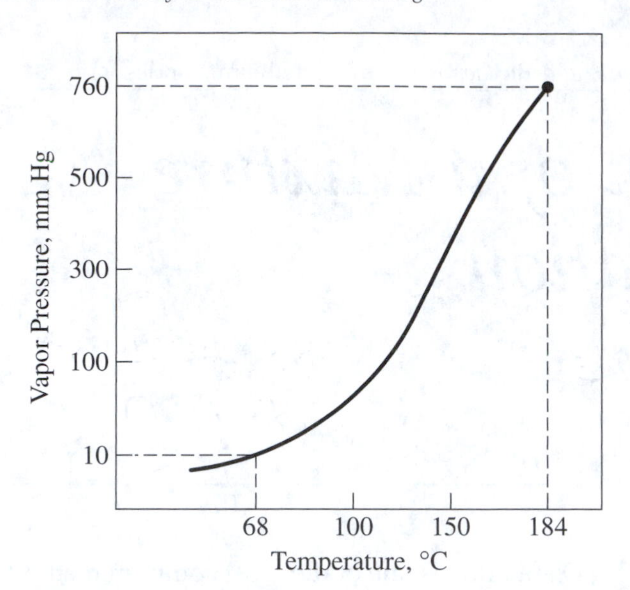

Figure 4.1
Temperature–vapor pressure curve for aniline.

decompose when boiled at atmospheric pressure, distill with minimum decomposition at reduced pressure and temperature. The relation of temperature to vapor pressure for the organic compound aniline can be shown by the curve in Figure 4.1. The organic liquid aniline, $C_6H_5NH_2$, a compound used to make synthetic dyes, can be distilled at 184°C (760 mm Hg) or at 68°C (10 mm Hg).

OBJECTIVES

1. To use distillation to separate a mixture.
2. To show that distillation can purify a liquid.

PROCEDURE

1. In this experiment a salt–water mixture will be separated by distillation. The volatile water will be separated from the nonvolatile salt (sodium chloride, NaCl). The purity of the collected distilled water will be demonstrated by chemical tests specific for sodium ions (Na^+) and chloride ions (Cl^-).
2. Assemble an apparatus as illustrated in Figure 4.2. A kit containing the necessary glassware can be obtained from your instructor. The glassware contains standard taper joints, which allow for quick assembly and disassembly. Before fitting the pieces together, apply a light coating of silicone grease to each joint to prevent the joints from sticking.
3. Use 100-mL round-bottom flasks for the boiling flask and the receiving flask. Fill the boiling flask with 50 mL of the prepared salt–water mixture. Add two boiling stones to the boiling flask to ensure smooth boiling of the mixture and to prevent bumping. Be sure that the rubber tubing to the condenser enters the lower opening and empties out of the upper opening. Turn on the water faucet and allow the water to fill

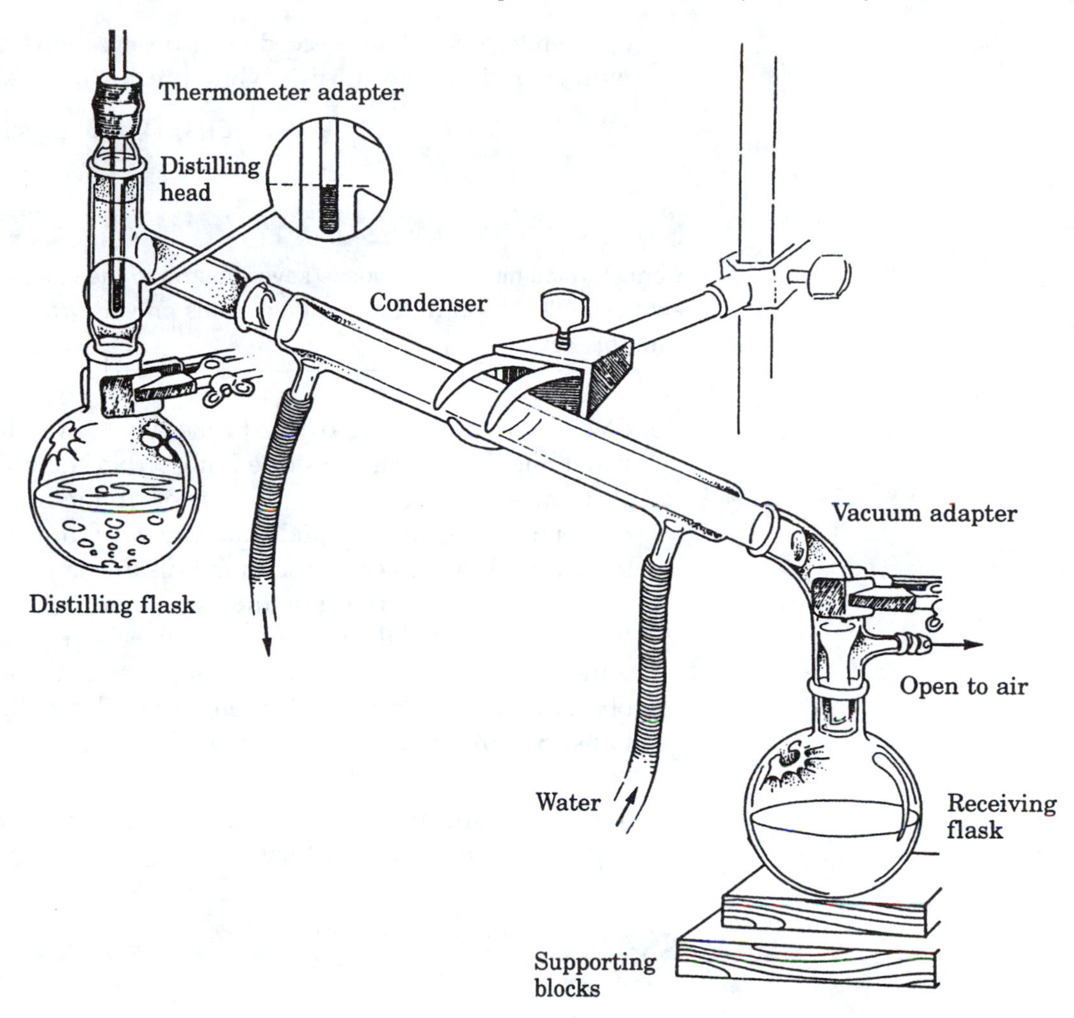

Figure 4.2
A distillation apparatus.

the jacket of the condenser slowly, so as not to trap air. Take care not to provide too much flow, otherwise the hoses will disconnect from the condenser. Adjust the bulb of the thermometer to below the junction of the condenser and the distillation column. **Be sure that the opening of the vacuum adapter is open to the atmosphere.**

4. Gently heat the boiling flask with a Bunsen burner. Eventually the liquid will boil, vapors will rise and enter the condenser, and liquid will recondense and be collected in the receiving flask.
5. Discard the first 1 mL of water collected. Record the temperature of the vapors as soon as the 1 mL of water has been collected. Continue collection of the distilled water until approximately one-half of the mixture has distilled. Record the temperature of the vapors at this point. Turn off the Bunsen burner and allow the system to return to room temperature.
6. The distilled water and the liquid in the boiling flask will be tested.
7. Place in separate clean, dry test tubes (100 × 13 mm) 2 mL of distilled water and 2 mL of the residue liquid from the boiling flask. Add to each sample 5 drops of silver nitrate solution. Look for the appearance

of a white precipitate. Record your observations. Silver ions combine with chloride ions to form a white precipitate of silver chloride.

$$Ag^{+} + Cl^{-} \longrightarrow AgCl(s)\ (\textbf{White precipitate})$$

CAUTION

Concentrated nitric acid causes severe burns to the skin. Handle this acid carefully. Flush with water if any spills on you. Wear gloves when working with this acid.

8. Place in separate clean, dry test tubes (100 × 13 mm) 2 mL of distilled water and 2 mL of the residue liquid from the boiling flask. Obtain a clean nickel wire from your instructor. In the hood, dip the wire into concentrated nitric acid and hold the wire in a Bunsen burner flame until the yellow color in the flame disappears. Dip the wire into the distilled water sample. Put the wire into the Bunsen burner flame. Record the color of the flame. Repeat the above procedure, cleaning the wire, dipping the wire into the liquid from the boiling flask, and observing the color of the Bunsen burner flame. Record your observations. Sodium ions produce a bright yellow flame with a Bunsen burner.
9. Make sure you wipe the grease from the joints before washing the glassware used in the distillation.

CHEMICALS AND EQUIPMENT

1. Boiling stones
2. Bunsen burner
3. Clamps
4. Distillation kit
5. Silicone grease
6. Thermometer
7. Nickel wire
8. Concentrated nitric acid, HNO_3
9. Salt–water mixture
10. 0.5 M silver nitrate, 0.5 M $AgNO_3$

The Empirical Formula of a Compound: The Law of Constant Composition

BACKGROUND

One of the most important fundamental observations in chemistry is summarized as the **Law of Constant (or Definite) Composition:** any pure chemical compound is made up of two or more elements in the same proportion by mass. In addition, it does not matter where the compound is found. Consider water from your kitchen tap and water from the Pacific Ocean: both are composed of the same elements, hydrogen and oxygen, and are found in exactly the same proportion—89% oxygen and 11% hydrogen—by weight. We also know that the compound water is composed of 2 atoms of hydrogen and 1 atom of oxygen and has the formula H_2O. If we consider that the mass of oxygen is 16 times the mass of hydrogen, water will always be found to contain 89% oxygen and 11% hydrogen.

For example, we can find the exact percentages for water and verify the above numbers. Using the **molecular formula** (the actual number of atoms in each molecule of a compound) of water, H_2O, the **gram molecular weight** or **molar mass** (the weight in grams of 1 mole of a compound) can be calculated:

$$\begin{aligned} \mathbf{2H} &= \mathbf{2 \times 1.008} = \mathbf{2.016} \\ \mathbf{1O} &= \mathbf{1 \times 15.999} = \underline{\mathbf{15.999}} \\ & \qquad\qquad\quad \mathbf{18.015 = 18.015\ g/mole} \end{aligned}$$

The percent composition of each element in water can then be calculated:

$$\%\mathbf{H} = \frac{\mathbf{2.016}}{\mathbf{18.015}} \times \mathbf{100 = 11.19 = 11\%}$$

$$\%\mathbf{O} = \frac{\mathbf{15.999}}{\mathbf{18.015}} \times \mathbf{100 = 88.81 = 89\%}$$

These values are constant and are never found in any other proportion!

The **empirical formula** (the simplest whole number ratio of atoms in a compound) is experimentally the simplest formula of a compound that can be found. For water, the formula, H_2O, is both the empirical and the molecular formula. Some other examples are carbon dioxide gas, CO_2; methane gas, CH_4; and hydrogen chloride gas, HCl. However, for the compound benzene, while the molecular formula is C_6H_6, the empirical formula is CH. Another example is the sugar found in honey, fructose: the molecular formula is $C_6H_{12}O_6$ and the empirical formula is CH_2O.

The empirical formula of a compound can be determined in a laboratory experiment by finding the ratio between the number of moles of the elements in the compound. The number of moles of each element can be calculated from the experimental values of the weights in which the elements combine by dividing by their corresponding atomic weights. If the molecular weight and the empirical formula of the compound are known, then the molecular formula of the compound can be determined.

Method

In this experiment we will verify that the empirical formula of copper(II) chloride is $CuCl_2$, and in so doing, demonstrate the Law of Constant Composition. We will do this by reducing a known weight of copper(II) chloride with aluminum to elemental copper. The reaction is shown by the following equation:

$$\mathbf{3CuCl_2 + 2Al \rightarrow 3Cu + 2AlCl_3} \qquad (1)$$

From the weight of $CuCl_2$, and the weight of Cu, subtraction will give the weight of Cl. From these weights, the mole ratio of copper to chlorine, the empirical formula, and the percent composition of $CuCl_2$ can then be calculated.

Example

Copper(II) chloride, 5.503 g, is reduced by excess aluminum and gives elemental copper, 2.603 g, according to equation (1). Using this data, the following calculations can be made:

1. Weight of chlorine in $CuCl_2$: (5.503 g $CuCl_2$) − (2.603 g Cu) = 2.900 g Cl.
2. Moles of Cu: (2.603 g Cu) $\times \left(\frac{1\text{ mole Cu}}{63.55\text{ g Cu}}\right)$ = 0.04100 mole Cu.
3. Moles of Cl: (2.900 g Cl) $\times \left(\frac{1\text{ mole Cl}}{35.45\text{ g Cl}}\right)$ = 0.08181 mole Cl.
4. Mole ratio of Cu to Cl: 0.04100 : 0.08181.
5. Simplest whole number ratio of Cu to Cl:

$$\frac{0.04100}{0.04100} : \frac{0.08181}{0.04100} = 1:2$$

6. The empirical formula for copper(II) chloride is $CuCl_2$.
7. %Cu in sample from data: $\frac{2.603\text{ g}}{5.503\text{ g}} \times 100 = 47.30\%$.

The theoretical calculated value of %Cu in $CuCl_2$, using the atomic masses, is 47.27%.

OBJECTIVES

1. To calculate the percent composition of an element in a compound.
2. To verify the empirical formula of copper(II) chloride.
3. To illustrate the Law of Constant Composition.

PROCEDURE

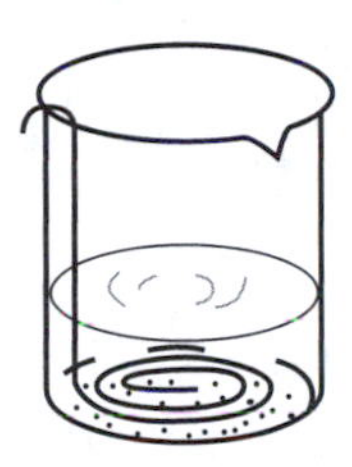

Figure 5.1
Suspension of aluminum wire in $CuCl_2$ solution.

1. Weigh out between 5 and 6 g of $CuCl_2$; record the weight to the nearest 0.001 g on your Report Sheet (1). Do not weigh directly on the balance pan, but be sure to use a container or weighing paper.
2. Transfer the $CuCl_2$ to a 250-mL beaker. Add 60 mL of distilled water and stir the contents with a glass stirring rod until the solid is completely dissolved.
3. Obtain a 45-cm length of aluminum wire (approx. 1.5 g). Make a flat coil on one end of the wire, and a handle at the other end. Make the handle long enough so that the wire can be hung over the side of the beaker. The coil must be covered by the solution and should reach the bottom of the beaker (Figure 5.1).
4. As the reaction proceeds, you will see flakes of brown copper accumulating on the wire. Occasionally shake the wire to loosen the copper. The disappearance of the initial blue color of the copper(II) ions indicates that the reaction is complete.
5. Test for the completion of the reaction.
   a. With a clean Pasteur pipet, place 10 drops of the supernatant solution into a clean test tube (100 × 13 mm).
   b. Add 3 drops of 6 M aqueous ammonia to the test tube. If a dark blue solution appears, copper(II) ions are still present, and the solution should be heated to 60°C for 15 min. (Use a hot plate.)
6. When the supernatant no longer tests for Cu^{2+} ions, the reaction is complete. Shake the aluminum wire so that all the copper clinging to it will fall into the solution. With a wash bottle filled with distilled water, wash the aluminum wire to remove any remaining residual copper. Remove the unreacted aluminum wire from the solution and discard into a solid waste container provided by your instructor.
7. Set up a vacuum filtration apparatus as shown in Figure 5.2.
8. Weigh a filter paper that fits into the Büchner funnel to the nearest 0.001 g; record on your Report Sheet (2).
9. Moisten the filter paper with distilled water, turn on the water aspirator, and filter the copper through the Büchner funnel. With a rubber policeman move any residue left in the beaker to the Büchner funnel; then rinse down all the copper in the beaker with water from a wash bottle and transfer to the Büchner funnel. If filtrate is cloudy, refilter, slowly. Finally, wash the copper in the funnel with 30 mL of acetone (to speed up the drying process). Let the copper remain on the filter paper for 10 min. with the water running to further the drying process.

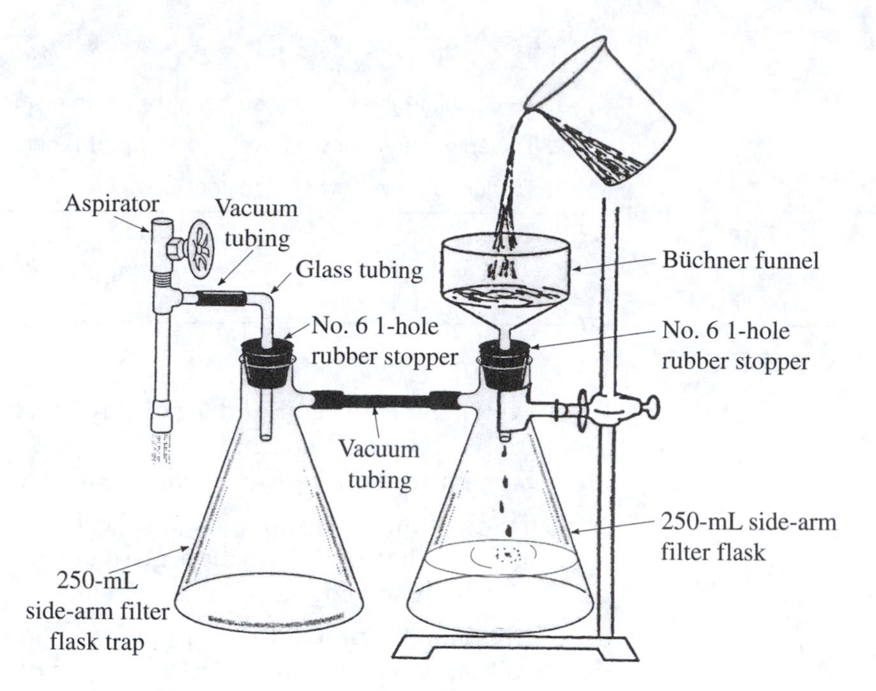

Figure 5.2
Vacuum filtration setup using the Büchner funnel.

10. Carefully remove the filter paper from the Büchner funnel so as not to tear the paper or lose any copper. Weigh the filter paper and the copper to the nearest 0.001 g and record on your Report Sheet (3). By subtraction obtain the weight of copper (4). From the weight of copper(II) chloride (1) and the weight of copper (4), the weight of chlorine can be calculated in the sample by subtraction (5).
11. From the experimental data, determine the empirical formula of copper(II) chloride, and the error in determining the percent of copper.

CHEMICALS AND EQUIPMENT

1. Aluminum wire (no. 18)
2. Acetone
3. 6 M aqueous ammonia, 6 M NH_3
4. Copper(II) chloride
5. Filter paper (Whatman no. 2, 7.0 cm)
6. Hot plate
7. Rubber policeman
8. Test tube (100 × 13 mm)
9. Pasteur pipets
10. Vacuum filtration setup
11. Wash bottle

name

section

date

partner

grade

5 EXPERIMENT 5

Pre-Lab Questions

1. Carbon dioxide, CO_2, is a common gas. It is found in the atmosphere, and we expel it when we breathe out in respiration. Determine: (a) the empirical formula; (b) the molecular formula; (c) the percent composition of carbon; (d) the percent composition of oxygen.

2. The molecular formula of *para*-dichlorobenzene, a compound found in some commercial mothballs, is $C_6H_4Cl_2$. What is the simple whole number ratio of the elements? What is the empirical formula?

3. Given the following molecular formulas, write the empirical formulas.

Molecular Formula	Empirical Formula
C_4H_{10} (butane)	________
$C_{10}H_8$ (naphthalene)	________
$C_6H_{12}O_6$ (glucose)	________

4. Calculate the percentage by weight of chloride ion (Cl^-) in common salt, NaCl. Show your work.

name | section | date

partner | grade

5 EXPERIMENT 5

Report Sheet

1. Weight of copper(II) chloride $CuCl_2$ ________ g
2. Weight of filter paper ________ g
3. Weight of filter paper and copper, Cu ________ g
4. Weight of Cu: (3) − (2) ________ g
5. Weight of Cl in sample: (1) − (4) ________ g
6. Molar mass of Cu ________ g
7. Molar mass of Cl ________ g
8. Number of moles of Cu atoms in sample: (4)/(6) ________ moles
9. Number of moles of Cl atoms in sample: (5)/(7) ________ moles
10. Mole ratio of Cu atoms to Cl atoms: (8) : (9) ________
11. Simple whole number mole ratio of Cu atoms to Cl atoms ________
12. Empirical formula for copper(II) chloride ________
13. Percentage of Cu in sample: % = [(4)/(1)] × 100 ________ %
14. Actual percentage of Cu in $CuCl_2$:

    $$\% = \frac{(6)}{(6) + [2 \times (7)]} \times 100$$ ________ %

15. Percentage error:

    $$\% = \frac{(14) - (13)}{(14)} \times 100$$ ________ %

Post-Lab Questions

1. Consider the following deviations from the described experimental procedure. How would they affect the accuracy of the percentage composition of copper?
   a. A student did not wait for the blue color to disappear but went ahead and collected whatever copper there was by vacuum filtration.

   b. Not all of the copper was transferred from the beaker to the Büchner funnel.

   c. The copper on the filter paper was not dry when a final weight was recorded.

2. For every mole of aluminum used in this reaction, how many moles of copper are formed? [*Hint:* See equation (1).]

3. In an experiment a student isolated 3.178 g of pure copper from an initial sample of 4.951 g of copper chloride. Determine the empirical formula of this copper chloride compound. Show your work.

4. In an experiment, 6.527 g of zinc chloride, $ZnCl_2$, yielded 2.935 g of zinc metal. Assuming that the original sample was pure, how much zinc should be obtained? Was all zinc possible from the sample recovered? Calculate the percentage recovery. (Your answer should be to the correct number of significant figures.)

Determination of the Formula of a Metal Oxide

BACKGROUND

Through the use of chemical symbols and numerical subscripts, the formula of a compound can be written. The simplest formula that may be written is the *empirical formula*. In this formula, the subscripts are in the form of the simplest whole number ratio of the atoms in a molecule or of the ions in a formula unit. The *molecular formula*, however, represents the actual number of atoms in a molecule. For example, although CH_2O represents the empirical formula of the sugar, glucose, $C_6H_{12}O_6$ represents the molecular formula. For water, H_2O, and carbon dioxide, CO_2, the empirical and the molecular formulas are the same. Ionic compounds are generally written as empirical formulas only; for example, common table salt is NaCl.

The formation of a compound from pure components is independent of the source of the material or of the method of preparation. If elements chemically react to form a compound, they always combine in definite proportions by weight. This concept is known as the *Law of Constant Composition*.

If the weight of each element that combines in an experiment is known, then the number of moles of each element can be determined. The empirical formula of the compound formed is the ratio between the number of moles of elements in the compound. This can be illustrated by the following example. If 32.06 grams of sulfur is burned in the presence of 32.00 grams of oxygen, then 64.06 grams of sulfur dioxide results. Thus

$$\frac{\textbf{32.06 g S}}{\textbf{32.06 g/mole S}} = \textbf{1 mole of sulfur}$$

$$\frac{\textbf{32.00 g O}}{\textbf{16.00 g/mole O}} = \textbf{2 moles of oxygen}$$

and the mole ratio of sulfur : oxygen is 1 : 2. The empirical formula of sulfur dioxide is SO_2. This also is the molecular formula.

In this experiment, the moderately reactive metal, magnesium, is combined with oxygen. The oxide, magnesium oxide, is formed. The equation for this reaction, based on the known chemical behavior, is

$$2Mg(s) + O_2(g) \xrightarrow{\text{heat}} 2MgO(s)$$

If the mass of the magnesium is known and the mass of the oxide is found in the experiment, the mass of the oxygen in the oxide can be calculated:

$$\begin{array}{l} \textbf{mass of magnesium oxide} \\ -\ \textbf{mass of magnesium} \\ \hline \textbf{mass of oxygen} \end{array}$$

As soon as the masses are known, the moles of each component can be calculated. The moles can then be expressed in a simple whole number ratio and an empirical formula written.

Example 1

When 2.43 g of magnesium was burned in oxygen, 4.03 g of magnesium oxide was produced.

$$\begin{array}{ll} \textbf{mass of magnesium oxide} & = 4.03\ \text{g} \\ -\ \textbf{mass of magnesium} & = 2.43\ \text{g} \\ \hline \textbf{mass of oxygen} & = 1.60\ \text{g} \end{array}$$

$$\text{No. of moles of magnesium} = \frac{2.43\ \text{g}}{24.31\ \text{g/mole}} = 0.100\ \text{moles}$$

$$\text{No. of moles of oxygen} = \frac{1.60\ \text{g}}{16.00\ \text{g/mole}} = 0.100\ \text{moles}$$

The molar ratio is 0.100 : 0.100 = 1 : 1
The empirical formula is Mg_1O_1 or MgO.

$$\%Mg = \frac{2.43\ \text{g}}{4.03\ \text{g}} \times 100 = 60.3\%$$

In the present experiment, magnesium metal is heated in air. Air is composed of approximately 78% nitrogen and 21% oxygen. A side reaction occurs between some of the magnesium and the nitrogen gas:

$$3Mg(s) + N_2(g) \xrightarrow{\text{heat}} Mg_3N_2(s)$$

Not all of the magnesium is converted into magnesium oxide; some becomes magnesium nitride. However, the magnesium nitride can be converted to magnesium oxide by the addition of water:

$$Mg_3N_2(s) + 3H_2O(l) \xrightarrow{\text{heat}} 3MgO(s) + 2NH_3(g)$$

As a result, all of the magnesium is transformed into magnesium oxide.

OBJECTIVES

1. To prepare a metal oxide.
2. To verify the empirical formula of a metal oxide.
3. To demonstrate the Law of Constant Composition.

PROCEDURE

CAUTION

A hot crucible can cause severe burns if handled improperly. Be sure to allow the crucible to cool sufficiently before handling. Always handle a hot crucible with crucible tongs.

Cleaning the Crucible

1. Obtain a porcelain crucible and cover. Carefully clean the crucible *in the hood* by adding 10 mL of 6 M HCl to the crucible; allow the crucible to stand for 5 min. with the acid. With crucible tongs, pick up the crucible, discard the HCl, and rinse the crucible with distilled water from a plastic squeeze bottle.
2. Place the crucible in a clay triangle, which is mounted on an iron ring and attached to a ring stand. Be sure the crucible is firmly in place in the triangle. Place the crucible cover on the crucible slightly ajar (Figure 6.1a).
3. Begin to heat the crucible with the aid of a Bunsen burner in order to evaporate water. Increase the heat, and, with the most intense flame (the tip of the inner blue cone), heat the crucible and cover for 5 min.; a cherry red color should appear when the bottom is heated strongly. Remove the flame. With tongs, remove the crucible to a heat-resistant surface and allow the crucible and cover to reach room temperature.
4. When cool, weigh the crucible and cover to 0.001 g (1). (Be sure to handle them with tongs because fingerprints leave a residue.)
5. Place the crucible and cover in the clay triangle again. Reheat the crucible to the cherry red color for 5 min. Allow the crucible and cover to cool to room temperature. Reweigh when cool (2). Compare weight (1) and weight (2). If the weight differs by more than 0.005 g, heat the crucible and cover again for 5 min. and reweigh when cool. Continue heating, cooling, and weighing until the weight of the crucible and cover are constant to within 0.005 g.

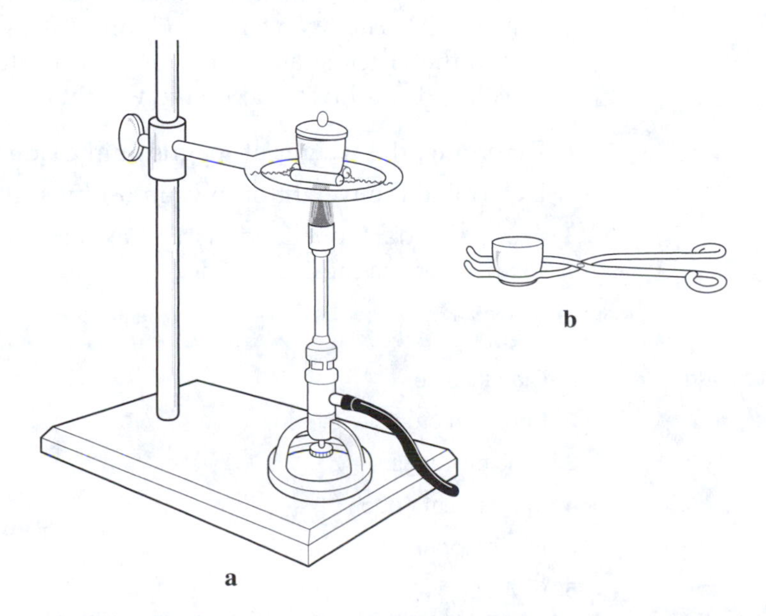

Figure 6.1
(a) Heating the crucible.
(b) Picking up the crucible with crucible tongs.

Forming the Oxide

1. Using forceps to handle the magnesium ribbon, cut a piece approximately 12 cm in length and fold the metal into a ball; transfer to the crucible. Weigh the crucible, cover, and magnesium to 0.001 g (3). Determine the weight of magnesium metal (4) by subtraction.
2. Transfer the crucible to the clay triangle; the cover should be slightly ajar (Figure 6.1a).
3. Using a small flame, gently apply heat to the crucible. Should fumes begin to appear, remove the heat and cover the crucible immediately. Again place the cover ajar and continue to gently heat for 10 min. (If fumes appear, cover as before.) Remove the flame and allow the assembly to cool for 2 min. With tongs, remove the cover. If the magnesium has been fully oxidized, the contents should be a dull gray. Shiny metal means there is still free metal present. The cover should be replaced as before and the crucible heated for an additional 5 min. Reexamine the metal and continue heating until no shiny metal surfaces are present.
4. When all the metal appears as the dull gray oxide, half-cover the crucible and gently heat with a small Bunsen flame. Over a period of 5 min., gradually adjust the intensity of the flame until it is at its hottest, then heat the crucible to the cherry red color for 5 min.

Completing the Reaction

1. Discontinue heating and allow the crucible assembly to cool to room temperature. Remove the cover and, with a glass rod, *carefully* break up the solid in the crucible. With 0.5 mL (10 drops) of distilled water dispensed from an eye dropper, wash the glass rod, adding the water to the crucible.
2. Set the cover ajar on the crucible and *gently* heat with a small Bunsen flame to evaporate the water. (Be careful to avoid spattering while heating; if spattering occurs, remove the heat and quickly cover the crucible completely.)
3. When all the water has been evaporated, half-cover the crucible and gradually increase to the hottest flame. Heat the crucible and the contents with the hottest flame for 10 min.
4. Allow the crucible assembly to cool to room temperature. Weigh the cool crucible, cover, and magnesium oxide to 0.001 g (5).
5. Return the crucible, cover, and magnesium oxide to the clay triangle. Heat at full heat of the Bunsen flame for 5 min. Allow to cool and then reweigh (6). The two weights, (5) and (6), must agree to within 0.005 g; if not, the crucible assembly must be heated for 5 min., cooled, and reweighed until two successive weights are within 0.005 g.

Calculations

1. Determine the weight of magnesium oxide (7) by subtraction.
2. Determine the weight of oxygen (8) by subtraction.
3. From the data obtained in the experiment, calculate the empirical formula of magnesium oxide.

CHEMICALS AND EQUIPMENT

1. Clay triangle
2. Porcelain crucible and cover
3. Crucible tongs
4. Magnesium ribbon
5. Eye dropper
6. 6 M HCl

name

section

date

partner

grade

EXPERIMENT 6

Pre-Lab Questions

1. Below are molecular formulas of selected organic compounds. Write the empirical formula for each:
   a. C_6H_6 (benzene)

   b. $C_2H_4O_2$ (acetic acid)

   c. $C_3H_6N_3O_9$ (nitroglycerin)

2. Calculate the molar mass of iron oxide, Fe_2O_3 (common rust), in grams. Show your work.

3. Calculate the percentage, by weight, of each of the elements (Fe, O) in iron oxide. Show your work.

name ______ section ______ date ______

partner ______ grade ______

6

EXPERIMENT 6

Report Sheet

1. Weight of crucible and cover (1) ______ g
2. Weight of crucible and cover (2) ______ g
3. Weight of crucible, cover, and Mg (3) ______ g
4. Weight of Mg metal (4): (3) – (2) ______ g
5. Weight of crucible, cover, and oxide (5) ______ g
6. Weight of crucible, cover, and oxide (6) ______ g
7. Weight of magnesium oxide (7): (6) – (2) ______ g
8. Weight of oxygen (8): (7) – (4) ______ g
9. Number of moles of magnesium
   (4)/24.31 g/mole ______ moles
10. Number of moles of oxygen
    (8)/16.00 g/mole ______ moles
11. Simplest whole number ratio of Mg atoms
    to O atoms = (9) : (10) ______ : ______
12. Empirical formula for magnesium oxide ______
13. % Mg in the oxide from data
    % = [(4)/(7)] × 100 ______ %
14. % Mg calculated from the formula MgO
    % = [24.31 g/40.31 g] × 100 ______ %
15. Error
    $$\% = \frac{(14) - (13)}{(14)} \times 100$$ ______ %

Post-Lab Questions

1. The white powder often found on objects made from aluminum is aluminum oxide, Al_2O_3. Write the balanced equation for the formation of aluminum oxide from the elements of aluminum, Al, and oxygen, O_2.

2. Why is water added to the crucible in order to complete the conversion to magnesium oxide?

3. During the experimental procedure you are cautioned not to allow "fumes" to escape from the crucible. What is the composition of the fumes? What error in calculation results if the fumes are allowed to escape?

4. Iron is obtained from the oxide in iron ore by heating with carbon in the form of coke. If a sample of the oxide, 10.78 g, is decomposed in this way, 8.378 g of iron is obtained. Determine the empirical formula for the iron oxide in this sample of ore. (*Hint:* First determine the number of moles of iron and oxygen in the sample.) Show your work.

Classes of Chemical Reactions

BACKGROUND

The Periodic Table shows over 100 elements. The chemical literature describes millions of compounds that are known—some isolated from natural sources, some synthesized by laboratory workers. The combination of chemicals, in the natural environment or the laboratory setting, involves chemical reactions. The change in the way that matter is composed is a *chemical reaction*, a process wherein reactants (or starting materials) are converted into products. The new products often have properties and characteristics that are entirely different from those of the starting materials.

Four ways in which chemical reactions may be classified are combination, decomposition, single replacement (substitution), and double replacement (metathesis).

Two elements reacting to form a compound is a *combination reaction*. This process may be described by the general formula:

$$\mathbf{A + B \rightarrow AB}$$

The rusting of iron or the combination of iron and sulfur are good examples.

$$\mathbf{4Fe(s) + 3O_2(g) \rightarrow 2Fe_2O_3(s)\ (rust)}$$
$$\mathbf{Fe(s) + S(s) \rightarrow FeS(s)}$$

Two compounds reacting together as in the example below also is a combination reaction.

$$\mathbf{CaO(s) + CO_2(g) \rightarrow CaCO_3(s)}$$

A compound that breaks down into elements or simpler components typifies the *decomposition reaction*. This reaction has the general formula:

$$\mathbf{AB \rightarrow A + B}$$

Table 7.1 *Activity Series of Common Metals*

K	(potassium)	Most active
Na	(sodium)	↑
Ca	(calcium)	
Mg	(magnesium)	
Al	(aluminum)	
Zn	(zinc)	
Fe	(iron)	Activity increases
Pb	(lead)	
H_2	(hydrogen)	
Cu	(copper)	
Hg	(mercury)	
Ag	(silver)	
Pt	(platinum)	
Au	(gold)	Least active

Some examples of this type of reaction are the electrolysis of water into hydrogen and oxygen:

$$\mathbf{2H_2O(l) \rightarrow 2H_2(g) + O_2(g)}$$

and the decomposition of potassium iodate into potassium iodide and oxygen:

$$\mathbf{2KIO_3(s) \rightarrow 2KI(s) + 3O_2(g)}$$

The replacement of one component in a compound by another describes the *single replacement* (or *substitution*) reaction. This reaction has the general formula:

$$\mathbf{AB + C \rightarrow CB + A}$$

Processes that involve oxidation (the loss of electrons or the gain of relative positive charge) and reduction (the gain of electrons or the loss of relative positive charge) are typical of these reactions. Use of Table 7.1, the activity series of common metals, enables chemists to predict which oxidation-reduction reactions are possible. A more active metal, one higher in the table, is able to displace a less active metal, one listed lower in the table, from its aqueous salt. Thus aluminum metal displaces copper metal from an aqueous solution of copper(II) chloride; but copper metal will not displace aluminum from an aqueous solution of aluminum(III) chloride.

$$\mathbf{2Al(s) + 3CuCl_2(aq) \rightarrow 3Cu(s) + 2AlCl_3(aq)}$$

$$\mathbf{Cu(s) + AlCl_3(aq) \rightarrow No\ Reaction}$$

(Note that Al is oxidized to Al^{3+} *and* Cu^{2+} *is reduced to Cu.)*

Hydrogen may be displaced from water by a very active metal. Alkali metals are particularly reactive with water, and the reaction of sodium with water often is exothermic enough to ignite the hydrogen gas released.

$$\mathbf{2Na(s) + 2HOH(l) \rightarrow 2NaOH(aq) + H_2(g) + heat}$$

(Note that Na is oxidized to Na^+ *and* H^+ *is reduced to* H_2*.)*

Active metals, those above hydrogen in the series, are capable of displacing hydrogen from aqueous mineral acids such as HCl or H_2SO_4; however, metals below hydrogen will not replace hydrogen. Thus, zinc reacts with aqueous solutions of HCl and H_2SO_4 to release hydrogen gas, but copper will not.

$$\mathbf{Zn(s) + 2HCl(aq) \rightarrow ZnCl_2(aq) + H_2(g)}$$
$$\mathbf{Cu(s) + H_2SO_4(aq) \rightarrow No\ Reaction}$$

Two compounds reacting with each other to form two different compounds describes *double replacement* (or *metathesis*). This process has the general formula:

$$\mathbf{AB + CD \rightarrow AD + CB}$$

There are two replacements in the sense that A replaces C in CD and C replaces A in AB. This type of reaction generally involves ions that form in solution either from the dissociation of ionic compounds or the ionization of molecular compounds. The reaction of an aqueous solution of silver nitrate with an aqueous solution of sodium chloride is a good example. The products are sodium nitrate and silver chloride. We know a reaction has taken place since the insoluble precipitate silver chloride forms and separates from solution.

$$\mathbf{AgNO_3(aq) + NaCl(aq) \rightarrow NaNO_3(aq) + AgCl(s)\ (White\ precipitate)}$$

In general, a double replacement results if one combination of ions leads to a precipitate, a gas or an un-ionized or very slightly ionized species such as water. In all of these reaction classes, it is very often possible to use your physical senses to observe whether a chemical reaction has occurred. The qualitative criteria may involve the formation of a gaseous product, the formation of a precipitate, a change in color, or a transfer of energy.

OBJECTIVES

1. To demonstrate the different types of chemical reactions.
2. To be able to observe whether a chemical reaction has taken place.
3. To use chemical equations to describe a chemical reaction.

PROCEDURE

Combination Reactions

1. Obtain a piece of aluminum foil approximately 2 × 0.5 in. Hold the foil at one end with a pair of forceps or crucible tongs and hold the other end in the hottest part of the flame of a Bunsen burner. Observe what happens to the foil. Record your observation and complete a balanced equation if you see that a reaction has occurred (1). Place the foil on a wire gauze to cool.
2. Obtain a piece of copper foil approximately 2 × 0.5 in. (A copper penny, one minted before 1982, may be substituted.) Hold the foil at one end with a pair of forceps or crucible tongs and hold the other end in the hottest part of the flame of a Bunsen burner. Observe what

happens to the metal. Record your observation and complete a balanced equation if you see that a reaction has occurred (2). Place the foil on a wire gauze to cool.

3. Scrape some of the gray solid from the surface of the aluminum obtained in step 1 into a test tube (100 × 13 mm). Add 1 mL of water and shake the test tube. Is the solid soluble? Record your observation (3).
4. Scrape some of the black solid from the surface of the copper obtained in step 2 into a test tube (100 × 13 mm). Add 1 mL of water and shake the test tube. Is the solid soluble? Record your observation (4).

Decomposition Reactions

1. *Decomposition of ammonium carbonate.* Place 0.5 g of ammonium carbonate into a clean, dry test tube (100 × 13 mm). Gently heat the test tube in the flame of a Bunsen burner (Figure 7.1). As you heat, hold a piece of wet red litmus paper with forceps at the mouth of the test tube. What happens to the solid? Are any gases produced? What happens to the color of the litmus paper? Ammonia gas acts as a base and turns moist red litmus paper blue. Record your observations and complete a balanced equation if you see that a reaction has occurred (5).

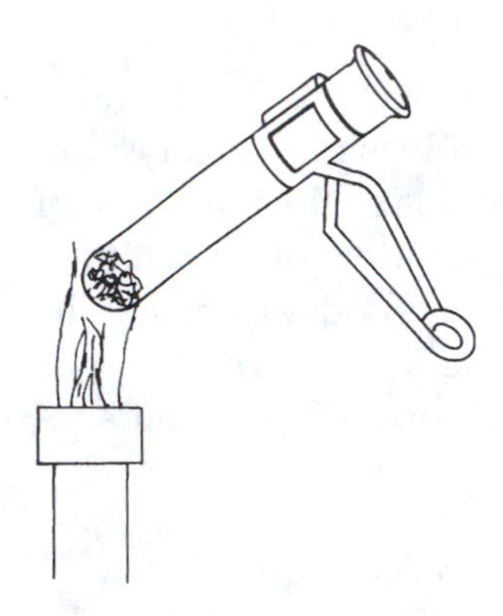

Figure 7.1
Position for holding a test tube in a Bunsen burner flame.

CAUTION

When heating the contents of a solid in a test tube, do not point the open end toward anyone.

2. *Decomposition of potassium iodate.*
   a. Obtain three clean, dry test tubes (100 × 13 mm). Label them and add 0.5 g of compound according to the table below.

Test Tube No.	*Compound*
1	KIO_3
2	KIO_3
3	KI

   b. Heat test tube no. 1 with the hottest part of the flame of the Bunsen burner as shown in Figure 7.2. Keep the test tube holder at the upper end of the test tube all the time. While test tube no. 1 is being

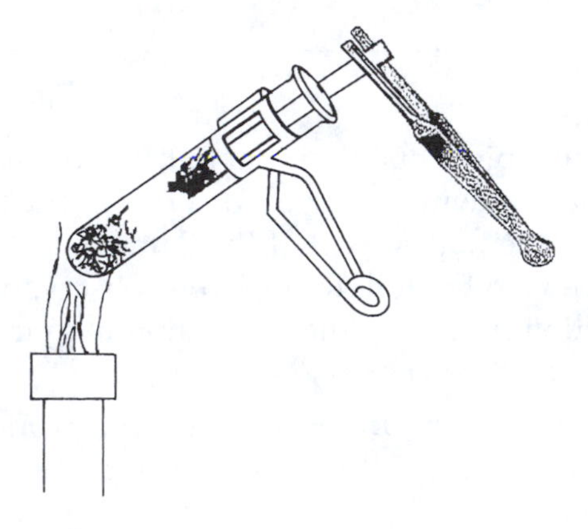

Figure 7.2
Testing for oxygen gas.

heated, thrust a glowing wooden splint about half way down the test tube (Figure 7.2). (The splint should not be burning but should be glowing with embers after the flame has been blown out. *Do not drop the glowing splint into the hot* KIO_3. Note that the wooden splint is held by forceps.) Oxygen supports combustion. The glowing splint should glow brighter or may burst into flame in the presence of oxygen. Record what happens to the glowing splint (6).

c. Remove the splint from the test tube. Remove the test tube from the flame and set it aside to cool.

d. Add 5 mL of distilled water to each of the three test tubes and mix thoroughly to ensure that the solids are completely dissolved. Add 10 drops of 0.1 M $AgNO_3$ solution to each test tube. Observe what happens to each solution. Record the colors of the precipitates and write complete balanced equations for the reactions taking place in test tube no. 2 and test tube no. 3. (The KIO_3 and KI solids can be distinguished by the test results with $AgNO_3$: AgI is a yellow precipitate; $AgIO_3$ is a white precipitate.) (7)

e. Does a precipitate result when the residue in test tube no. 1 is mixed with $AgNO_3$ solution (8)? What is the color of the precipitate? What compound is present in test tube no. 1 after heating KIO_3 (9)?

f. Write a complete balanced equation for the decomposition reaction (10).

Single Replacement Reactions

1. In a test tube rack, set up labeled test tubes (100 × 13 mm) numbered from 1 through 9. Place a small piece of metal into each test tube as outlined in the table below. Use forceps to handle the metal. Quickly add 1 mL (approx. 20 drops) of the appropriate solution to the test tubes, again as outlined in the table.

Test Tube No.	*Solution*	*Metal*
1	H_2O	Ca
2	H_2O	Mg
3	H_2O	Al
4	3 M HCl	Zn
5	6 M HCl	Pb
6	6 M HCl	Cu
7	0.1 M $NaNO_3$	Al
8	0.1 M $CuCl_2$	Al
9	0.1 M $AgNO_3$	Cu

2. Observe the mixtures over a 20-min. period of time. Note any color changes, any evolution of gases, any formation of precipitates, or any energy changes (hold each test tube in your hand and note whether the solution becomes warmer or colder) that occur during each

reaction; record your observations in the appropriate spaces on the Report Sheet (11). Write a complete and balanced equation for each reaction that occurred. For those cases where no reaction took place, write "No Reaction."

3. Dispose of the unreacted metals as directed by your instructor. *Do not discard them into the sink.*

Double Replacement Reaction

1. Each experiment in this part requires mixing equal volumes of two solutions in a test tube (100 × 13 mm). Use about 10 drops of each solution. Record your observation at the time of mixing (12). When there appears to be no evidence of a reaction, feel the test tube for an energy change (exothermic or endothermic). The solutions to be mixed are outlined in the table below.

Test Tube No.	*Solution No. 1*	*Solution No. 2*
1	0.1 M $NaCl$	0.1 M KNO_3
2	0.1 M $NaCl$	0.1 M $AgNO_3$
3	0.1 M Na_2CO_3	3 M HCl
4	3 M $NaOH$	3 M HCl
5	0.1 M $BaCl_2$	3 M H_2SO_4
6	0.1 M $Pb(NO_3)_2$	0.1 M K_2CrO_4
7	0.1 M $Fe(NO_3)_3$	3 M $NaOH$
8	0.1 M $Cu(NO_3)_2$	3 M $NaOH$

2. For those cases where a reaction occurred, write a complete and balanced equation. Indicate precipitates, gases, and color changes. Table 7.2 lists some insoluble salts. For those cases where no reaction took place, write "No Reaction."
3. Discard the solutions as directed by your instructor. *Do not discard them into the sink.*

Table 7.2 *Some Insoluble Salts*

$AgCl$	Silver chloride (white)
Ag_2CrO_4	Silver chromate (red)
$AgIO_3$	Silver iodate (white)
AgI	Silver iodide (yellow)
$BaSO_4$	Barium sulfate (white)
$Cu(OH)_2$	Copper(II) hydroxide (blue)
$Fe(OH)_3$	Iron(III) hydroxide (red)
$PbCrO_4$	Lead(II) chromate (yellow)
PbI_2	Lead(II) iodide (yellow)
$PbSO_4$	Lead(II) sulfate (white)

CHEMICALS AND EQUIPMENT

1. Aluminum foil
2. Aluminum wire
3. Copper foil
4. Copper wire
5. Ammonium carbonate, $(NH_4)_2CO_3$
6. Potassium iodate, KIO_3
7. Potassium iodide, KI
8. Calcium turnings
9. Magnesium ribbon
10. Mossy zinc
11. Lead shot
12. 3 M HCl
13. 6 M HCl
14. 3 M H_2SO_4
15. 3 M NaOH
16. 0.1 M $AgNO_3$
17. 0.1 M NaCl
18. 0.1 M $NaNO_3$
19. 0.1 M Na_2CO_3
20. 0.1 M KNO_3
21. 0.1 M K_2CrO_4
22. 0.1 M $BaCl_2$
23. 0.1 M $Cu(NO_3)_2$
24. 0.1 M $CuCl_2$
25. 0.1 M $Pb(NO_3)_2$
26. 0.1 M $Fe(NO_3)_3$

name ______ section ______ date ______

partner ______ grade ______

7 EXPERIMENT 7

Pre-Lab Questions

For each of the reactions below, classify as a combination, decomposition, single replacement, or double replacement.

1. $Ca(s) + Cl_2(g) \rightarrow CaCl_2(s)$ ______
2. $2Cu(s) + O_2(g) \rightarrow 2CuO(s)$ ______
3. $Ca(NO_3)_2(aq) + H_2SO_4(aq) \rightarrow 2HNO_3(aq) + CaSO_4(s)$ ______
4. $NH_3(aq) + HCl(aq) \rightarrow NH_4Cl(aq)$ ______
5. $Hg(NO_3)_2(aq) + 2NaI(aq) \rightarrow HgI_2(s) + 2NaNO_3(aq)$ ______
6. $AgNO_3(aq) + NaCl(aq) \rightarrow AgCl(s) + NaNO_3(aq)$ ______
7. $Zn(s) + H_2SO_4(aq) \rightarrow ZnSO_4(aq) + H_2(g)$ ______
8. $H_2CO_3(aq) \rightarrow CO_2(g) + H_2O(l)$ ______
9. $2H_2O(l) \rightarrow 2H_2(g) + 2O_2(g)$ ______
10. $2Li(s) + 2H_2O(l) \rightarrow 2LiOH(aq) + H_2(g)$ ______

name ______ section ______ date ______

partner ______ grade ______

7 EXPERIMENT 7

Report Sheet

Write *complete, balanced equations* for all cases in which a reaction takes place. Your observation that a reaction occurred would be by a color change, by the formation of a gas, by the formation of a precipitate, or by an energy change (exothermic or endothermic). In cases showing no evidence of a reaction, write "No Reaction."

Classes of chemical reactions

Combination reactions | *Observation*

1. ______ Al(s) + ______ $O_2(g) \rightarrow$ ______

2. ______ Cu(s) + ______ $O_2(g) \rightarrow$ ______

3. Solubility of aluminum oxide

4. Solubility of copper oxide

Decomposition reactions

5. ______ $(NH_4)_2CO_3(s) \rightarrow$ ______

6. What happens to the glowing splint? ______

7. ______ $KIO_3(aq)$ + ______ $AgNO_3(aq) \rightarrow$ ______

______ KI(aq) + ______ $AgNO_3(aq) \rightarrow$ ______

8. Results when the residue of KIO_3 is mixed with $AgNO_3$ solution ______

9. The formula of the residue present after heating KIO_3 ______

10. ______ $KIO_3(s) \xrightarrow{\text{heat}}$

Single replacement reactions *Observation*

11. *Test tube number*

1. _____ Ca(s) + _____ $H_2O(l)$ _______________

2. _____ Mg(s) + _____ $H_2O(l)$ _______________

3. _____ Al(s) + _____ $H_2O(l)$ _______________

4. _____ Zn(s) + _____ HCl(l) _______________

5. _____ Pb(s) + _____ HCl(l) _______________

6. _____ Cu(s) + _____ HCl(l) _______________

7. _____ Al(s) + _____ $NaNO_3(aq)$ _______________

8. _____ Al(s) + _____ $CuCl_2(aq)$ _______________

9. _____ Cu(s) + _____ $AgNO_3(aq)$ _______________

Double replacement reactions

12. *Test tube number*

1. _____ NaCl(aq) + _____ $KNO_3(aq)$ _______________

2. _____ NaCl(aq) + _____ $AgNO_3(aq)$ _______________

3. _____ $Na_2CO_3(aq)$ + _____ HCl(aq) _______________

4. _____ NaOH(aq) + _____ HCl(aq) _______________

5. _____ $BaCl_2(aq)$ + _____ $H_2SO_4(aq)$ _______________

6. _____ $Pb(NO_3)_2(aq)$ + _____ $K_2CrO_4(aq)$ _______________

7. _____ $Fe(NO_3)_3(aq)$ + _____ NaOH(aq) _______________

8. _____ $Cu(NO_3)_2(aq)$ + _____ NaOH(aq) _______________

Post-Lab Questions

1. Consider the following reactions. Determine whether a reaction will take place (yes) or not take place (no). Write balanced equations for those that will occur.

   a. $Ag + CuCl_2$

   b. $Mg + PbCl_2$

   c. $Hg + H_2SO_4$

   d. $Fe + MgCl_2$

2. For the list of chemicals below, write complete, balanced equations for all possible reactions that would lead to an insoluble salt. (Refer to Table 7.2.)

   KNO_3 $NaCl$ $AgNO_3$ $PbCl_2$ K_2SO_4 HNO_3

3. When ammonium carbonate [$(NH_4)_2CO_3$] decomposes, what would happen to a glowing splint in the presence of the gases that are evolved? Explain your answer.

4. What observations would indicate that a chemical reaction had occurred?

Chemical Properties of Consumer Products

BACKGROUND

Concern for the environment has placed considerable attention on the identification of chemicals that enter our everyday world. Analytical chemistry deals with these concerns in both a quantitative and qualitative sense. In *quantitative analysis*, the concern is for exact amounts of certain chemicals present in a sample; experiments in this manual will deal with this problem (for example, see Experiments 19, 20, 47, and 48). *Qualitative analysis* is limited to establishing the presence or absence of certain chemicals in detectable amounts in a sample. This experiment will focus on the qualitative determination of inorganic chemicals. Subsequent experiments in this manual will deal with organic chemicals.

The simplest approach to the detection of inorganic chemicals is to use tests that will identify the ions that make up the inorganic sample. These ions are cations and anions. *Cations* are ions that carry positive charges; Na^+, NH_4^+, Ca^{2+}, Cu^{2+}, and Al^{3+} are representative examples. *Anions* are ions that carry negative charges; Cl^-, HCO_3^-, CO_3^{2-}, SO_4^{2-}, and PO_4^{3-} are examples of this type. Because each ion has unique properties, each will give a characteristic reaction or test result. By examining an aqueous solution of the chemical, qualitative spot tests often will identify the cation and anion present. The tests used will bring about some chemical change. This change will be seen in the form of a solid precipitate, gas bubbles, or a color change.

This experiment will use chemicals commonly found around the house, so-called consumer chemical products. You may not think of these products as chemicals or refer to them by their inorganic chemical names. Nevertheless, they are chemicals, and simple qualitative analytical techniques can be used to identify the ions found in their makeup.

Table salt, NaCl. Table salt is most commonly used as a flavoring agent. Individuals with high blood pressure (hypertension) are advised to restrict salt intake in order to reduce the amount of sodium ion, Na^+, absorbed. When dissolved in water, table salt releases the sodium cation, Na^+, and the chloride anion, Cl^-. Chloride ion is detected by silver nitrate, $AgNO_3$; a characteristic white precipitate of silver chloride forms.

$$\mathbf{Ag^+(aq) + Cl^-(aq) \rightarrow AgCl(s)\ (White\ precipitate)}$$

Sodium ions produce a characteristic bright yellow color in a flame.

Ammonia, NH_3. Ammonia is a gas with a strong irritating odor. The gas dissolves readily in water, giving an aqueous ammonia solution; the solution is commonly referred to as ammonium hydroxide. Aqueous ammonia solutions are used as cleaning agents because of their ability to solubilize grease, oils, and waxes. Ammonia solutions are basic and will change moistened red litmus paper to blue. Ammonium salts (for example, ammonium chloride, NH_4Cl) react with strong bases to form ammonia gas.

$$\mathbf{NH_4^+(aq) + OH^-(aq) \rightarrow NH_3(g) + H_2O(l)}$$

Baking soda, sodium bicarbonate, $NaHCO_3$. Baking soda, sodium bicarbonate, $NaHCO_3$, acts as an antacid in some commercial products (e.g., Alka Seltzer) and as a leavening agent, helping to "raise" a cake. When sodium bicarbonate reacts with acids, carbon dioxide, a colorless, odorless gas, is released.

$$\mathbf{HCO_3^-(aq) + H^+(aq) \rightarrow CO_2(g) + H_2O(l)}$$

The presence of CO_2 can be confirmed with barium hydroxide solution, $Ba(OH)_2$; a white precipitate of barium carbonate results.

$$\mathbf{CO_2(g) + Ba(OH)_2(aq) \rightarrow H_2O(l) + BaCO_3(s)\ (White\ precipitate)}$$

Epsom salt, $MgSO_4 \cdot 7H_2O$. Epsom salt has several uses; it may be taken internally as a laxative or purgative, or it may be used externally as a solution for soaking one's feet. When dissolved in water, Epsom salt releases magnesium cations, Mg^{2+}, and sulfate anions, SO_4^{2-}. The magnesium cation may be detected by first treating with a strong base, such as NaOH, and then with the organic dye *p*-nitrobenzene-azoresorcinol. The magnesium hydroxide, $Mg(OH)_2$, which initially forms, combines with the dye to give a blue color. This behavior is specific for the magnesium cation.

$$\mathbf{Mg^{2+}(aq) + 2OH^-(aq) \rightarrow Mg(OH)_2(s) \xrightarrow{dye} Blue\ complex}$$

The sulfate anion, SO_4^{2-}, reacts with barium chloride, $BaCl_2$, to form a white precipitate of barium sulfate, $BaSO_4$.

$$\mathbf{Ba^{2+}(aq) + SO_4^{2-}(aq) \rightarrow BaSO_4(s)\ (White\ precipitate)}$$

Bleach, sodium hypochlorite, NaOCl. Bleach sold commercially is a dilute solution of sodium hypochlorite, NaOCl, usually 5% in concentration. The active agent is the hypochlorite anion. In solution, it behaves as if free chlorine, Cl_2, were present. Chlorine is an effective oxidizing agent. Thus in the presence of iodide salts, such as potassium iodide, KI, iodide anions are oxidized to iodine, I_2; chlorine is reduced to chloride anions, Cl^-.

$$\mathbf{Cl_2(aq) + 2I^-(aq) \rightarrow I_2(aq) + 2Cl^-(aq)}$$

The iodine gives a reddish-brown color to water. However, because iodine is more soluble in organic solvents, such as hexane, C_6H_{14}, the iodine dissolves in the organic solvent. The organic solvent separates from the water, and the iodine colors the organic solvent violet.

Sodium phosphate, Na_3PO_4. In some communities that use well water for their water supply, dissolved calcium and magnesium salts make the water "hard." Normal soaps do not work well as a result. In order to increase the efficiency of their products, especially in hard water areas, some commercial soap preparations, or detergents, contain sodium phosphate, Na_3PO_4. The phosphate anion is the active ingredient and keeps the calcium and magnesium ions from interfering with the soap's cleaning action. Other products containing phosphate salts are plant fertilizers; here, ammonium phosphate, $(NH_4)_3PO_4$, serves as the source of phosphorus. The presence of the phosphate anion can be detected with ammonium molybdate, $(NH_4)_2MoO_4$. In acid solution, phosphate anions combine with the molybdate reagent to form a bright yellow precipitate.

$$\mathbf{PO_4^{3-}(aq) + 12MoO_4^{2-}(aq) + 3NH_4^+(aq) + 24H^+(aq) \rightarrow (NH_4)_3PO_4(MoO_3)_{12}(s) + 12H_2O(l)}$$

(Yellow precipitate)

OBJECTIVES

1. To examine the chemical properties of some common substances found around the house.
2. To use spot tests to learn which inorganic cations and anions are found in these products.

PROCEDURE

CAUTION

Although we are using chemical substances common to our everyday life, conduct this experiment as you would any other. Wear safety glasses; do not taste anything; mix only those substances as directed.

Analysis of Table Salt, NaCl

1. Place a small amount (covering the tip of a small spatula) of table salt in a test tube (100 × 13 mm). Add 1 mL (approx. 20 drops) of distilled water and mix to dissolve. Add 2 drops of 0.1 M $AgNO_3$. Record your observation (1).
2. Take a small spatula and clean the tip by holding it in a Bunsen burner flame until the yellow color disappears. Allow to cool but do not let the tip touch anything. Place a few crystals of table salt on the clean spatula tip and heat in the flame of the Bunsen burner. Record your observation (2).

Analysis of Household Ammonia, NH_3, and Ammonium Ions, NH_4^+

1. Place 1 mL of household ammonia in a test tube (100 × 13 mm). Hold a piece of dry red litmus paper over the mouth of the test tube (be careful not to touch the glass with the paper). Record your observation (3). Moisten the red litmus paper with distilled water and hold it over the mouth of the test tube. Record your observation (4).
2. Place a small amount (covering the tip of a small spatula) of ammonium chloride, NH_4Cl, in a test tube (100 × 13 mm). Add 0.5 mL (about 10 drops) of 6 M NaOH to the test tube. Hold a moist piece of red litmus inside the mouth of the test tube (be careful not to touch the glass with

the paper). Does the litmus change color? If the litmus paper does not change color, gently warm the test tube (**do not boil the solution**). Record your observation (5).

3. Place a small amount (covering the tip of a small spatula) of commercial fertilizer in a test tube (100 × 13 mm). Add 0.5 mL (about 10 drops) of 6 M NaOH to the test tube. Test as above with moist red litmus paper. Record your observation and conclusion (6).

Analysis of Baking Soda, $NaHCO_3$

1. Place a small amount (covering the tip of a small spatula) of baking soda in a test tube (100 × 13 mm). Dissolve the solid in 1 mL (about 20 drops) of distilled water. Add 5 drops of 6 M H_2SO_4 and tap the test tube to mix. Record your observation (7).
2. We will test the escaping gas for CO_2. Prepare a mixture as in step 1, above. Place a small amount (covering the tip of a small spatula) of baking soda in a test tube (100 × 13 mm). Dissolve the solid in 1 mL (about 20 drops) of distilled water. Have the 5 drops of 6 M H_2SO_4 ready for use, but do not add it yet. Make a loop in a wire; the loop should be about 5 mm in diameter. Dip the wire loop into 5% barium hydroxide, $Ba(OH)_2$, solution; a drop should cling to the loop. Now you are ready to add the acid to the test tube. Add the acid to the test tube and then carefully lower the wire loop, with the drop of barium hydroxide on it, down into the mouth of the test tube. Avoid touching the walls and the solution. Record what happens to the drop (8).

Analysis of Epsom Salt, $MgSO_4 \cdot 7H_2O$

1. Place a small amount (covering the tip of a small spatula) of Epsom salt into a test tube (100 × 13 mm). Dissolve in 1 mL (about 20 drops) of distilled water. Add 5 drops of 6 M NaOH. Then add 5 drops of the "organic dye" solution (0.01% *p*-nitrobenzene-azoresorcinol). Record your observation (9).
2. Place a small amount (covering the tip of a small spatula) of Epsom salt into a test tube (100 × 13 mm). Dissolve in 1 mL (about 20 drops) of distilled water. Add 1 drop of 3 M HNO_3, followed by 2 drops of 1 M $BaCl_2$ solution. Record your observation (10).

Analysis of Bleach, NaOCl

Place a small amount (covering the tip of a small spatula) of potassium iodide, KI, in a test tube (100 × 13 mm). Dissolve in 1 mL (about 20 drops) of distilled water. Add 1 mL of bleach to the solution, followed by 10 drops of hexane, C_6H_{14}. Cork the test tube and shake vigorously. Set aside and allow the layers to separate. Note the color of the upper organic layer and record your observation (11).

Analysis of Sodium Phosphate, Na_3PO_4

Label three clean test tubes (100 × 13 mm) no. 1, no. 2, and no. 3. In test tube no. 1, place 2 mL of 1 M Na_3PO_4; in test tube no. 2, place a small amount (covering the tip of a small spatula) of a detergent; in test tube no. 3, place a small amount (covering the tip of a small spatula) of a fertilizer. Add 2 mL of distilled water to the solids in test tubes no. 2 and no. 3 and mix. Add 6 M HNO_3 dropwise to all three test tubes until the solutions test acid to litmus paper (blue litmus turns red when treated with acid). Mix each solution well and then add 10 drops of the $(NH_4)_2MoO_4$ reagent to each test tube. Warm the test tube in a water bath maintained at 60–70°C. Compare the three solutions and record your observations (12).

CHEMICALS AND EQUIPMENT

1. Bunsen burner
2. Copper wire
3. Litmus paper, blue
4. Litmus paper, red
5. Commercial ammonia solution, NH_3
6. Ammonium chloride, NH_4Cl
7. Commercial baking soda, $NaHCO_3$
8. Commercial bleach, $NaOCl$
9. Detergent, Na_3PO_4
10. Epsom salt, $MgSO_4 \cdot 7H_2O$
11. Garden fertilizer, $(NH_4)_3PO_4$
12. Table salt, $NaCl$
13. Ammonium molybdate reagent, $(NH_4)_2MoO_4$
14. 1 M $BaCl_2$
15. 5% $Ba(OH)_2$
16. 3 M HNO_3
17. 6 M HNO_3
18. Potassium iodide, KI
19. 0.1 M $AgNO_3$
20. 6 M $NaOH$
21. 1 M Na_3PO_4
22. 6 M H_2SO_4
23. 0.01% *p*-nitrobenzene-azoresorcinol ("organic dye" solution)
24. Hexane, C_6H_{14}

Angela Kerth

name ______________________ section ____________ date ____________

partner ______________________ grade ____________

EXPERIMENT 8

Pre-Lab Questions

1. Explain the difference between quantitative analysis and qualitative analysis. Are the tests carried out in this experiment quantitative or qualitative? Quantitative is the concern for exact amounts of certain chemicals present in a sample. Qualitative is establishing the presence/absence of certain chemicals in samples.

2. Below are some observations that were made when tests were carried out on solutions for specific ions. Based on the observed result, what is the most likely ion present in the solution? Write the chemical equation for the reaction of the reagent with the ion.

a. A white precipitate formed with barium chloride, $BaCl_2$.

$Ba^{2+}(aq) + SO_4^{2-}(aq) \rightarrow BaSO_4(s)$ white precip.

b. A white precipitate formed with silver nitrate, $AgNO_3$.

$Ag^+(aq) + Cl^-(aq) \rightarrow AgCl(s)$ (white precip.)

c. A solution treated with potassium iodide, KI, turned reddish-brown and produced a violet-colored layer with hexane.

$Cl_2(aq) + 2I^-(aq) \rightarrow I_2(aq) + 2Cl^-(aq)$

3. Below is a list of the materials to be analyzed. Complete the table by providing the name and formula of the salt found in each product, the name and formula of the cation, and the name and formula of the anion.

Product	Salt	Cation	Anion
1. Baking soda	$NaHCO_3$	Na^-	HCO^-
2. Bleach	$NaOCl$	Na^-	Cl^-
3. Detergent	NH_3	N^+	H^{-3}
4. Epsom salt	$MgSO_4 \cdot 7H_2O$	Mg^{2+}	SO_4^{2-}
5. Fertilizer	~~Na_3PO_4~~ NH_4PO_4	NH_4^+	PO_4^-
6. Table salt	$NaCl$	Na^+	Cl^-

✓

a.Ventn

name

section

date

megan Karnes

partner

grade

8

EXPERIMENT 8

Report Sheet

Analysis of table salt, NaCl

1. $AgNO_3 + NaCl$ white precipitate

2. Color of flame orange

Analysis of household ammonia, NH_3, and ammonium ions, NH_4^+

3. Color of dry litmus with ammonia fumes no change

4. Color of wet litmus with ammonia fumes turns blue

5. Color of wet litmus with $NH_4Cl + NaOH$ turns blue

6. Presence of ammonium ions in fertilizer turns slightly blue

Analysis of baking soda, $NaHCO_3$

7. $H_2SO_4 + NaHCO_3$ bubbles, fizzes foams - some precipitate

8. Presence of CO_2 gas fizzed, solid formed @ bottom

Analysis of Epsom salt, $MgSO_4 \cdot 7H_2O$

9. Presence of magnesium cation

cloudy, white percipitate forms @ bottom

organic dye- purple top
dark blue middle
bottom-white perc. solid

10. Presence of sulfate anion

turned white- cloudy cloudy

Analysis of bleach, NaOCl

11. Color of hexane layer

top layer- clear

Analysis of sodium phosphate, Na_3PO_4

12. Presence of phosphate

no. 1 clear

no. 2 cloudy w) white perc

no. 3 extremely yellow

Post-Lab Questions

1. How could you tell whether the white crystals you have are sodium chloride (table salt), $NaCl$, or sodium iodide (a poison), NaI?

2. Eggshells, oyster shells, limestone, and blackboard chalk contain calcium carbonate, $CaCO_3$. What test would determine the presence of carbonate in these substances? Write the chemical equation for the reaction.

3. A sample of tap water produces a white precipitate when treated with silver nitrate, $AgNO_3$. What ion is most likely responsible for this effect? What is the white precipitate?

4. Directions in some cake recipes call for the use of lemon juice or sour milk along with baking soda. Why do the directions call for these substances? What is the expected action when everything is mixed together?

5. In some communities, products that contain phosphates are banned from sale because of pollution of well water, streams, and coastal bays. How could these salts enter the environment? How could the authorities test for the presence of phosphates in the water?

Calorimetry: The Determination of the Specific Heat of a Metal

BACKGROUND

Any chemical or physical change involves a change in energy. Heat is a form of energy that can be observed as a flow of energy. Heat can pass spontaneously from an object at a high temperature to an object at a lower temperature. Two objects in contact at different temperatures, given enough time, will eventually reach the same temperature. The flow of heat energy can also be either into or out of a system under study.

The amount of heat can be measured in a device called a *calorimeter*. A calorimeter is a container with insulated walls. The insulation prevents rapid heat exchange between the contents of the calorimeter and the surroundings. In the closed environment of the system, there is no loss or gain of heat. Because the change in temperature of the contents of the calorimeter is used to measure the magnitude of the heat flow, a thermometer is included with the calorimeter.

The specific heat of any substance can be determined in a calorimeter. The *specific heat* is an intensive physical property of a substance and is the quantity of heat (in calories) necessary to raise the temperature of one gram of substance by one degree Celsius. The specific heats for some common substances are listed in Table 9.1. Notice that specific heat has the units calories per gram per degree Celsius. From Table 9.1, the specific heat of water is 1.00 cal/g °C; this means that it would take one calorie to raise the temperature of one gram of water by 1°C. In contrast, iron has a specific heat of 0.11 cal/g °C; it would take only 0.11 calorie to raise the temperature of one gram of iron by 1°C. Just by comparing these two substances, you can see that water is a convenient coolant and explains its use in the internal combustion engine of automobiles. A small quantity of water is capable of absorbing a relatively large amount of heat, yet shows only a modest rise in temperature.

Table 9.1 *Specific Heat Values for Some Common Substances*

Substance	*Specific Heat (cal/g °C)*	*Substance*	*Specific Heat (cal/g °C)*
Lead (Pb)	0.038	Glass (Flint)	0.12
Tin (Sn)	0.052	Table salt (NaCl)	0.21
Silver (Ag)	0.056	Aluminum (Al)	0.22
Copper (Cu)	0.092	Wood	0.42
Zinc (Zn)	0.093	Ethyl alcohol (C_2H_6O)	0.59
Iron (Fe)	0.11	Water (H_2O)	1.00

In general, when a given mass of a substance undergoes a temperature change, the heat energy required for the change is given by the equation

$$\textbf{Amount of Heat} = \mathbf{Q} = \mathbf{SH \times m \times \Delta T = SH \times m \times (T_{\mathit{final}} - T_{\mathit{initial}})}$$

where **Q** is the change in heat energy (amount of heat), **SH** is the specific heat of the substance, **m** is the mass of the substance in grams, and **ΔT** is the change in temperature (the difference between the *final* and *initial* temperatures); thus

$$\textbf{calories} = (\textbf{cal/g °C}) \times \textbf{g} \times \textbf{°C}$$

The specific heat of a metal can be found with a water calorimeter. This can be conveniently done by using the Principle of Conservation of Energy: *Energy can neither be created nor destroyed in any process, but can be transferred from one part of a system to another.* Experimentally, the amount of heat absorbed by a known mass of water can be measured when a known mass of hot metal is placed in the water. The temperature of the water will rise as the temperature of the metal falls. Using the known heat capacity of water, the amount of heat added to the water can be calculated, just as in Example 1. This is exactly the amount of heat given up by the metal.

Example 1

If 20 g of water is heated so that its temperature rises from 20° to 25°C, then we know that 100 cal have been absorbed.

$$\mathbf{Q = SH \times m \times \Delta T}$$

$$\mathbf{1.0 \frac{cal}{g\ °C} \times 20\ g \times (25 - 20)°C = 100\ cal}$$

Heat(cal) lost by metal = Heat(cal) gained by water

$$Q_{metal} = Q_{water}$$

$$\mathbf{m_m \times SH_m \times (\Delta T)_m = m_w \times SH_w \times (\Delta T)_w}$$

All the terms in the above equation are either known or can be determined experimentally, except for the value SH_m, the specific heat of the metal. The unknown can then be calculated.

$$\mathbf{SH_m = \frac{m_w \times SH_w \times (\Delta T)_w}{m_m \times (\Delta T)_m}}$$

Example 2

An unknown hot metal at 100.0°C with a mass of 50.03 g was mixed with 40.11 g of water at a temperature of 21.5°C. A final temperature of 30.6°C was reached. The heat gained by the water is calculated by

$$Q_w = 1.00 \frac{\text{cal}}{\text{g °C}} \times (40.11 \text{ g}) \times (30.6 - 21.5)\text{°C} = 365 \text{ cal}$$

The heat lost by the metal is equal to the heat gained by the water.

$$Q_m = Q_w = 365 \text{ cal}$$

The specific heat of the unknown metal is calculated to be

$$SH_m = \frac{365 \text{ cal}}{(50.03 \text{ g}) \times (100.0 - 30.6)\text{°C}} = 0.105 \frac{\text{cal}}{\text{g °C}}$$

The specific heat of iron is 0.11 cal/g °C; thus, from the value of SH_m determined experimentally, the unknown metal is iron.

If the specific heat of the metal is known, an approximate atomic mass can be determined. This can be done using the relationship between the specific heat of solid metallic objects and their atomic mass observed by Pierre Dulong and Alexis Petit in 1819; it is known as the Law of Dulong and Petit.

$$SH_m \times \text{Atomic Mass} = 6.3 \text{ cal/mole °C}$$

Example 3

The specific heat from Example 2 is 0.11 cal/g °C (to two significant figures). The approximate atomic mass is calculated to be

$$\text{Atomic Mass} = \frac{6.3 \text{ cal/mole °C}}{0.11 \text{ cal/g °C}} = 57 \text{ g/mole}$$

The atomic mass of iron is 56 g/mole (to two significant figures).

The calculations assume no heat is lost from the calorimeter to the surroundings and that the calorimeter absorbs a negligible amount of heat. However, this is not entirely correct. The calorimeter consists of the container, the stirrer, and the thermometer. All three get heated along with the water. As a result, the calorimeter absorbs heat. Therefore, the heat capacity for the calorimeter will be obtained experimentally, and the value derived applied whenever the calorimeter is used.

$$Q_{\text{calorimeter}} = C_{\text{calorimeter}} \Delta T$$

Example 4

The temperature of 50.0 mL of warm water is 36.9°C. The temperature of 50.0 mL of cold water in a calorimeter is 19.9°C. When the two were mixed together in the calorimeter, the temperature after mixing was 28.1°C. The heat capacity of the calorimeter is calculated as follows (assume the density of water is 1.00 g/mL):

The heat lost by the warm water is

$$(28.1 - 36.9)\text{°C} \times 50.0 \text{ g} \times 1.00 \text{ cal/g °C} = -440 \text{ cal}$$

The heat gained by the cold water is

$$(28.1 - 19.9)^\circ\text{C} \times 50.0 \text{ g} \times 1.00 \text{ cal/g }^\circ\text{C} = 410 \text{ cal}$$

The heat lost to the calorimeter is

$$-440 \text{ cal} + 410 \text{ cal} = -30 \text{ cal}$$

The heat capacity of the calorimeter is

$$\frac{30 \text{ cal}}{(28.1 - 19.9)^\circ\text{C}} = 3.7 \text{ cal/}^\circ\text{C}$$

In this experiment, you also will plot the water temperature in the calorimeter versus time. As the calorimeter walls and cover are not perfect insulators, some heat will be lost to the surroundings. In fact, when the hot water (or hot metal) is added to the colder water in the calorimeter, some heat will be lost before the maximum temperature is reached. In order to compensate for this loss, the maximum temperature is obtained by extrapolation of the curve as shown in Figure 9.2. This gives the maximum temperature rise that would have been recorded had there been no heat loss through the calorimeter walls. Once T_{max} is found, ΔT can be determined.

OBJECTIVES

1. To construct a simple calorimeter.
2. To measure the heat capacity of the calorimeter.
3. To measure the specific heat of a metal.

PROCEDURE

Determination of the Heat Capacity of the Calorimeter

1. Construct a calorimeter as shown in Figure 9.1. The two dry 8-oz. Styrofoam cups are inserted one into the other, supported in a 250-mL beaker. The plastic lid should fit tightly on the cup. With a suitable-sized cork borer, make two holes in the lid; one hole should be near the center for the thermometer and one hole to the side for the stirring wire. In order to keep the thermometer bulb 2 cm above the bottom

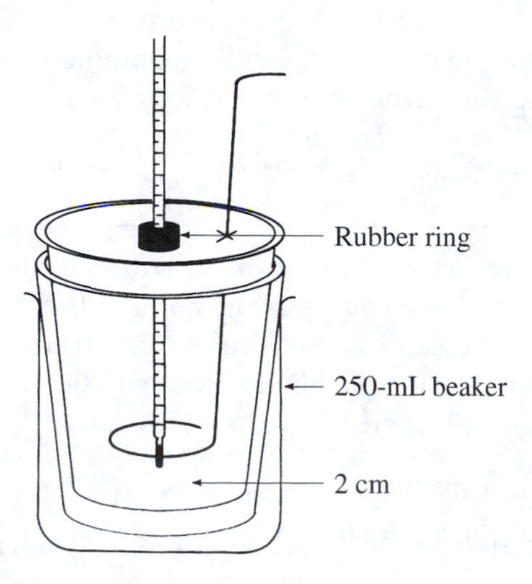

Figure 9.1
The Styrofoam calorimeter.

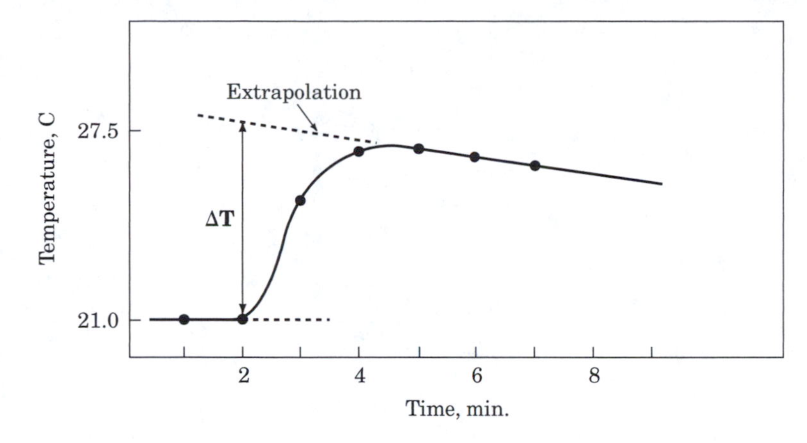

Figure 9.2
Plot of temperature vs. time.

of the inner cup, fit a rubber ring (cut from latex rubber tubing) around the thermometer and adjust the ring by moving it up or down the thermometer.

2. Because the density of water is nearly 1.00 g/mL over the temperature range for this experiment, the amount of water used in the calorimeter will be measured by volume. With a volumetric pipet, place 50.0 mL of cold water in the calorimeter cup; determine and record the mass (1). Cover the cup with the lid-thermometer-stirrer assembly. Stir the water for 5 min., observing the temperature during the time; record the temperature at 1-min. intervals on the Data Sheet. When the system is at equilibrium, record the temperature to the nearest 0.2°C (3).
3. With a volumetric pipet, place 50.0 mL of water in a clean, dry 150-mL beaker; determine and record the mass (2). Heat the water with a low flame until the temperature of the water is about 70°C. Allow the hot water to stand for a few minutes, stirring occasionally during this time period. Quickly record the temperature to the nearest 0.2°C (4) and pour the water completely into the calorimeter that has been assembled and has reached equilibrium (Figure 9.1).
4. Replace the cover assembly and stir the contents gently. Observe the temperature for 5 min. and record the temperature on the Data Sheet (p. 112) every 30 sec. during that 5-min. period. Plot the temperature as a function of time, as shown in Figure 9.2. (Use the graph paper on p. 113.) Determine from your curve the maximum temperature by extrapolation and record it (5). Determine the ΔT. From the data, calculate the heat capacity of the calorimeter according to the calculations on the Report Sheet (p. 109).

Determination of the Specific Heat of a Metal

1. Dry the Styrofoam cups used for the calorimeter calibration. Reassemble the apparatus as in Figure 9.1.
2. With a volumetric pipet, place 50.0 mL of cold water in the calorimeter cup; record the mass (1).
3. Obtain an unknown metal sample from your instructor. Record the number of the unknown on the Report Sheet (p. 110).
4. Weigh a clean, dry 50-mL beaker to the nearest 0.01 g (2). Place about 40 g of your unknown sample in the beaker and reweigh to the nearest 0.01 g (3). Determine the mass of the metal by subtraction (4). Pour the sample into a 150 × 25 mm clean, dry test tube.

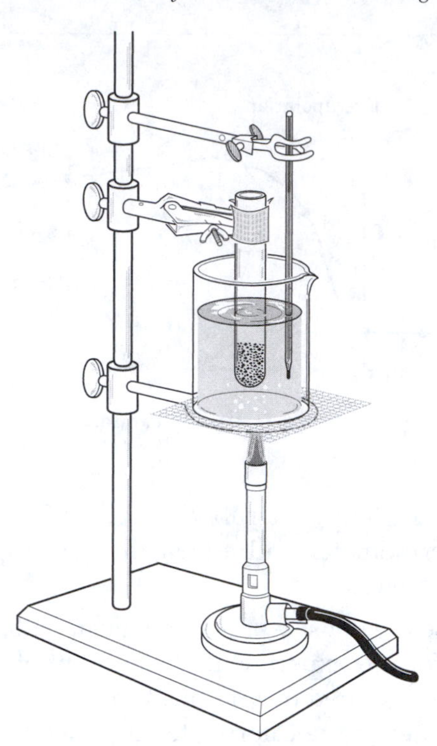

Figure 9.3
Assembly for heating the metal.

5. Place the test tube in the water bath as shown in Figure 9.3. Be sure that all of the metal in the test tube is below the surface of the water. Heat the water to a gentle boil and keep the test tube in the bath for 10 min. Make certain that water does not splash into the test tube.
6. While the metal is heating, follow the temperature of the cold water in the calorimeter for 5 min.; record the temperature on the Data Sheet at 1-min. intervals. After 5 min., record the temperature on the Report Sheet of the cold water to the nearest 0.2°C (5).
7. After 10 min. of heating the metal, observe and record the temperature on the Report Sheet of the boiling water in the beaker to the nearest 0.2°C (6). Obtain and use another thermometer for the calorimeter.
8. All steps must be done *quickly* and *carefully* at this point. Remove the test tube from the boiling water; dry the outside glass with a paper towel; remove the lid on the calorimeter; add the hot metal to the calorimeter. Be careful no water is added to nor lost from the calorimeter on the transfer.
9. Record the calorimeter temperature on the Data Sheet as soon as the apparatus has been reassembled. Note the time when the temperature is determined. Stir the water. Continue to follow the temperature, recording the temperature on the Data Sheet every 30 sec. for the next 5 min.
10. Plot the temperature as a function of time, as shown in Figure 9.2. (Use the graph paper on page 114.) Determine from your curve the maximum temperature; record the temperature on the Report Sheet (7). Determine the ΔT. From the data, determine the specific heat and the atomic mass of the metal.

CHEMICALS AND EQUIPMENT

1. Metal pellets
2. Styrofoam cups (2)
3. Lid for Styrofoam cups
4. Metal stirring loop
5. Thermometers, 110°C (2)
6. Latex rubber ring
7. Volumetric pipet, 50 mL
8. Thermometer clamp
9. Test tube, 150 × 25 mm

name section date

partner grade

9 EXPERIMENT 9

Pre-Lab Questions

1. Why can you use a calorimeter to study the exchange of heat?

2. Why is specific heat an intensive property?

3. Is specific heat a dimensionless quantity? If there are dimensions, what are they?

4. Walkways on a beach are usually made of wood rather than of a metal such as iron. Why (forget about cost)?

5. Which would be a better coolant in your car engine: pure antifreeze (ethylene glycol, SH = 0.57 cal/g °C) or water (SH = 1.00 cal/g °C)? Why?

name ______ section ______ date ______

partner ______ grade ______

9 EXPERIMENT 9

Report Sheet

Determination of the heat capacity of the calorimeter

1. Mass of the cold water

   50.0 mL × 1.00 g/mL ______ g

2. Mass of the warm water

   50.0 mL × 1.00 g/mL ______ g

3. Temperature of the equilibrated system: cold water and calorimeter ______ °C

4. Temperature of the warm water ______ °C

5. Maximum temperature from the graph ______ °C

6. ΔT of cold water and calorimeter

   (5) – (3) ______ °C

7. ΔT of warm water

   (5) – (4) ______ °C

8. Heat lost by warm water

   (2) × 1.00 cal/g °C × (7) ______ cal

9. Heat gained by cold water and the calorimeter: –(8) ______ cal

10. Heat gained by cold water

    (1) × 1.00 cal/g °C × (6) ______ cal

11. Heat gained by the calorimeter

    (9) – (10) ______ cal

12. Heat capacity of calorimeter, C_{cal}

    (11)/(6) ______ cal/°C

Determination of the specific heat of a metal

1. Mass of cold water

   50.0 mL × 1.00 g/mL ____________ g

2. Mass of 50-mL beaker ____________ g

3. Mass of the beaker plus metal ____________ g

4. Mass of metal: (3) – (2) ____________ g

5. Temperature of the equilibrated system ____________ °C

6. Temperature of hot metal
   (Temperature of boiling water) ____________ °C

7. Maximum temperature from the graph ____________ °C

8. ΔT of cold water and calorimeter

   (7) – (5) ____________ °C

9. Heat gained by the calorimeter and water

   $[(1) \times 1.00\ \text{cal/g °C} \times (8)] + [C_{cal} \times (8)]$ ____________ cal

10. ΔT of the metal

    (7) – (6) ____________ °C

11. Heat lost by the metal: –(9) ____________ cal

12. Specific heat of the metal

    $\frac{(11)}{(4) \times (10)}$ ____________ cal/g °C

13. Atomic mass

    $\frac{6.3\ \text{cal/mole °C}}{(12)}$ ____________ g/mole

Unknown number ____________ Metal unknown ____________

Post-Lab Questions

1. Would it be a good idea to use glass beakers for the calorimeter rather than the Styrofoam cups? Explain your answer (think about holding a cup of hot coffee from your favorite fast-food chain in a glass versus a Styrofoam cup).

2. In this experiment, why do you use the extrapolated value for the maximum temperature from the plot of temperature as a function of time? If, instead, you used the value of the water just after mixing with the metal as the maximum temperature, would the heat capacity of the metal, Q, be affected? How?

3. A 26.2-g sample of ethyl alcohol was heated from an initial temperature of 20.0°C (room temperature) to a final temperature of 37.5°C (body temperature). How many calories has the alcohol absorbed? If the same size sample of water was heated in the same temperature range, how many calories would the water absorb? Which absorbs more energy?

4. A 40.5-g sample of metal beads was heated in a water bath to 99.5°C. The metal was then added to a sample of water (31.6 g) and the temperature of the water changed from an initial temperature of 20.0°C to a final temperature of 37.5°C.

   a. What was the temperature of the metal at equilibrium?

   b. What was the temperature change of the metal?

   c. Calculate the specific heat and the atomic mass of the metal. What is the metal?

Temperature Data Sheet

Calorimeter Calibration		*Specific Heat Determination*	
Time (min.)	*Temp (°C)*	*Time (min.)*	*Temp (°C)*
0	______	0	______
1.0	______	1.0	______
2.0	______	2.0	______
3.0	______	3.0	______
4.0	______	4.0	______
5.0	______	5.0	______
5.5	______	5.5	______
6.0	______	6.0	______
6.5	______	6.5	______
7.0	______	7.0	______
7.5	______	7.5	______
8.0	______	8.0	______
8.5	______	8.5	______
9.0	______	9.0	______
9.5	______	9.5	______
10.0	______	10.0	______

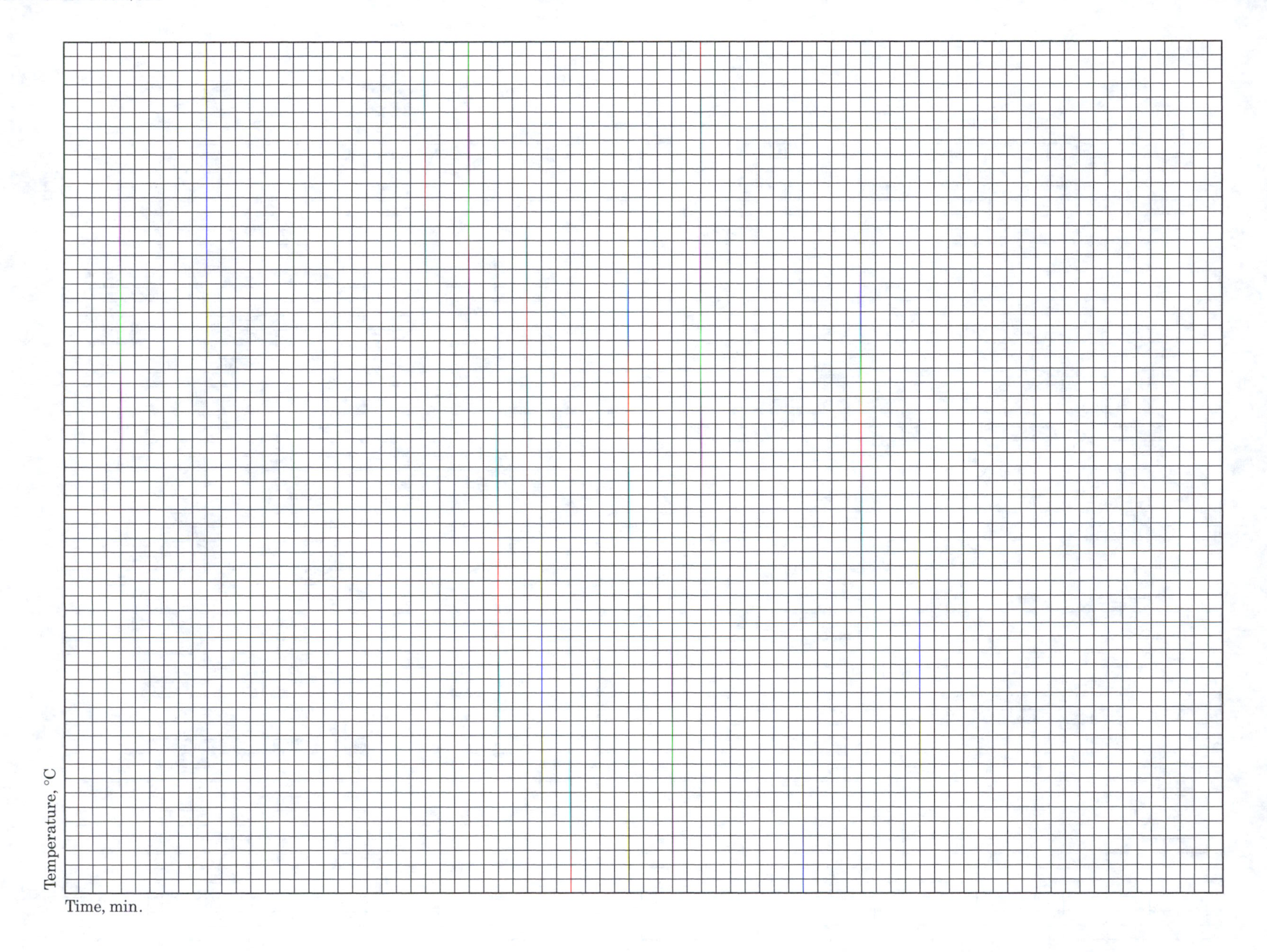
Temperature, °C
Time, min.

Time, min.
Temperature, °C

Boyle's Law: The Pressure–Volume Relationship of a Gas

BACKGROUND

The British scientist Robert Boyle made many contributions in the fields of medicine, astronomy, physics, and chemistry. However, he is best known for his work on the behavior of gases. In 1662, Boyle found that when the temperature is held constant, the pressure of a trapped amount of gas (any gas) is inversely proportional to its volume. That is, when the pressure of the gas increases, the volume of the gas decreases; when the pressure of the gas decreases, the volume of the gas increases. Boyle's Law can be written mathematically as follows:

$$\mathbf{V} = \mathbf{k} \times \frac{\mathbf{1}}{\mathbf{P}} \quad \text{or} \quad \mathbf{V} = \frac{\mathbf{k}}{\mathbf{P}} \quad \text{or} \quad \mathbf{PV} = \mathbf{k}$$

where **V** is the volume of the gas, **P** is the pressure of the gas, and **k** is a constant that depends on the temperature and amount of the gas. By looking at these equations, it is easy to see the inverse relationship. For example, if pressure on a sample of trapped gas is *doubled*, the volume of the sample will be reduced by *half* of the value it had been before the increase in pressure. On the other hand, if the pressure is reduced by *half*, the volume will *double*.

If there is a pressure change, for example from $\mathbf{P_1}$ to $\mathbf{P_2}$, then the volume also changes from $\mathbf{V_1}$ to $\mathbf{V_2}$. The relationship between the initial pressure and volume to the new pressure and volume can be expressed as follows.

$$\mathbf{P_1V_1} = \mathbf{P_2V_2} \quad \text{or} \quad \frac{\mathbf{P_1}}{\mathbf{P_2}} = \frac{\mathbf{V_2}}{\mathbf{V_1}}$$

Volume may be expressed in a variety of units—liter (L), milliliter (mL), cubic meter (m^3), or cubic centimeter (cc or cm^3). Similarly, pressure can be expressed in a variety of units, but the standard unit of pressure is the atmosphere (atm); one atmosphere (1 atm) is defined as the pressure needed to support a column of mercury 760 mm in height at 0°C at sea level. In honor of Evangelista Torricelli, the Italian inventor of the barometer, the unit **torr** is used and is equal to 1 mm Hg. Thus

$$\mathbf{1\ mm\ Hg = 1\ torr}$$
$$\mathbf{1\ atm = 760\ mm\ Hg = 760\ torr}$$

Breathing is a good example of how Boyle's Law works. We breathe as a result of the movements that take place in the diaphragm and the rib cage. When the diaphragm contracts (moves downward) and the rib cage is raised (expands), the volume of the chest cavity increases. This action decreases the pressure in the lungs and this pressure is lower than the outside pressure. The result: air flows from the outside, higher-pressure area into the lungs and they expand—we inhale. When we breathe out, the process is reversed: the diaphragm is relaxed (moves upward) and the rib cage is lowered (contracts). This decreases the volume of the chest cavity and increases the pressure inside the lungs. With pressure greater inside the lungs than the outside, air flows out and the lungs contract—we exhale.

In this experiment, a volume of air is trapped in a capillary tube by a column of mercury. The mercury acts as a movable piston. Depending on how the capillary tube is tilted, the mercury column moves and thus causes the volume of trapped air to change. The pressure of the trapped air supports not only the pressure exerted by the atmosphere but also the pressure of the mercury column. The pressure exerted by the mercury column varies depending on the angle of the tilt. If θ is the angle of the tilt, the pressure of the mercury column can be calculated by the following equation:

$$\mathbf{P_{Hg} = (length\ of\ Hg\ column) \times \sin\theta}$$

The total pressure of the trapped gas is the sum of the atmospheric pressure and the pressure due to the mercury column.

You need only measure the length of the column of air, L_{air}, because the length is directly related to the volume. The column of air is geometrically a regular cylinder. The radius of the cylinder, in this case the capillary tube, remains the same. The volume of a regular cylinder is a constant (πr^2) times the height of the cylinder (L_{air}); the only quantity that varies in this experiment is the value L_{air}.

OBJECTIVES

1. To show the validity of Boyle's Law.
2. To measure the volume of a fixed quantity of air as the pressure changes at constant temperature.

PROCEDURE*

CAUTION

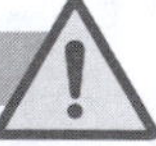

Mercury can be spilled easily. Although mercury has a low vapor pressure, its vapor is extremely toxic. Mercury can also be absorbed through the skin. If any mercury is spilled, notify the instructor immediately for proper clean-up.

*Adapted from R. A. Hermens, *J. Chem. Educ.* **60** (1983), 764.

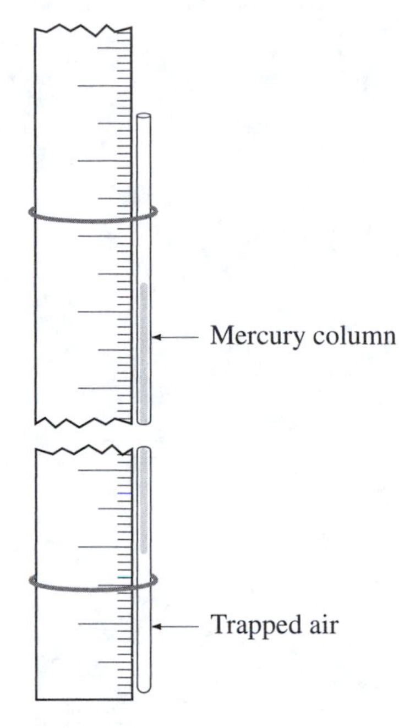

Figure 10.1
The Boyle's Law apparatus.

1. Obtain a Boyle's Law apparatus and a 30°-60°-90° plastic triangle. The Boyle's Law apparatus consists of a piece of glass tubing that contains a column of mercury and is attached to a ruler by means of rubber bands (Figure 10.1).
2. Record the temperature on the Report Sheet (1).
3. Record the barometric pressure, P_{at}, in mm Hg on the Report Sheet (2).
4. Measure the length of the column of mercury, L_{Hg}, to the nearest 0.5 mm; record this length on the Report Sheet (3).
5. The length of the column of trapped air is to be measured when the tube is at various angles to the bench top, as outlined in the following table.

Angle of Tube	*Position of Open End of Tube*
0°	Horizontal
90°	Up
90°	Down
60°	Up
60°	Down
30°	Up
30°	Down

The column length of the trapped air, L_{air}, is measured (from the glass seal to the mercury) to the nearest 0.5 mm and is recorded in the table on the Report Sheet (4). The correct angle can be obtained with the aid of the 30°-60°-90° triangle by placing the Boyle's Law apparatus along the appropriate edge.

CAUTION

Do not touch the glass tube during the measurements, to avoid any temperature changes. Do not jar the tube at any time. This will avoid separation or displacement of the mercury column when measurements have begun.

6. Calculate the reciprocal, $\frac{1}{L_{air}}$, and enter on the Report Sheet (5).
7. Using the appropriate formula from the table on the Report Sheet (6), calculate the pressure, P, of the column of air (7).
8. Plot the data on graph paper as follows: y-axis, the calculated pressure, P; x-axis, the reciprocal of the length of the trapped air, $\frac{1}{L_{air}}$.
9. Replot the data with P (y-axis) versus L_{air} (x-axis).

CHEMICALS AND EQUIPMENT

1. Boyle's Law apparatus
2. 30°-60°-90° plastic triangle

name

section

date

partner

grade

10 EXPERIMENT 10

Pre-Lab Questions

1. Write the mathematical expression for Boyle's Law and explain it in words.

2. As discussed in the **Background** section, as the pressure for a trapped gas changes from P_1 to P_2, the volume changes from V_1 to V_2. We then can write a formula for the results as follows: $P_1V_1 = P_2V_2$. Why can we write this equality?

3. Weather reports usually quote the barometric pressure as rising, falling, or steady. What causes the pressure and what is the standard unit of pressure?

4. Annapurna is a mountain in Nepal of height 26,504 ft. How does the barometric pressure at that height compare to the barometric pressure at your desk now?

5. A bicycle pump has a volume of air of 200 mL at 1 atm. When you push in the plunger, the volume decreases to 50 mL. What will be the pressure exerted by the trapped air (temperature is constant)?

name

partner

section

date

grade

10

EXPERIMENT 10

Report Sheet

1. Temperature ____________ °C

2. Barometric pressure, P_{at} ____________ mm Hg

3. Length of mercury column, L_{Hg} ____________ mm

Boyle's Law Data

Angle of Tube (opening)*	*(4) Length of Trapped Air, L_{air}, mm*	*(5)* $\frac{1}{L_{air}}$	*(6) Pressure (calculation)*	*(7) Pressure P*
0°			P_{at}	
90° (U)			$P_{at} + L_{Hg}$	
90° (D)			$P_{at} - L_{Hg}$	
60° (U)			$P_{at} + (L_{Hg} \sin 60°)$	
60° (D)			$P_{at} - (L_{Hg} \sin 60°)$	
30° (U)			$P_{at} + (L_{Hg} \sin 30°)$	
30° (D)			$P_{at} - (L_{Hg} \sin 30°)$	

*U = up; D = down

Post-Lab Questions

1. Make the following conversions:
   a. 2.0 atm = ___________ torr
   b. 720 mm Hg = ___________ atm
   c. 700 torr = ___________ mm Hg = ___________ atm

2. During the experiment, as measurements were being taken, you held the tube containing the mercury in your fingers. How would this affect the results?

3. Explain what happens when muscles cause the diaphragm in our chest cavity to be raised.

4. The sequence of diagrams below shows what is taking place in an automobile cylinder when the motor is running.
   a. How does this illustrate Boyle's Law?
   b. If in (a) the mixture of air and gasoline vapor occupies 200 mL at 1 atm, what would the pressure be in (c) if the volume is reduced to 10 mL?

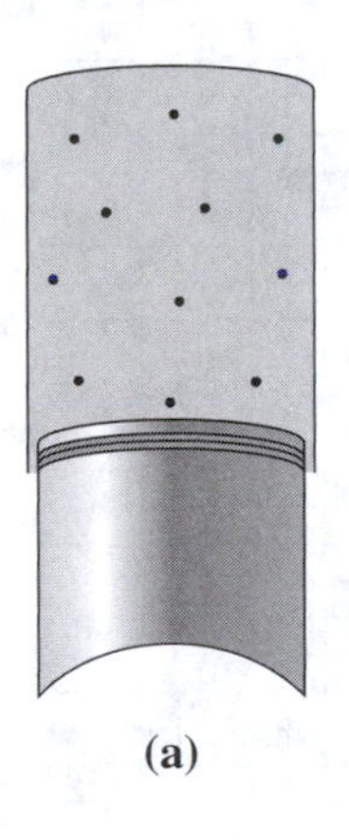

(a)

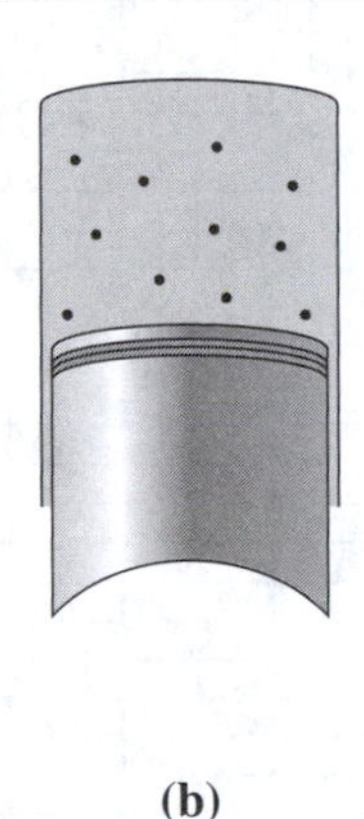

(b)

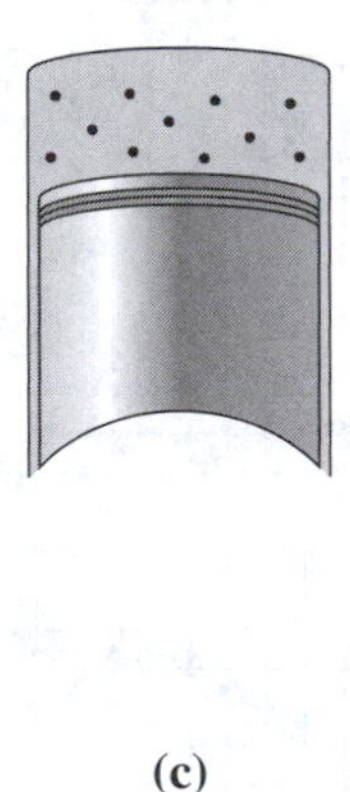

(c)

P (mm Hg)

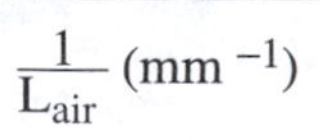

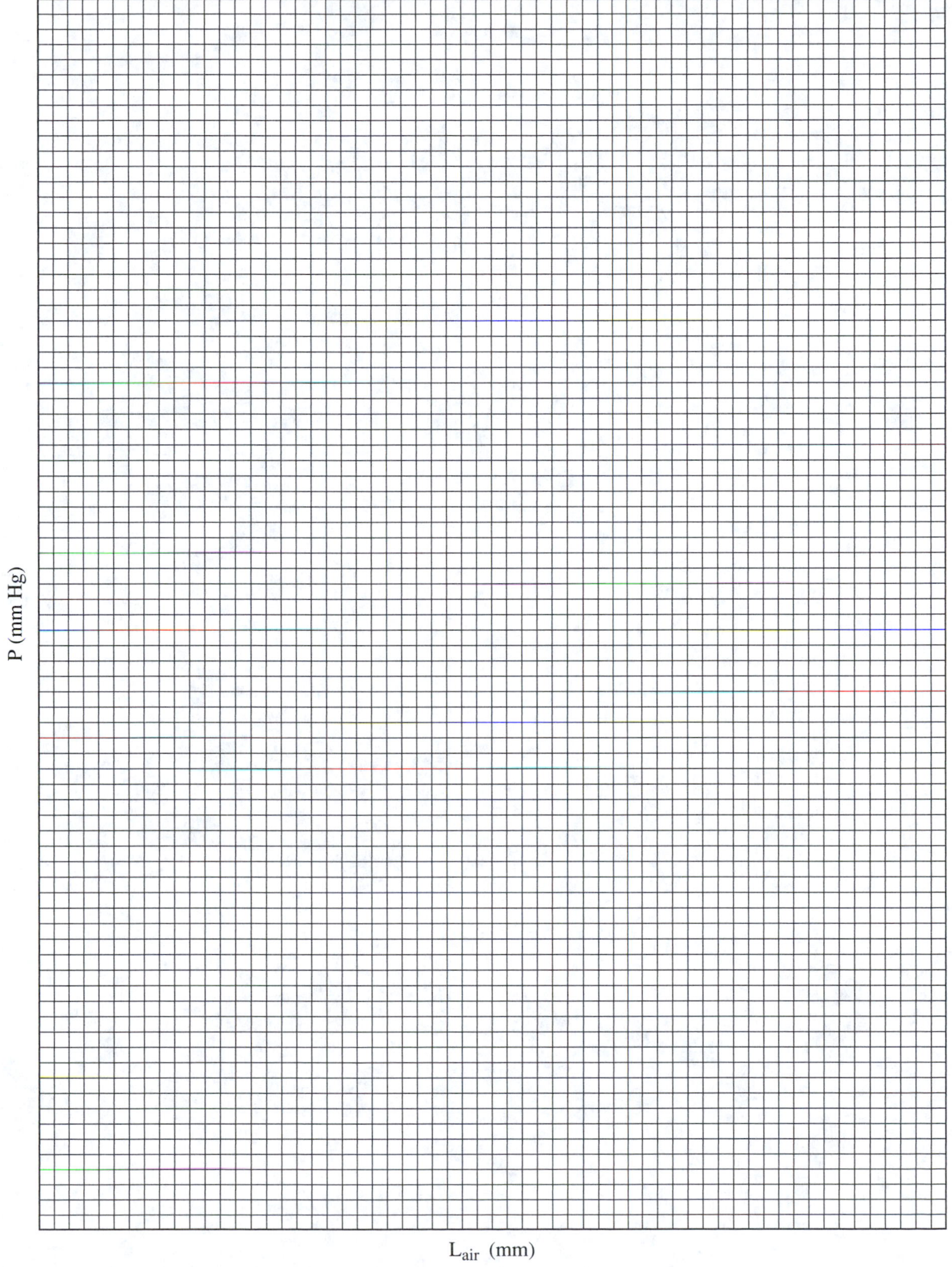
P (mm Hg)
L_{air} (mm)

Charles's Law: The Volume–Temperature Relationship of a Gas

BACKGROUND

Jacques Charles observed that for a fixed quantity of gas, the volume at constant pressure changes when temperature changes: the volume increases (V↑) when the temperature increases (T↑); the volume decreases (V↓) when the temperature decreases (T↓). Although first described by Charles in 1787, it was not until 1802 that Joseph Gay-Lussac expressed the relationship mathematically.

Charles's Law states that when the pressure is held constant, the volume of a fixed mass of ideal gas is in direct proportion to the temperature in degrees Kelvin. Charles's Law can be written mathematically as follows:

$$\mathbf{V} = \mathbf{k} \times \mathbf{T} \quad \text{or} \quad \frac{\mathbf{V}}{\mathbf{T}} = \mathbf{k} \qquad (1)$$

where **V** is the volume of the gas, **T** is the temperature in degrees Kelvin, and **k** is a constant that depends on the pressure and amount of gas. The direct relationship is clear by looking at the equations. If a sample of gas at a fixed pressure has its temperature *doubled,* the volume in turn is *doubled.* Conversely, decreasing the temperature by *one-half* brings about a decrease in volume of *one-half.*

The law applies, for a given pressure and quantity of gas, at all sets of conditions. Thus for two sets of **T** and **V**, the following can be written:

$$\frac{\mathbf{V_1}}{\mathbf{T_1}} = \frac{\mathbf{V_2}}{\mathbf{T_2}} \quad \text{or} \quad \mathbf{V_1T_2} = \mathbf{V_2T_1} \quad \text{or} \quad \frac{\mathbf{V_1T_2}}{\mathbf{V_2T_1}} = 1 \qquad (2)$$

where at constant pressure, $\mathbf{V_1}$ and $\mathbf{T_1}$ refer to the set of conditions at the beginning of the experiment, and $\mathbf{V_2}$ and $\mathbf{T_2}$ refer to the set of conditions at the end of the experiment.

Charles's Law can be illustrated by a hot-air balloon. The material that the balloon is made from is stretchable, so the pressure of the air inside is constant. As the air inside is heated (**T**↑), the volume of the air increases (expands; V↑) and the balloon fills out. With the mass the same but the volume larger, the density decreases (see Experiment 2). Because the air inside is less dense than the air outside, the balloon rises.

This experiment determines the volume of a sample of air when measured at two different temperatures with the pressure held constant.

OBJECTIVES

1. To measure the volume of a fixed quantity of air as the temperature changes at constant pressure.
2. To verify Charles's Law.

PROCEDURE

1. Use a clean and dry 250-mL Erlenmeyer flask (Flask no. 1). Fit the flask with a prepared stopper assembly, consisting of a no. 6 one-hole rubber stopper with a 5-cm to 8-cm length of glass tubing inserted through the hole. If an assembly needs to be constructed, use the following procedure:
   a. Select a sharpened brass cork borer with a diameter that just allows the glass tubing to pass through it easily.
   b. Lubricate the outside of the cork borer with glycerine and push it through the rubber stopper from the bottom.
   c. Once the cork borer is through the stopper, pass the glass tubing through the cork borer so that the tubing is flush with the bottom of the stopper.
   d. Grasp the tubing and the stopper with one hand to hold these two pieces stationary; with the other hand carefully remove the borer. The glass tubing stays in the stopper. (Check to be certain that the end of the glass tubing is flush with the bottom of the rubber stopper.)
2. Mark the position of the bottom of the rubber stopper on Flask no. 1 with a marking pencil. Connect a 2-ft. piece of latex rubber tubing to the glass tubing.
3. Place 300 mL of water and three (3) boiling stones in an 800-mL beaker. Support the beaker on a ring stand using a ring support and wire gauze, and heat the water with a Bunsen burner to boiling (Figure 11.1) (or place the beaker on a hot plate and heat to boiling). Keep the water at a gentle boil. Record the temperature of the boiling water on the Report Sheet (1).
4. Prepare an ice-water bath using a second 800-mL beaker half-filled with a mixture of ice and water. Record the temperature of the bath on the Report Sheet (3). Set aside for use in step 8.
5. Put about 200 mL of water into a second 250-mL Erlenmeyer flask (Flask no. 2) and place the end of the rubber tubing into the water. Make sure that the end of the rubber tubing reaches to the bottom of the flask and stays submerged at all times. (You may wish to hold it in place with a clamp attached to a ring stand.)

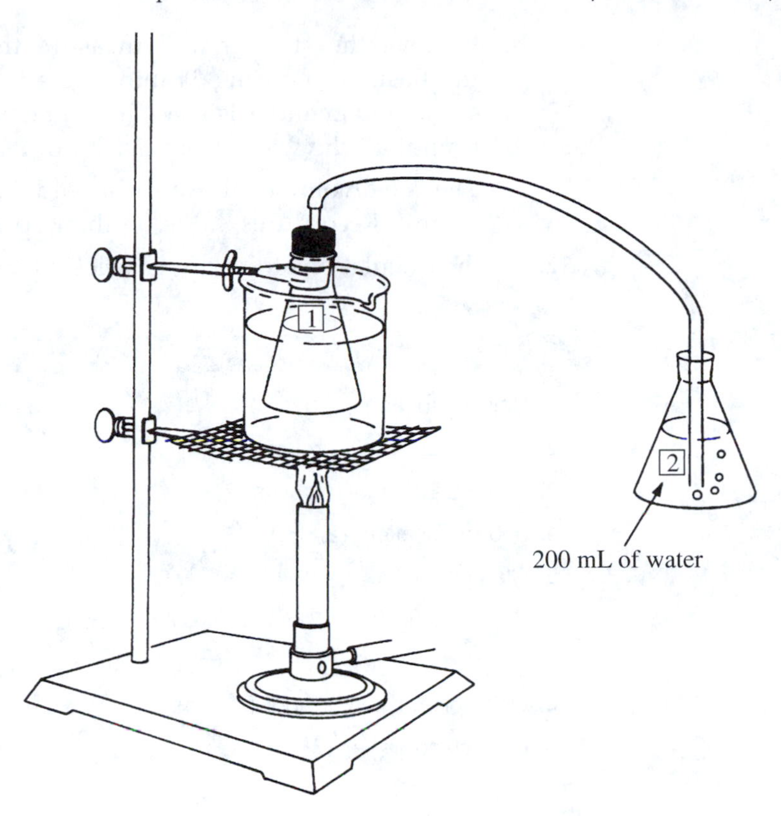

Figure 11.1
Equipment to study Charles's Law.

6. With a clamp holding the neck of Erlenmeyer Flask no. 1, lower the flask as far as it will go into the boiling water. Secure onto the ring stand (Figure 11.1). Adjust the water level in the beaker to cover as much of the Erlenmeyer flask as possible.
7. Boil gently for 5 min. Air bubbles should emerge from the rubber tubing submerged in Flask no. 2. Add water to the beaker if boiling causes the water level to go down.
8. When bubbles no longer emerge from the end of the submerged tubing (after 5 min.), carefully lift Flask no. 1 from the boiling water bath and quickly place it into the ice-water bath. Record what you observe happening as Flask no. 1 cools (2). **Be sure to keep the end of the rubber tubing always submerged in the water in Flask no. 2.**

CAUTION

The water, the glassware, and the ironware are hot.

9. When no more water is drawn into Flask no. 1, raise the flask until the level of water inside the flask is at the same height as the water in the ice-water bath. Then remove the stopper from Flask no. 1.
10. Using a graduated cylinder, measure the water in Flask no. 1. Record the volume to the nearest 0.1 mL on the Report Sheet (4).
11. Determine the volume of Erlenmeyer Flask no. 1 as follows:
    a. First, fill it with water to the level marked by the marking pencil. Insert the stopper with the glass tubing into the flask to be sure the bottom of the stopper touches the water with no air space present. Adjust the water level if necessary.

b. Remove the stopper and measure the volume of the water in the flask by pouring it into a graduated cylinder. If a 100-mL graduated cylinder is used, it will be necessary to empty and refill it until all the water from Flask no. 1 has been measured.

c. The *total* volume of water should be measured to the nearest 0.1 mL. Record this value on the Report Sheet (5).

12. Do the calculations to verify Charles's Law.

CHEMICALS AND EQUIPMENT

1. Boiling stones
2. Bunsen burner (or hot plate)
3. 250-mL Erlenmeyer flasks (2)
4. 800-mL beakers (2)
5. Clamps
6. Glass tubing (6- to 8-cm length; 7-mm OD)
7. Marking pencil
8. One-hole rubber stopper (size no. 6)
9. Ring stand
10. Ring support
11. Rubber tubing (2-ft. length)
12. Thermometer, 110°C
13. Wire gauze

name section date

partner grade

EXPERIMENT 11

Pre-Lab Questions

1. What did Jacques Charles describe in 1787?

2. In the mathematical relation that Gay-Lussac developed, how must temperature be expressed?

3. Circle the underlined word or phrase that correctly completes the following sentences:

a. With pressure constant, as temperature increases, the volume decreases or increases.

b. With pressure constant, as the volume decreases by one-half, the temperature decreases by one-half or increases by 2 times.

c. Volume is inversely or directly related to temperature at constant pressure.

d. The mathematical equation for Charles's Law is $V \times T = k$ or $T \times k = V$.

4. Helium in a balloon occupies a volume of 25 L at a temperature of 27°C. What temperature must be reached in order for the balloon to expand to a volume of 50 L, at constant pressure?

name ____________________ section __________ date __________

partner ____________________ grade __________

11 EXPERIMENT 11

Report Sheet

1. Temperature, boiling water (T_2) __________ °C __________ K

2. Observation as Flask no. 1 cools:

3. Temperature, ice water (T_1) __________ °C __________ K

4. Volume of water sucked into Flask no. 1 (V_w) __________ mL

5. Volume of Flask no. 1 __________ mL

6. Volume of air at the temperature of boiling water (5)

(V_2) __________ mL

7. Volume of the air at the temperature of ice water (V_1)

($V_1 = V_2 - V_w$) __________ mL

8. Verify Charles's Law

$$\frac{V_2 \times T_1}{V_1 \times T_2} =$$ __________

9. Percent deviation from Charles's Law

$$\% = \frac{1.00 - (8)}{1.00} \times 100 =$$ __________ %

Post-Lab Questions

1. There are a number of steps in the **Procedure,** which, if not followed, could introduce errors. Consider the variations below and speculate whether the results would be affected.
   a. All the water in the beaker was allowed to boil away (refer to steps 3 and 7).

   b. The experiment was stopped while bubbles continued to emerge from the rubber tubing (refer to step 8).

   c. The volume of Flask no. 1 was assumed to be 250 mL (refer to step 11).

   d. The temperatures were recorded in degrees Kelvin (refer to Report Sheet no. 1 and no. 3).

2. A student carried out the procedure described in this experiment to verify Charles's Law. The following data were obtained:
   a. Temperature of boiling water, $T = 99.5°C$

   b. Room temperature, $T = 25.0°C$

   c. Volume of water drawn into the flask, $V = 60.0$ mL

   d. Total volume of the flask, $V = 271.5$ mL

Using these data, verify Charles's Law according to equation (2) (see **Background**), and determine any percent deviation. Show all your work.

Properties of Gases: Determination of the Molar Mass of a Volatile Liquid

BACKGROUND

In the world in which we live, the gases with which we are familiar (for example, O_2, N_2, CO_2, H_2) possess a molecular volume and show interactions between molecules. When cooled or compressed, these real gases eventually condense into the liquid phase. In the hypothetical world, there are hypothetical gases, which we refer to as *ideal* gases. The molecules of these gases are presumed to possess negligible volume. There are no attractions between molecules of ideal gases, and, as a result, molecules do not stick together upon collision. The gases possess the volume of the container. Volumes of all gases decrease linearly with decreasing temperature. When we extrapolate these lines, they all converge to one temperature, −273.15°C, where the volume of the gas is presumably zero. Obviously this is an imaginary situation because gases cannot exist at such low temperatures; they condense and become liquids and then solids. But −273.15°C is the lowest possible temperature, and although we can never reach it, scientists have come pretty close, to one millionth of a degree Celsius above it. For that reason it is rational to start a temperature scale at the lowest possible temperature and call that zero. This is the absolute, or Kelvin, scale. On the Kelvin scale, this low temperature is termed *absolute zero* and is −273.15 degrees below zero on the Celsius scale.

The relationship that unites pressure, **P**, volume, **V**, and temperature, **T**, for a given quantity of gas, **n**, in the sample is the ideal gas equation:

$$\mathbf{P} \times \mathbf{V} = \mathbf{n} \times \mathbf{R} \times \mathbf{T}$$

This equation expresses pressure in atmospheres, volume in liters, temperature in degrees Kelvin, and the quantity of gas in moles. These four quantities are related exactly through the use of the ideal gas constant, **R**; the value for R is 0.0821 L atm/K mole. (Notice that the units for R are in terms of V, P, T, and n.)

A container of fixed volume at a given temperature and pressure holds only one possible quantity of gas. This quantity can be calculated by using the ideal gas equation:

$$n = \frac{P \times V}{R \times T}$$

The number of moles, **n**, can be determined from the mass of the gas sample, **m**, and the molar mass of the gas, **M**:

$$n = \frac{m}{M}$$

Substituting this expression into the ideal gas equation gives

$$P \times V = \frac{m}{M} \times R \times T$$

Solving for the molar mass, **M**:

$$M = \frac{m \times R \times T}{P \times V}$$

This equation gives us the means for determining the molar mass of a gas sample from the measurements of mass, temperature, pressure, and volume. Most real gases at low pressures (1 atm or less) and high temperatures (300 K or more) behave as an ideal gas and thus, under such conditions, the ideal gas law is applicable to real gases as well.

If a volatile liquid (one of low boiling point and low molar mass) is used, the method of Dumas (John Dumas, 1800–1884) can give a reasonably accurate value of the molar mass. In this experiment, a small quantity of a volatile liquid will be vaporized in a pre-weighed flask of known volume at a known temperature and barometric pressure. Because the boiling point of the liquid will be below that of boiling water when the flask is submerged in a boiling-water bath, the liquid will vaporize completely. If the density of the sample gas is greater than the density of the air, the gas will drive out the air and fill the flask with the gaseous sample. The gas pressure in the flask is in equilibrium with atmospheric pressure and, therefore, can be determined from a barometer. The temperature of the gas is the same as the temperature of the boiling water. Cooling the flask condenses the vapor. The weight of the vapor can be measured by weighing the liquid, and M, the molar mass of the gas, can be calculated.

Example

A sample of an unknown liquid is vaporized in an Erlenmeyer flask with a volume of 250 mL. At 100°C the vapor exerts a pressure of 0.975 atm. The condensed vapor weighs 0.685 g. Determine the molar mass of the unknown liquid.

Using the equation

$$M = \frac{m \times R \times T}{P \times V}$$

we can substitute all the known values:

$$M = \frac{(0.685\ g)(0.0821\ \cancel{L}\ \cancel{atm}/mole\ \cancel{K})(373\ \cancel{K})}{(0.975\ \cancel{atm})(0.250\ \cancel{L})} = 86.1\ g/mole$$

Hexane, with the molecular formula C_6H_{14}, has a molar mass of 86.17 g/mole.

OBJECTIVES

1. Experimentally determine the mass of the vapor of a volatile liquid.
2. Calculate the molar mass of the liquid by applying the ideal gas equation to the vapor.

PROCEDURE

1. Obtain a sample of unknown liquid from your instructor. Record the code number of the unknown liquid on the Report Sheet (1). The unknown liquid will be one of the liquids found in Table 12.1.
2. Weigh together a clean, dry 125-mL Erlenmeyer flask, a 2.5- by 2.5-in. square of aluminum foil, and a 4-in. piece of copper wire. Record the total weight to the nearest 0.001 g (2).
3. Pour approximately 3 mL of the liquid into the flask. Cover the mouth of the flask with the aluminum foil square and crimp the edges tightly over the neck of the flask. Secure the foil by wrapping the wire around the neck and twisting the ends by hand.
4. With a larger square of aluminum foil (3 × 3 in.), secure a second cover on the mouth of the flask with a rubber band.
5. Carefully punch a *single, small* hole in the foil covers with a needle or pin. The assembly is now prepared for heating (Figure 12.1).

Table 12.1 *Volatile Liquid Unknowns*

Liquid	*Formula*	*Molar Mass*	*Boiling Point (°C at 1 atm)*
Pentane	C_5H_{12}	72.2	36.2
Acetone	C_3H_6O	58.1	56.5
Methanol	CH_4O	32.0	64.7
Hexane	C_6H_{14}	86.2	69.0
Ethanol	C_2H_6O	46.1	78.5
2-Propanol	C_3H_8O	60.1	82.3

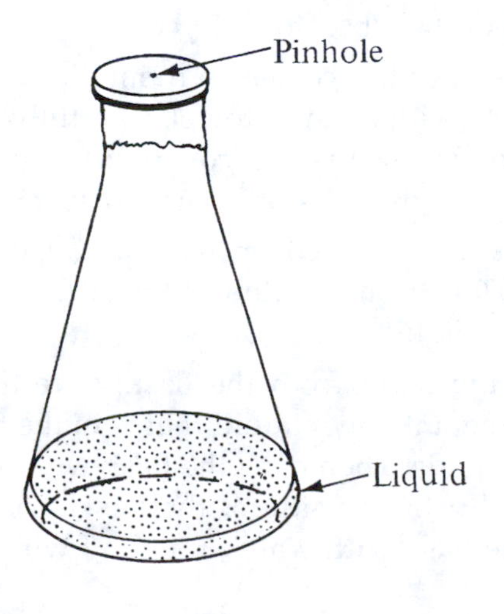

Figure 12.1
Assembly for vaporization of the liquid.

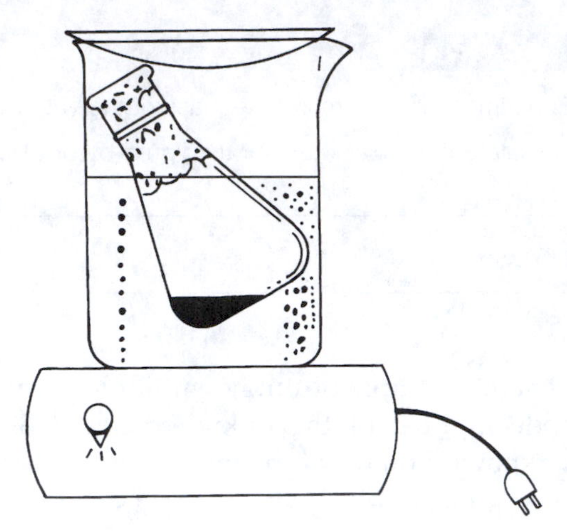

Figure 12.2
Assembly for the determination of molar mass.

6. Take a 1000-mL beaker and add 300 mL of water and a few boiling stones. Heat the water to boiling using a hot plate. Regulate the heat so the boiling does not cause the water to splash.

CAUTION

Use a hot plate and not a Bunsen burner, because the liquids listed in Table 12.1 are flammable.

7. Immerse the flask containing the volatile unknown liquid in the boiling water so that most of the flask is beneath the hot water as shown in Figure 12.2. (You may need to weigh down the flask with a test tube clamp or a lead sinker.)
8. Observe the unknown liquid. There is more liquid than is required to fill the flask with vapor. As the liquid evaporates, the level will decrease and excess vapor will escape through the pinhole. When all the liquid appears to be gone, continue to heat for an additional 5 min.
9. Record the temperature of the boiling water (5). Record the temperature of the vapor in the flask (6).
10. Using tongs, carefully remove the flask from the water and set it on the laboratory bench.
11. While the flask cools to room temperature, record the barometric pressure in the laboratory (7).
12. When the flask has cooled to room temperature, wipe dry the outside of the flask with a paper towel. Carefully remove only the second foil cover; blot dry the first foil cover with a paper towel. Take a look inside the flask. Are droplets of liquid present?
13. Weigh the flask, foil cover, wire, and condensed liquid. Record to the nearest 0.001 g on the Report Sheet (3). Determine the weight of the condensed liquid (4).
14. Remove the cover from the flask; save the aluminum foil and wire. Discard the unknown liquid that is in the flask in a container for waste organic liquids. Then rinse the flask with water and refill with water to the rim; dry the outside of the flask. Carefully and without spilling, weigh the flask with water, foil, and wire to the nearest 0.01 g (8).

15. Determine the weight of the water (9). Determine the temperature of the water in the flask (10). Look up the density of water at that temperature in the *Handbook of Chemistry and Physics*, and, using this value, calculate the volume of water in the flask (11). (You may check the volume in the flask by measuring the water with a graduated cylinder.)

16. Calculate the molar mass of the unknown (12). Identify the unknown (13).

CHEMICALS AND EQUIPMENT

1. Aluminum foil
2. Tongs
3. Hot plate
4. Copper wire
5. Rubber bands
6. Boiling stones
7. Absorbent paper towels
8. Unknown liquid chosen from Table 12.1
9. Lead sinkers

name | section | date

partner | grade

12

EXPERIMENT 12

Pre-Lab Questions

1. What happens to an ideal gas as it is cooled to −273°C? How does this differ from a real gas?

2. What are the three variable quantities that determine gas volume?

3. What is meant by *absolute zero*?

4. Which is the coldest temperature: 0°C or 0°F or 0 K?

5. A gas weighing 22.40 g occupies a 10.00-L flask at 27°C and 722 torr of pressure. What is the molar mass of the gas? (Remember to use proper units: V in liters, T in Kelvin, P in atmospheres.) Show your work. Identify the gas from Table 12.1.

name | section | date

partner | grade

12

EXPERIMENT 12

Report Sheet

1. Code number of unknown liquid ______
2. Weight of 125-mL Erlenmeyer flask, aluminum foil, and copper wire ______ g
3. Weight of cooled flask, foil, wire, and condensed liquid ______ g
4. Weight of condensed liquid: (3) − (2) ______ g
5. Temperature of boiling water ______ °C ______ K
6. Temperature of vapor in flask ______ °C ______ K
7. Barometric pressure ______ atm
8. Weight of 125-mL Erlenmeyer flask, aluminum foil, copper wire, and water ______ g
9. Weight of water: (8) − (2) ______ g
10. Temperature of water ______ °C
11. Volume of the flask: (9)/(density of water) ______ mL ______ L
12. Molar mass of the unknown:

$$M = \frac{(4) \times (0.0821 \text{ L atm/K mole}) \times (6)}{(7) \times (11)}$$ ______ g/mole

13. The unknown is ______

Post-Lab Questions

1. Under what conditions do real gases behave most like an ideal gas? How does this experiment meet those conditions?

2. The following errors in procedure will affect the experimental result. Explain how each action causes an incorrect result.

   a. Heating was stopped before all of the liquid in the flask had evaporated (step 8).

   b. Water remained on the outside of the flask and on the foil when the final weight was determined (steps 12 and 13).

   c. The volume of the Erlenmeyer flask was assumed to be 125 mL and not actually measured as directed in steps 14 and 15.

3. A student had an unknown liquid and determined its molar mass by collecting data as described in this experiment.

   a. Weight of the 125-mL Erlenmeyer flask and cover (aluminum foil and copper wire), 79.621 g

   b. Weight of cooled flask and cover containing condensed liquid, 79.927 g

   c. Temperature of boiling water, 99.5°C

   d. Volume of the 125-mL Erlenmeyer flask, 131.5 mL

   e. Atmospheric pressure, 750 torr

Using these data, calculate the liquid's molar mass and identify the unknown from Table 12.1. Show all of your work.

Physical Properties of Chemicals: Melting Point, Sublimation, and Boiling Point

BACKGROUND

If you were asked to describe a friend, most likely you would start by identifying particular physical characteristics. You might begin by giving your friend's height, weight, hair color, eye color, or facial features. These characteristics would allow you to single out the individual from a group.

Chemicals also possess distinguishing physical properties which enable their identification. In many circumstances, a thorough determination of the physical properties of a given chemical can be used for its identification. If faced with an unknown sample, a chemist may compare the physical properties of the unknown to properties of known substances that are tabulated in the chemical literature; if a match can be made, an identification can be assumed (unless chemical evidence suggests otherwise).

The physical properties most commonly listed in handbooks of chemical data are color, crystal form (if a solid), refractive index (if a liquid), density (discussed in Experiment 2), solubility in various solvents (discussed in Experiment 14), melting point, sublimation characteristics, and boiling point. When a new compound is isolated or synthesized, these properties almost always accompany the report in the literature.

The transition of a substance from a solid to a liquid to a gas, and the reversal, represent physical changes. In a physical change there is only a change in the form or state of the substance. No chemical bonds break; no alteration in the chemical composition occurs. Water undergoes state changes from ice to liquid water to steam; however, the composition of molecules in all three states remains H_2O.

$$\underset{\textbf{Ice}}{H_2O(s)} \rightleftharpoons \underset{\textbf{Liquid}}{H_2O(l)} \rightleftharpoons \underset{\textbf{Steam}}{H_2O(g)}$$

The *melting* or *freezing point* of a substance refers to the temperature at which the solid and liquid states are in equilibrium. The terms are interchangeable and correspond to the same temperature; how the terms are applied depends upon the state the substance is in originally. The melting point is the temperature at equilibrium when starting in the solid state and going to the liquid state. The freezing point is the temperature at equilibrium when starting in the liquid state and going to the solid state.

Melting points of pure substances occur over a very narrow range and are usually quite sharp. The criteria for purity of a solid is the narrowness of the melting point range and the correspondence to the value found in the literature. Impurities will lower the melting point and cause a broadening of the range. For example, pure benzoic acid has a reported melting point of 122.13°C; benzoic acid with a melting point range of 121–122°C is considered to be quite pure.

The *boiling point* or *condensation* point of a liquid refers to the temperature when its vapor pressure is equal to the external pressure. If a beaker of liquid is brought to a boil in your laboratory, bubbles of vapor form throughout the liquid. These bubbles rise rapidly to the surface, burst, and release vapor to the space above the liquid. In this case, the liquid is in contact with the atmosphere; the normal boiling point of the liquid will be the temperature when the pressure of the vapor is equal to the atmospheric pressure (1 atm or 760 mm Hg). Should the external pressure vary, so will the boiling point. A liquid will boil at a higher temperature when the external pressure is higher and will boil at a lower temperature when the external pressure is reduced. The change in state from a gas to a liquid represents condensation and is the reverse of boiling. The temperature for this change of state is the same as the boiling temperature but is concerned with the approach from the gas phase.

Just as a solid has a characteristic melting point, a liquid has a characteristic boiling point. At one atmosphere, pure water boils at 100°C, pure ethanol (ethyl alcohol) boils at 78.5°C, and pure diethyl ether boils at 34.6°C. The vapor pressure curves shown in Figure 13.1 illustrate the variation of the vapor pressure of these liquids with temperature. One can use these curves to predict the boiling point at a reduced pressure. For example, diethyl ether has a vapor pressure of 422 mm Hg at 20°C. If the external pressure were reduced to 422 mm Hg, diethyl ether would boil at 20°C.

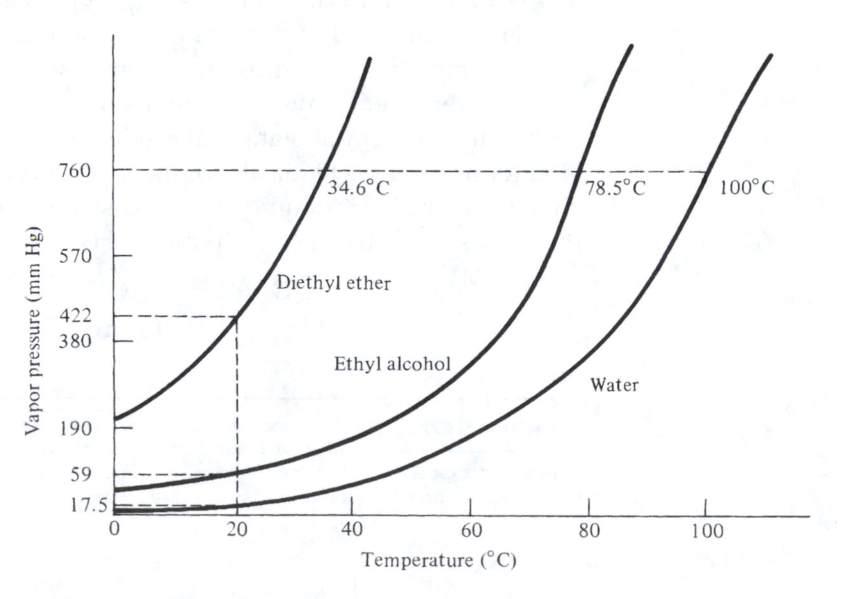

Figure 13.1
Diethyl ether, ethyl alcohol (ethanol), and water vapor pressure curves.

Sublimation is a process that involves the direct conversion of a solid to a gas without passing through the liquid state. Relatively few solids do this at atmospheric pressure. Some examples are the solid compounds naphthalene (mothballs), caffeine, iodine, and solid carbon dioxide (commercial Dry Ice). Water, on the other hand, sublimes at −10°C and at 0.001 atm. Sublimation temperatures are not as easily obtained as melting points or boiling points.

OBJECTIVES

1. To use melting points and boiling points in identifying substances.
2. To use sublimation as a means of purification.

PROCEDURE

Melting Point Determination

1. Unknowns are provided by the instructor. Obtain approximately 0.1 g of unknown solid and place it on a small watch glass. Record the code number of the unknown on the Report Sheet (1). (The instructor will weigh out a 0.1-g sample as a demonstration; take approximately that amount with your spatula.) Carefully crush the solid on a watch glass into a powder with the flat portion of a spatula.
2. Obtain a melting-point capillary tube. One end of the tube will be sealed. The tube is packed with solid in the following way:

   Step A Press the open end of the capillary tube vertically into the solid sample (Figure 13.2 A). A small amount of sample will be forced into the open end of the capillary tube.

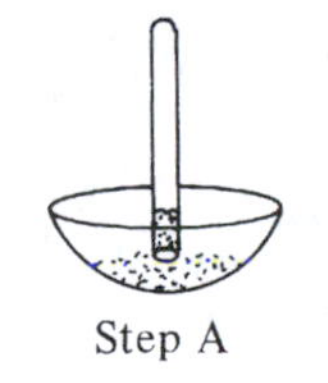

A. Forcing solid into the capillary tube.

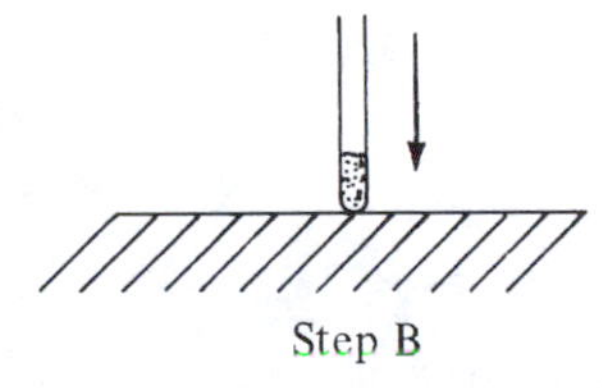

B. Tapping to force down solid.

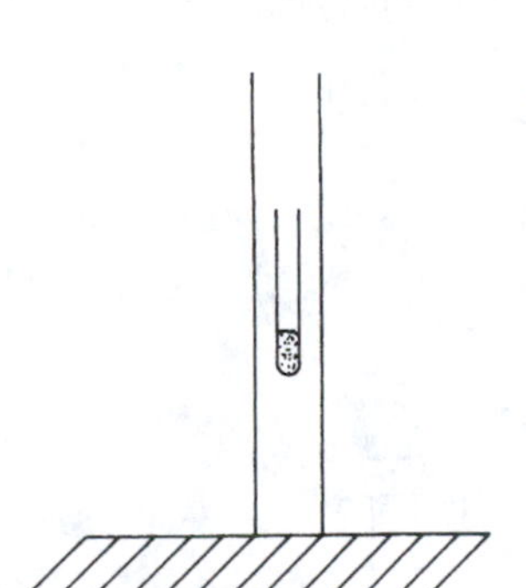

C. Alternative method for bringing the solid down.

Figure 13.2
Packing a capillary tube.

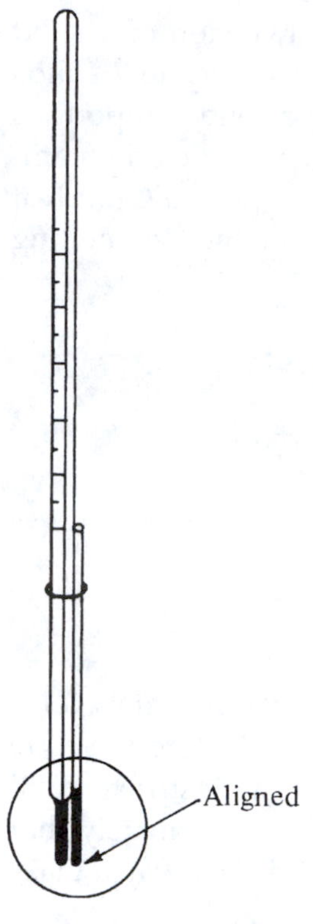

Figure 13.3
Proper alignment of the capillary tube and the mercury bulb.

Step B Invert the capillary tube so that the closed end is pointing toward the bench top. Gently tap the end of the tube against the laboratory bench top (Figure 13.2 B). Continue tapping until the solid is forced down to the closed end. A sample depth of 5–10 mm is sufficient.

Step C An alternative method for bringing the solid sample to the closed end uses a piece of glass tubing of approximately 20 to 30 cm. Hold the capillary tube, closed end down, at the top of the glass tubing, held vertically; let the capillary tube drop through the tubing so that it hits the laboratory bench top. The capillary tube will bounce and bring the solid down. Repeat if necessary (Figure 13.2 C).

3. The melting point may be determined using either a commercial melting-point apparatus or a Thiele tube.

   a. The use of the Thiele tube is as follows:

      - Attach the melting-point capillary tube to the thermometer by means of a rubber ring. Align the mercury bulb of the thermometer so that the tip of the melting-point capillary containing the solid is next to it (Figure 13.3).
      - Use an extension clamp to support the Thiele tube on a ring stand. Add mineral oil or silicone oil to the Thiele tube, and fill to a level above the top of the side arm. Use a thermometer clamp to support the thermometer with the attached melting point capillary tube in the oil. The bulb and capillary tube should be immersed in the oil; keep the rubber ring and open end of the capillary tube out of the oil (Figure 13.4).

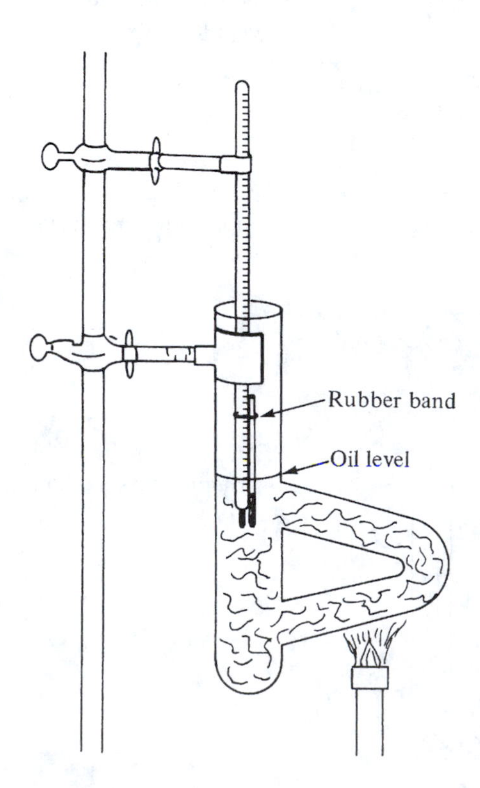

Figure 13.4
Thiele tube apparatus.

Table 13.1 *Melting Points of Selected Solids*

Solid	*Melting Point (°C)*	*Use*
Acetamide	82	Plasticizer; stabilizer
Acetanilide	114	Manufacture of other medicinals and dyes
Adipic acid	152	Manufacture of nylon
Benzophenone	48	Manufacture of antihistamines, hypnotics, insecticides
Benzoic acid	122	Preserving foods; antifungal agent
p-Dichlorobenzene	54	Moth repellent; insecticidal fumigant
Naphthalene	80	Moth repellent; insecticide
Stearic acid	70	Suppositories; ointments

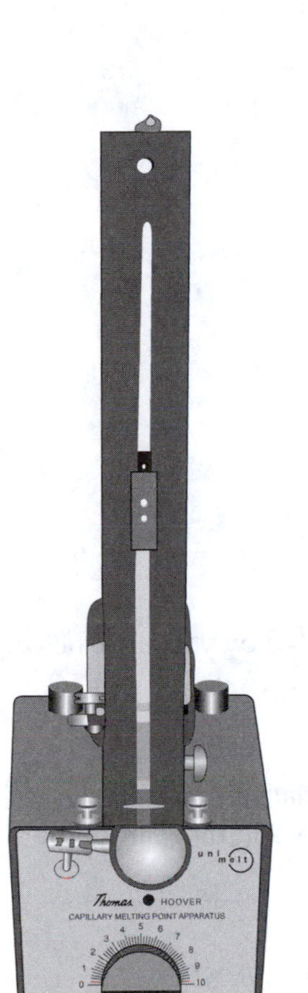

Figure 13.5
A Thomas-Hoover Unimelt®.

- Heat the arm of the Thiele tube very slowly with a Bunsen burner flame. Use a small flame and gently move the burner along the arm of the Thiele tube.
- You should position yourself so that you can follow the rise of the mercury in the thermometer as well as observe the solid in the capillary tube. Record the temperature when the solid begins to liquefy (2) (the solid will appear to shrink). Record the temperature when the solid is completely a liquid (3). Express these readings as a melting point range (4).
- Identify the solid by comparing the melting point with those listed in Table 13.1 for different solids (5).

b. A commercial melting-point apparatus, such as the Thomas-Hoover Unimelt® (Figure 13.5), will be demonstrated by your instructor.

4. Do as many melting-point determinations as your instructor may require. Just remember to use a new melting point capillary tube for each melting point determination.

5. Dispose of the solids as directed by your instructor.

Purification of Naphthalene by Sublimation

1. Place approximately 0.5 g of impure naphthalene into a 100-mL beaker. (Your instructor will weigh out 0.5 g of sample as a demonstration; with a spatula take an amount that approximates this quantity.)

2. Into the 100-mL beaker place a smaller 50-mL beaker. Fill the smaller one halfway with ice cubes or ice chips. Place the assembled beakers on a wire gauze supported by a ring clamp (Figure 13.6).

3. Using a small Bunsen burner flame, gently heat the bottom of the 100-mL beaker by passing the flame back and forth beneath the beaker.

4. You will see solid flakes of naphthalene collect on the bottom of the 50-mL beaker. When a sufficient amount of solid has been collected, turn off the burner.

5. Pour off the ice water from the 50-mL beaker and carefully scrape the flakes of naphthalene onto a piece of filter paper with a spatula.

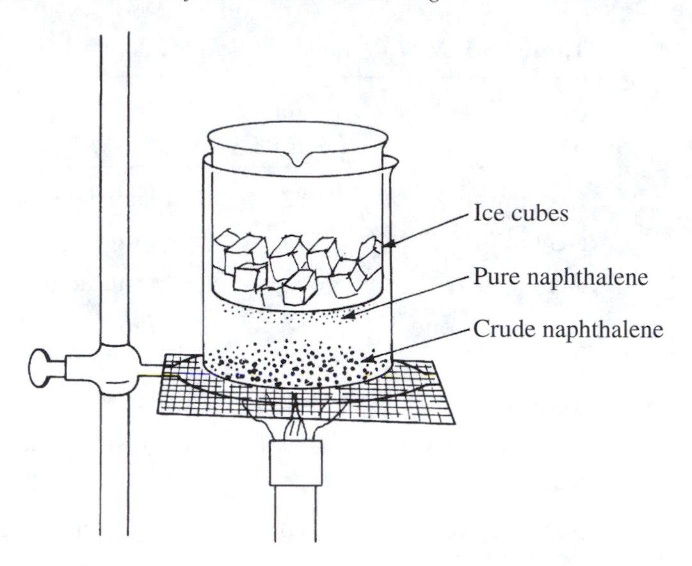

Figure 13.6
Setup for sublimation of naphthalene.

6. Take the melting point of the pure naphthalene and compare it to the value listed in Table 13.1 (6).
7. Dispose of the crude and pure naphthalene as directed by your instructor.

Boiling Point Determination

CAUTION

The chemicals used for boiling point determinations are flammable. Be sure all Bunsen burner flames are extinguished before starting this part of the experiment.

1. Obtain an unknown liquid from your instructor and record its code number on the Report Sheet (7).
2. Clamp a clean, dry test tube (100 × 13 mm) onto a ring stand. Add to the test tube approximately 3 mL of the unknown liquid and two small boiling stones. Lower the test tube into a 250-mL beaker which contains 100 mL of water and two boiling stones. Adjust the depth of the test tube so that the unknown liquid is below the water level of the water bath (Figure 13.7).
3. With a thermometer clamp, securely clamp a thermometer and lower it into the test tube through a neoprene adapter. Adjust the thermometer so that it is approximately 1 cm above the surface of the unknown liquid.
4. Use a piece of aluminum foil to cover the mouth of the test tube. (Be certain that the test tube mouth has an opening; the system should not be closed.)
5. Gradually heat the water in the beaker with a hot plate and watch for changes in temperature. As the liquid begins to boil, the temperature above the liquid will rise. When the temperature no longer rises but remains constant, record the temperature to the nearest 0.1°C (8). This is the observed boiling point. From the list in Table 13.2, identify your unknown liquid by matching your observed boiling point with the compound whose boiling point best corresponds (9).
6. Do as many boiling point determinations as required by your instructor.
7. Dispose of the liquid as directed by your instructor.

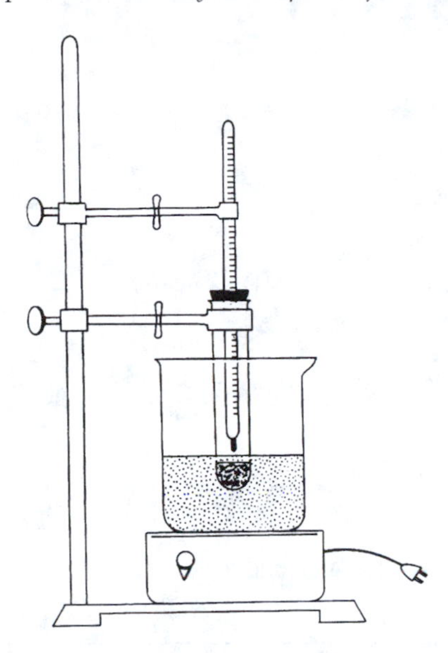

Figure 13.7
Setup for determining the boiling point.

Table 13.2 *Boiling Points of Selected Liquids*

Liquid	*Boiling Point (°C at 1 atm)*	*Use*
Acetone	56	Solvent; paint remover
Cyclohexane	81	Solvent for lacquers and resins
Ethyl acetate	77	Solvent for airplane dopes; artificial fruit essence
Hexane	69	Liquid in thermometers with blue or red dye
Methanol (methyl alcohol)	65	Solvent; radiator antifreeze
1-Propanol	97	Solvent
2-Propanol (isopropyl alcohol)	83	Solvent for shellac; essential oils; body rubs

CHEMICALS AND EQUIPMENT

1. Aluminum foil
2. Boiling stones
3. Bunsen burner
4. Hot plate
5. Commercial melting-point apparatus (if available)
6. Melting-point capillary tubes
7. Rubber rings
8. Thiele tube melting-point apparatus
9. Thermometer clamp
10. Glass tubing
11. Acetamide
12. Acetanilide
13. Acetone

14. Adipic acid

15. Benzophenone

16. Benzoic acid

17. Cyclohexane

18. *p*-Dichlorobenzene

19. Ethyl acetate

20. Hexane

21. Methanol (methyl alcohol)

22. Naphthalene (pure)

23. Naphthalene (impure)

24. 1-Propanol

25. 2-Propanol (isopropyl alcohol)

26. Stearic acid

Angela Kerth

name

section

date

Megan Karnes

partner

grade

8.0

13

EXPERIMENT 13

Pre-Lab Questions

1. When liquid water boils and is converted into the gas steam, is this a physical or chemical change? What is the chemical formula of the compound in liquid water and in the gas?

$H_2O(l) - H_2O(g)$ the water (liquid) is changed (forms) into a gas - physical change

2. When ethanol condenses and goes from the gaseous phase into the liquid phase, is this a physical change or a chemical change? Explain your answer. (The chemical formula for ethanol is CH_3CH_2OH.) physical change. No chemical bonds are broken.

3. The melting point of water is 0°C; the freezing point of water is 0°C. How is it that the value for temperature is the same for the two different terms? They are at an equilibrium thats why the solid pt + the liquid pt are the same temp.

4. Would you expect the boiling point for water to be the same at your laboratory bench in Garden City and laboratory at the top of Mt. Everest (elevation 29,028 ft.)? Explain your answer.

No because of the pressure and elevation the boiling pt would be different. Mt Everest would be a ~~higher~~ lower caused by the ~~higher pressure~~.

-0.5

5. What criteria are applied to define a "pure" solid?

melting pt range and the correct pendence to the value found.

6. Describe all the state changes that solid naphthalene passes through in order to become a gas.

sublimation - when a solid ~~becau~~ becomes a gas w/out passing through a liquid stage.

-0.5

name Angela Yerth | section ____ | date ____

partner Megan Yarnes | grade ____

13 EXPERIMENT 13

Report Sheet

Melting point determination

	Trial No. 1	Trial No. 2
1. Code number of unknown	C	C
2. Temperature melting begins	120 °C	120 °C
3. Temperature melting ends	135 °C	135 °C
4. Melting-point range	120-135 °C	120-130 °C
5. Identification of unknown	Benzonic	Acid

Purification of naphthalene by sublimation

6. Melting-point range	____ °C	____ °C

Boiling point determination

7. Unknown number	2	2
8. Observed boiling point	75 °C	75 °C
9. Identification of unknown	Ethyl Acetate	

Angela Roth

Post-Lab Questions

1. Student A found that a sample of adipic acid, used to make nylon, had a melting point of 151–153°C. Student B had a sample that melted over a range, 135–149°C. Are both samples of the same purity? Which student has the better sample, and why do you make that conclusion?

   Student A has a more pure sample because the range of temp for that sample is more accurate then Student B's sample.

2. A student in Denver, Colorado (the city often nicknamed the "Mile-High City"), did this laboratory at her school. All of her boiling points were noticeably lower than the ones listed in Table 13.2. She concluded that all were impure. Assuming everything was done as directed with a properly calibrated thermometer, was her conclusion correct? Explain your answer.

   The pressure in denver is different than over here so she could have done the lab correctly. Boiling pts vary due to pressure.

3. A bottle containing solid iodine crystals fills with a violet haze upon standing at room temperature. What is the violet haze due to and why does it form?

   Sublimation causes the haze its going from a solid to a gas.

4. Cocaine is a white solid that melts at 98°C when pure. Sucrose (table sugar) also is a white solid, but it melts at 185–186°C. A forensic chemist working for the New York Police Department has a white solid believed to be cocaine. What can the chemist do to determine quickly whether the sample is pure cocaine, simply table sugar, or a mixture of the two? (A good chemist never tastes any unknown chemical.)

   If the sample is heated then the melting pt of the sample can be found.

   1-0.5 what would you look for? -0.5

Solubility and Solutions

BACKGROUND

Most materials encountered in every day life are *mixtures*. This means that more than one component is found together in a system. Think back to your morning breakfast beverage; orange juice, coffee, tea, and milk are examples of mixtures.

Some mixtures have special characteristics. A mixture that is uniform throughout, with no phase boundaries, is called a *homogeneous mixture*. If you were to sample any part of the system, the same components in the same proportions would be found in each sample. The most familiar of these homogeneous mixtures is the liquid *solution;* here a *solute* (either a solid or a liquid) is thoroughly and uniformly dispersed into a *solvent* (a liquid). If the solution were allowed to remain standing, the components would not separate, no matter how much time was allowed to pass.

There are limits as to how much solute may be dispersed or dissolved in a given amount of solvent. This limit is the *solubility* and is defined as the *maximum weight of solute that dissolves in 100 g of a given solvent at a given temperature*. For example, sucrose (or table sugar) is soluble to the extent of 203.9 g per 100 g of water at 20°C. This means that if you have 100 g of water, you can dissolve up to 203.9 g of table sugar, but no more, in that quantity of water at 20°C. If more is added, the extra amount sinks to the bottom undissolved. A solution in this state is referred to as *saturated*. A solution with less than the maximum at the same temperature is called *unsaturated*. Solubility also varies with temperature (Figure 14.1).

Liquids dissolved in liquids similarly may form homogeneous solutions. Some liquids have limited solubility in water. Diethyl ether, $CH_3CH_2OCH_2CH_3$ (an organic liquid), is soluble to the extent of 4 g per 100 g of water at 25°C; an excess of the diethyl ether will result in a separation of phases with the less dense organic liquid floating on the water. Some liquids mix in all proportions; these liquids are completely *miscible*. The mixture of commercial antifreeze, ethylene glycol, $HOCH_2CH_2OH$, and water, used as a coolant in automobile radiators, is such a solution.

The solubility of a given solute in a particular solvent depends on a number of factors. One generalization for determining solubility is "like dissolves like." This means that the more similar the polarity of a solute

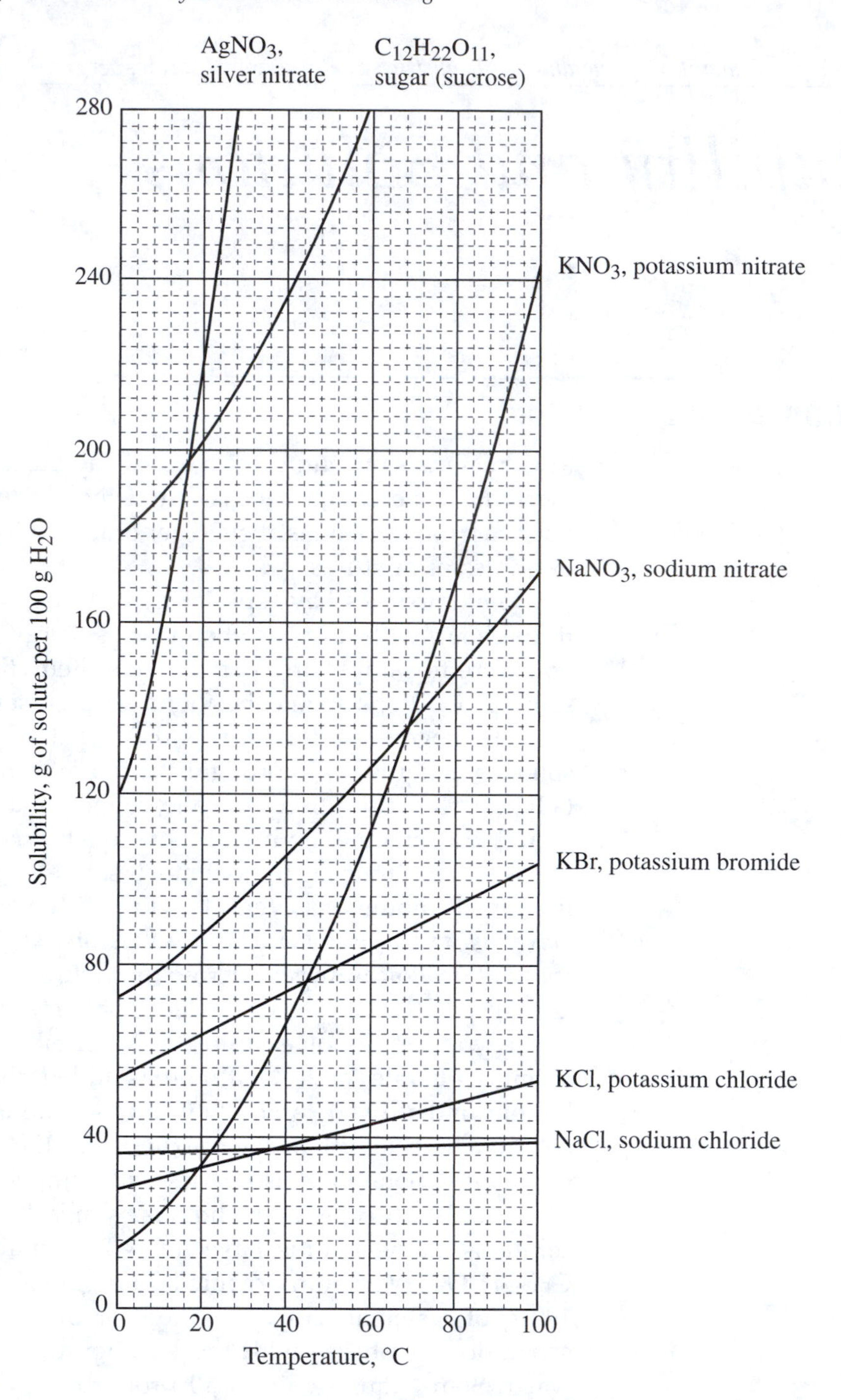

Figure 14.1
The effect of temperature on the solubility of some solutes in water.

is to the polarity of the solvent, the more likely the two will form a homogeneous solution. A polar solvent, such as water, will dissolve a polar compound: an ionic salt like common table salt, NaCl, will dissolve in water; a polar covalent solid like table sugar, sucrose, will dissolve in water. Nonpolar solvents such as naphtha or turpentine will dissolve nonpolar material, such as grease or oil. On the other hand, oil and water do not mix because of their different polar characteristics.

When ionic salts dissolve in water, the individual ions separate. These positively and negatively charged particles in the water medium are mobile and can move from one part of a solution to another. Because of this

Table 14.1 *Selected Electrolytes and Nonelectrolytes*

Strong Electrolytes	*Weak Electrolytes*	*Nonelectrolytes*
Sodium chloride, $NaCl$	Acetic acid, CH_3CO_2H	Methanol, CH_3OH
Sulfuric acid, H_2SO_4	Carbonic acid, H_2CO_3	Benzene, C_6H_6
Hydrochloric acid, HCl	Ammonia, NH_3	Acetone, $(CH_3)_2CO$
Sodium hydroxide, $NaOH$		Sucrose, $C_{12}H_{22}O_{11}$

movement, solutions of ions can conduct electricity. *Electrolytes* are substances that can form ions when dissolved in water and can conduct an electric current. These substances are also capable of conducting an electric current in the molten state. *Nonelectrolytes* are substances that do not conduct an electric current. Electrolytes may be further characterized as either strong or weak. A strong electrolyte dissociates almost completely when in a water solution; it is a good conductor of electricity. A weak electrolyte has only a small fraction of its particles dissociated into ions in water; it is a poor conductor of electricity. Table 14.1 lists examples of compounds behaving as electrolytes or nonelectrolytes in a water solution.

OBJECTIVES

1. To show how temperature affects solubility.
2. To demonstrate the difference between electrolytes and nonelectrolytes.
3. To show how the nature of the solute and the solvent affects solubility.

PROCEDURE

Saturated Solutions

1. Prepare a warm water bath. Place 300 mL of water into a 400-mL beaker and heat with a hot plate to 50°C.
2. Place 10 mL of distilled water into a 150 × 25 mm test tube; record the temperature of the water (1).
3. While stirring with a glass rod, add solid potassium nitrate, KNO_3, in 2-g portions; keep adding until no more potassium nitrate dissolves. The solution should be saturated. Record the mass of potassium nitrate added (2).
4. Place the test tube into the warm water bath and keep the temperature at 50°C. Again add to the solution, with stirring, potassium nitrate in 2-g portions until no more potassium nitrate dissolves. Record the mass of potassium nitrate added (3).
5. Slowly heat the solution above 50°C (to no more than 60°C) until all of the solid dissolves. With a test tube holder, remove the test tube from the water bath and set it into a 250-mL beaker (for support) to cool. Observe what happens as the solution cools.
6. When the first crystals appear, record the temperature. Fill a 250-mL beaker with ice and place the test tube into the ice. Observe what happened and offer an explanation for what has taken place (5). (If no crystals have formed, stir the solution with a stirring rod.)

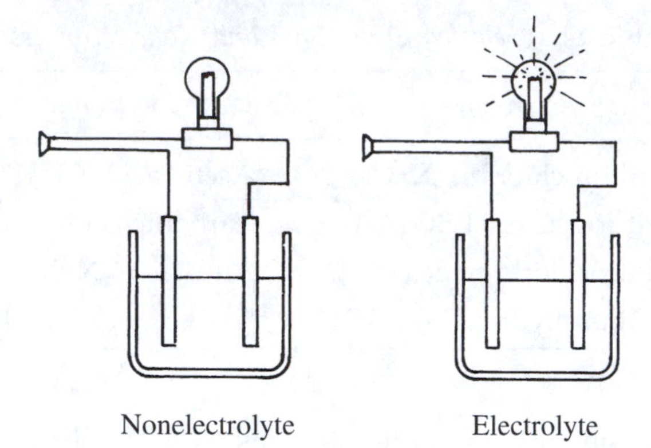

Figure 14.2
Conductivity apparatus.

Electrical Conductivity

This part of the experiment can be done in pairs. Obtain and set up a conductivity apparatus (Figure 14.2). It consists of two terminals connected to a light bulb and a plug for connection to a 110-volt electrical wall outlet.

CAUTION

To avoid a shock, do not touch the terminals when the apparatus is plugged in. Be sure to unplug the apparatus between tests and while rinsing and drying. Do not let the terminals touch each other.

The following solutions are to be tested with the conductivity apparatus:

- **a.** distilled water
- **b.** tap water
- **c.** 1 M NaCl
- **d.** 0.1 M NaCl
- **e.** 1 M sucrose, $C_{12}H_{22}O_{11}$
- **f.** 0.1 M sucrose, $C_{12}H_{22}O_{11}$
- **g.** 1 M HCl
- **h.** 0.1 M HCl
- **i.** glacial acetic acid, CH_3CO_2H
- **j.** 0.1 M acetic acid, CH_3CO_2H

1. For each solution follow steps 2, 3, 4, and 5.
2. Place about 20 mL of the solution to be tested into a 30-mL beaker that has been rinsed with distilled water. A convenient way to rinse the beaker is with a squeezable plastic wash bottle. Direct a stream of water from the wash bottle into the beaker, swirl the water about, and discard the water into the sink.
3. Lower the terminals into the beaker so that the solution covers the terminals. For each test solution, try to keep the same distance between the terminals and the terminals submerged to the same depth.
4. Plug the apparatus into the wall socket. Observe the effect on the light bulb. A solution containing an electrolyte conducts electricity—the circuit is completed and the bulb will light. Strong electrolytes will

give a bright light; weak electrolytes will give a dim light; nonelectrolytes will give no light. Note the effect of concentration. Record your observations on the Report Sheet.

5. Between each test, disconnect the conductivity apparatus from the wall socket, raise the terminals from the solution, and rinse the terminals with distilled water from the wash bottle.

Solubility: Solute and Solvent Characteristics

1. Clean and dry 16 test tubes (100 × 13 mm).
2. Place approximately 0.1 g of the following solids into test tubes numbered as indicated (your instructor will weigh exactly 0.1 g of solid as a demonstration; use your spatula to estimate the 0.1-g sample):
   a. No. 1: table salt, NaCl
   b. No. 2: table sugar, sucrose, $C_{12}H_{22}O_{11}$
   c. No. 3: naphthalene, $C_{10}H_8$
   d. No. 4: iodine, I_2
3. Add 4 mL of distilled water to each test tube and shake the mixture (sharp tapping of the test tube with your fingers will agitate the contents enough).
4. Record on the Report Sheet whether the solid dissolved completely (soluble), partially (slightly soluble), or not at all (insoluble).
5. With new sets of labeled test tubes containing the solids listed above, repeat the solubility tests using the solvents ethanol (ethyl alcohol), C_2H_5OH, acetone, $(CH_3)_2CO$, and petroleum ether in place of the water. Record your observations.
6. Discard your solutions in waste containers provided. *Do not discard them into the sink.*

CHEMICALS AND EQUIPMENT

1. Potassium nitrate, KNO_3
2. Sucrose (solid and solutions)
3. NaCl (solid and solutions)
4. Naphthalene
5. Iodine
6. HCl solutions
7. Acetic acid (glacial and solutions)
8. Ethanol (ethyl alcohol)
9. Acetone
10. Petroleum ether
11. Conductivity apparatus
12. Hot plate
13. Wash bottle
14. Beakers, 30-mL
15. Test tubes, 150 × 25 mm

name | section | date

partner | grade

EXPERIMENT 14

Pre-Lab Questions

1. If you were to draw samples from the top of a homogeneous solution and then the bottom, would you find differences between the properties of the samples? Explain your answer.

2. Name two factors that limit the solubility of a solute in a given volume of solvent.

3. Refer to Figure 14.1. Look at the solubility curves for potassium bromide, KBr, and potassium nitrate, KNO_3. Which salt is more soluble: (a) at 20°C; (b) at 80°C? Is there a temperature at which both salts are equally soluble?

4. Potassium bromide, KBr, is soluble to the extent of 64 g per 100 g of H_2O at 20°C. Is the solution saturated or unsaturated? What would happen if: (a) more KBr was added to the solution; (b) the temperature was lowered to 0°C?

5. Why is table salt, NaCl, classed as a strong electrolyte, and table sugar, $C_{12}H_{22}O_{11}$, classed as a nonelectrolyte? Does solubility have anything to do with this?

name ______ section ______ date ______

partner ______ grade ______

14 EXPERIMENT 14

Report Sheet

Saturated solution

1. Temperature of distilled water ______ °C
2. Mass of potassium nitrate ______ g/10 mL
3. Mass of additional potassium nitrate ______ g
4. Total mass of potassium nitrate: (2) + (3) ______ g/10 mL at 50°C
5. Observations and explanation

Electrical conductivity

Rate the brightness of the light bulb on a scale from 0 to 5: 0 for no light to 5 for very bright light.

Substance	*Observation*
Distilled water	______
Tap water	______
1 M NaCl	______
0.1 M NaCl	______
1 M sucrose	______
0.1 M sucrose	______
1 M HCl	______
0.1 M HCl	______
Glacial acetic acid	______
0.1 M acetic acid	______

Solubility: solute and solvent characteristics

Record the solubility as soluble (s), slightly soluble (ss), or insoluble (i). (A change in color of the solvent would indicate some solubility.)

	Solvent			
Solute	*Water*	*Ethanol*	*Acetone*	*Petroleum Ether*
Table salt, NaCl				
Table sugar, sucrose				
Naphthalene				
Iodine				

Post-Lab Questions

1. How do you know when a solution is saturated?

2. If you have a solution that is saturated, how you dissolve the excess solute?

3. Table sugar, sucrose, is soluble in water, but is insoluble in petroleum ether (an organic solvent similar to gasoline). What would account for this difference?

4. What are the most likely particles present in NaCl solutions and in sucrose solutions? From the observed brightness of the light bulb in the electrical conductivity experiment, would this account for the results?

Water of Hydration

BACKGROUND

Some compounds do not melt when heated but undergo decomposition. In decomposing, the compound can break down irreversibly or reversibly into two or more substances. If it is reversible, recombination leads to reformation of the original material. Hydrates are examples of compounds that do not melt but that decompose upon heating. The decomposition products are an anhydrous salt and water. The original hydrates can be regenerated by addition of water to the anhydrous salt.

Hydrates contain water as an integral part of the crystalline structure of the salt. When a hydrate crystallizes from an aqueous solution, water molecules are bound to the metal ion. A characteristic of the metal is the number of water molecules that bind to the metal ion. These water molecules are called *waters of hydration* and are in a definite proportion. Thus when copper sulfate crystallizes from water, the blue salt copper(II) sulfate pentahydrate, $CuSO_4 \cdot 5H_2O$, forms. As indicated by the formula, 5 waters of hydration are bound to the copper(II) ion in copper sulfate. Notice how the formula is written—the waters of hydration are separated from the formula of the salt by a *dot*.

Heat can transform a hydrate into an anhydrous salt. The water can often be seen escaping as steam. For example, the blue crystals of copper(II) sulfate pentahydrate can be changed into a white powder, the anhydrous salt, by heating to approximately 250°C.

$$\underset{\text{Blue}}{\mathbf{CuSO_4 \cdot 5H_2O(s)}} \xrightarrow[250^\circ\mathrm{C}]{} \underset{\text{White}}{\mathbf{CuSO_4(s)}} + \mathbf{5H_2O(g)}$$

This process is reversible; adding water to the white anhydrous copper sulfate salt will rehydrate the salt and regenerate the blue pentahydrate. Not all hydrates behave in this way. For example, the waters of hydration that are bound to the iron(III) ion, Fe^{3+}, in the salt iron(III) chloride hexahydrate, $FeCl_3 \cdot 6H_2O$, are held very strongly; intense heat will not drive off the water.

Some anhydrous salts are capable of becoming hydrated upon exposure to the moisture in their surroundings. These salts are called *hygroscopic salts* and can be used as chemical drying agents or *desiccants*. Some salts are such excellent desiccants and are able to absorb so much moisture from their surroundings that they can eventually dissolve themselves! Calcium chloride, $CaCl_2$, is such a salt and is said to be *deliquescent*.

These salts and their hydrates are used in commercial applications. Containers holding pharmaceutical pills often have small packets of desiccant to control moisture so the pills last longer. Unless a desiccant is present, fertilizers will become wet and sticky as they absorb moisture from the air; some will even "turn to liquid" after some time as they absorb so much water that they dissolve. Some humidity indicators use cobalt or copper salts and vary in color as the moisture in the air varies.

Because many hydrates contain water in a stoichiometric quantity, it is possible to determine the molar ratio of water to salt. First, you would determine the weight of the water lost from the hydrate by heating a weighed sample. From the weight of the water lost, you then can calculate the percent of water in the hydrate. From the weight of the water lost you can also determine the number of water molecules in the hydrate salt and thus the molar ratio.

Example

A sample of Epsom salt, the hydrate of magnesium sulfate, 5.320 g, lost water on heating; the anhydrous salt, which remained, weighed 2.598 g.

a. The weight of the water lost:

Weight of hydrate sample (g)	**5.320 g**
−Weight of the anhydrous salt (g)	**−2.598 g**
Weight of the water lost (g)	**2.722 g**

b. The percent by mass of water:

$$\frac{\textbf{Weight of water lost (g)}}{\textbf{Weight of hydrate sample (g)}} \times 100 = \frac{2.722\,\text{g}}{5.320\,\text{g}} \times 100 = 51.17\%$$

c. The number of moles of water lost:

$$\frac{\textbf{Weight of water lost (g)}}{\textbf{MW of water (g/mole)}} = \frac{2.722\,\text{g}}{18.02\,\text{g/mole}} = 0.1511\ \text{mole}$$

d. The number of moles of $MgSO_4$:

$$\frac{\textbf{Weight of MgSO}_4\textbf{ anhydrous (g)}}{\textbf{MW of MgSO}_4\textbf{ (g/mole)}} = \frac{2.598\,\text{g}}{120.4\,\text{g/mole}} = 0.02158\ \text{mole}$$

e. The mole ratio of H_2O to anhydrous $MgSO_4$:

$$\frac{\textbf{Moles of H}_2\textbf{O}}{\textbf{Moles of MgSO}_4} = \frac{0.1511}{0.02158} = 7$$

Therefore, the formula of the hydrate of magnesium sulfate is $MgSO_4 \cdot 7H_2O$.

OBJECTIVES

1. To learn some properties and characteristics of hydrates.
2. To verify the percent of water in the hydrate of copper sulfate.
3. To verify that the mole ratio of water to salt in the hydrate of copper sulfate is fixed.

PROCEDURE

Properties of Anhydrous $CaCl_2$

1. Take a small spatula full of anhydrous $CaCl_2$ and place it on a watch glass.
2. Set the watch glass to the side, out of the way, and continue the rest of the experiment. From time to time during the period, examine the solid and record your observations.
3. Did anything happen to the solid $CaCl_2$ by the end of the period?

Composition of a Hydrate

1. Obtain from your instructor a porcelain crucible and cover. Clean with soap and water and dry thoroughly with paper towels.
2. Place the crucible and cover in a clay triangle supported by a metal ring on a ring stand (Figure 15.1). Heat the crucible with a Bunsen burner to red heat for 5 min. Using crucible tongs, place the crucible and cover on a wire gauze and allow it to cool to room temperature.
3. Weigh the crucible and cover to the nearest 0.001 g (1).
4. Repeat this procedure (heating, cooling, weighing) until two successive weights of the covered crucible agree to within 0.005 g or less (2).

CAUTION

Handle the crucible and cover with the crucible tongs from this point on. This will avoid possible burns and will avoid transfer of moisture and oils from your fingers to the porcelain.

5. Add between 3 and 4 g of the hydrate of copper sulfate to the crucible. Weigh the covered crucible to the nearest 0.001 g (3). Determine the exact weight of the hydrate (4) by subtraction.
6. Place the covered crucible and contents in the clay triangle. Move the cover so that it is slightly ajar (Figure 15.1a). Begin heating the crucible with a small flame for 5 min. (If any spattering of the solid occurs, remove the heat and completely cover the crucible.) Gradually increase the flame until the blue inner cone touches the bottom of the crucible. Heat to red hot for an additional 5 min.

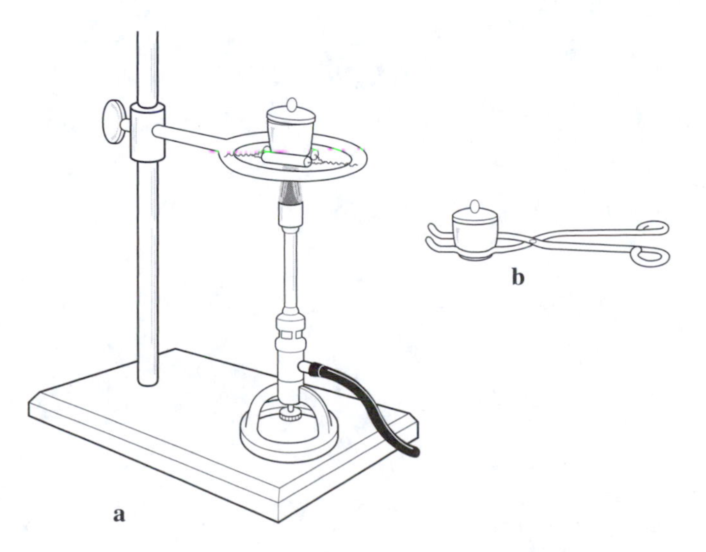

Figure 15.1
(a) Heating the crucible. (b) Moving the crucible with crucible tongs.

7. Remove the heat. Using crucible tongs, place the covered crucible on a wire gauze. Have the crucible completely covered by the lid. Allow to cool to room temperature. Weigh to the nearest 0.001 g (5).
8. Repeat the procedure (heating, cooling, weighing) until two successive weights of the covered crucible and contents agree to within 0.005 g or less (6).
9. Determine the weight of the anhydrous copper sulfate (7) and the weight of the water lost (8).
10. Carry out the calculations indicated on the Report Sheet.
11. If time permits, repeat the procedure again for a Trial 2.
12. Before you discard the white anhydrous salt (as directed by the instructor), add a few drops of water to the salt. What happens (14)?

CHEMICALS AND EQUIPMENT

1. Crucible and cover
2. Ring stand
3. Clay triangle
4. Crucible tongs
5. Calcium chloride, $CaCl_2$
6. Copper(II) sulfate pentahydrate, $CuSO_4 \cdot 5H_2O$

name | section | date

partner | grade

15 EXPERIMENT 15

Pre-Lab Questions

1. How does the Law of Constant Composition apply to hydrate salts?

2. Give an example of a salt that is
   a. hygroscopic
   b. a desiccant
   c. deliquescent

3. Natural gypsum is a hydrate of calcium sulfate, $CaSO_4 \cdot 2H_2O$.
   a. How many total atoms of all kinds are present?
   b. On heating the hydrate, how many moles of water should be driven off per mole of hydrate?
   c. Calculate the percent water in the hydrate. Show your work.
   d. If you heat 20.00 g of hydrate and drive off the water, what is the weight of the anhydrous salt remaining? Show your work.

name　　section　　date

partner　　grade

15

EXPERIMENT 15

Report Sheet

Observations on the properties of anhydrous $CaCl_2$

Composition of a hydrate	*Trial 1*	*Trial 2*
1. Weight of crucible and cover, first heating	________ g	________ g
2. Weight of crucible and cover, second heating	________ g	________ g
3. Weight of covered crucible plus sample	________ g	________ g
4. Weight of sample (hydrate): (3) − (2)	________ g	________ g
5. Weight of covered crucible plus sample, after first heating	________ g	________ g
6. Weight of covered crucible plus sample, after second heating	________ g	________ g
7. Weight of anhydrous salt: (6) − (2)	________ g	________ g
8. Weight of water lost: (4) − (7)	________ g	________ g
9. Percent of water in hydrate: % = [(8)/(4)] × 100	________ %	________ %
10. Moles of water lost: (8)/18.02 g/mole	________ mole	________ mole

11. Moles of anhydrous $CuSO_4$:

(7)/159.6 g/mole ____________ mole ____________ mole

12. Moles of water per mole of $CuSO_4$:

(10)/(11) ____________ ____________

13. The formula for the hydrated copper(II) sulfate ____________ ____________

14. Observation: water added to the anhydrous copper(II) sulfate:

Post-Lab Questions

1. During the heating of the hydrate, some solid was lost due to "spattering." How would this affect the experimentally determined percent of water in the hydrate?

2. Suppose your sample contained a volatile impurity. Would your determination of the percent of water in the hydrate (9) be too high or too low? Explain your answer.

3. Plaster of Paris is a hydrate of calcium sulfate, $(CaSO_4)_2 \cdot H_2O$.

   a. How many atoms of each kind are present?

   b. What is the mole ratio of salt to water?

   c. Calculate the percent water in the hydrate. Show your work.

4. A student found the percent of water in $ZnSO_4 \cdot 7H_2O$ to be 45.5%. Determine the experimental error. Show your work.

Factors Affecting Reaction Rates

BACKGROUND

Some chemical reactions take place rapidly; others are very slow. For example, antacid neutralizes stomach acid (HCl) rapidly but hydrogen and oxygen react with each other to form water very slowly. A tank containing a mixture of H_2 and O_2 shows no measurable change even after many years. The study of reaction rates is called *chemical kinetics*. The *rate of reaction* is the change in concentration of a reactant (or product) per unit time. For example, in the reaction

$$\mathbf{2HCl(aq) + CaCO_3(s) \rightleftharpoons CaCl_2(aq) + H_2O(l) + CO_2(g)}$$

we monitor the evolution of CO_2, and we find that 4.4 g of carbon dioxide gas was produced in 10 min. Because 4.4 g corresponds to 0.1 moles of CO_2, the rate of the reaction is 0.01 moles CO_2/min (0.1 mole/10 min.). On the other hand, if we monitor the HCl concentration, we may find that at the beginning we had 0.6 M HCl and after 10 min. the concentration of HCl was 0.4 M. This means that we used up 0.2 M HCl in 10 min. Thus the rate of reaction is 0.02 moles HCl/L-min (0.2 M/10 min.). From the above we can see that when describing the rate of reaction it is not sufficient to give a number. We have to specify the units and also the reactant (or product) we monitored.

In order for a reaction to take place, molecules or ions must first collide. Not every collision yields a reaction. In many collisions, the molecules simply bounce apart without reacting. A collision that results in a reaction is called an *effective collision*. The minimum energy necessary for the reaction to happen is called the *activation energy* (Figure 16.1). In this energy diagram, we see that the rate of reaction depends on this activation energy.

The lower the activation energy the faster the rate of reaction; the higher the activation energy the slower the reaction. This is true for both exothermic and endothermic reactions.

A number of factors affect the rates of reactions. Our experiments will demonstrate how these factors affect reaction rates.

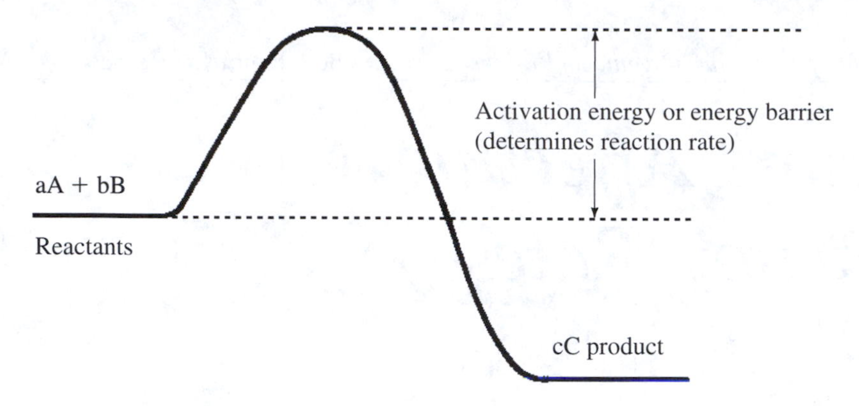

Figure 16.1
Energy diagram for a typical reaction.

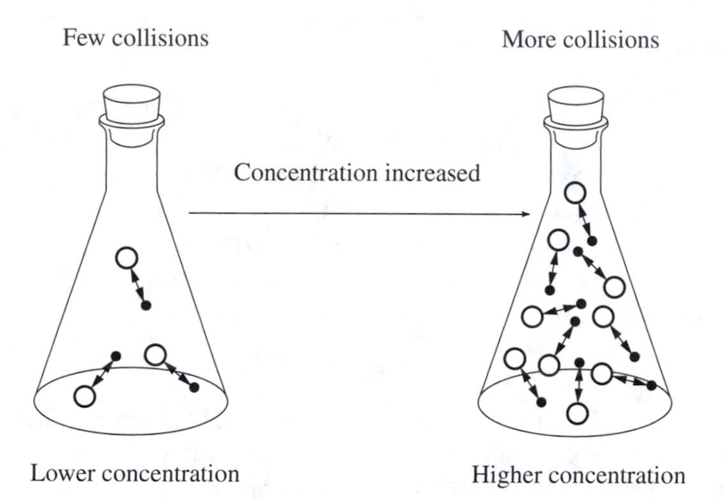

Figure 16.2
Concentration affects the rate of reaction.

1. **Nature of reactants.** Some compounds are more reactive than others. In general, reactions that take place between ions in aqueous solutions are rapid. Reactions between covalent molecules are much slower.
2. **Concentration.** In most reactions, the rate increases when the concentration of either or both reactants is increased. This is understandable on the basis of the collision theory. If we double the concentration of one reactant, it will collide in each second twice as many times with the second reactant as before. Because the rate of reaction depends on the number of effective collisions per second, the rate is doubled (Figure 16.2).
3. **Surface area.** If one of the reactants is a solid, the molecules of the second reactant can collide only with the surface of the solid. Thus the surface area of the solid is in effect its concentration. An increase in the surface area of the solid (by grinding to a powder in a mortar) will increase the rate of reaction.
4. **Temperature.** Increasing the temperature makes the reactants more energetic than before. This means that more molecules will have energy equal to or greater than the activation energy. Thus one expects an increase in the rate of reaction with increasing temperature. As a rule of thumb, every time the temperature goes up by 10°C, the rate of reaction doubles. This rule is far from exact, but it applies to many reactions.

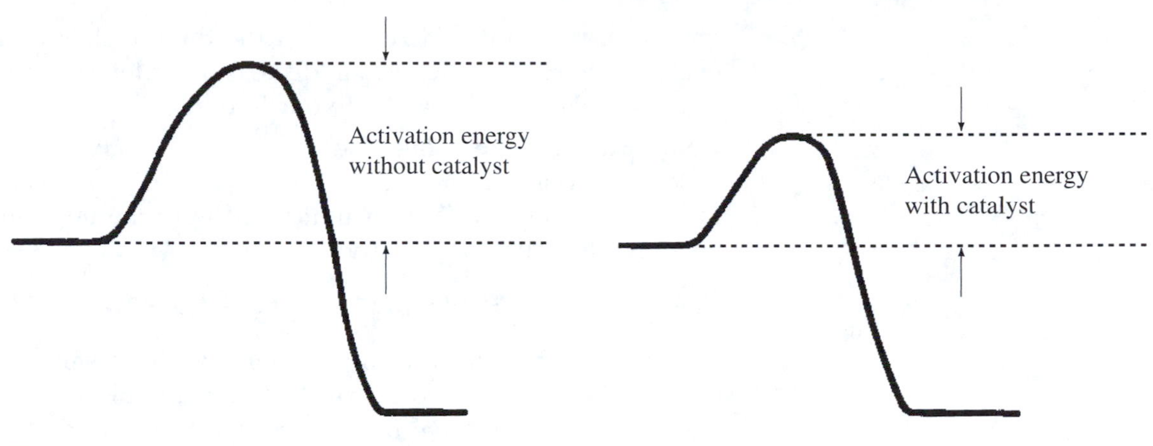

Figure 16.3
Energy diagrams of reactions with and without a catalyst.

5. **Catalyst.** Any substance that increases the rate of reaction without itself being used up in the process is called a *catalyst*. A catalyst increases the rate of reaction by lowering the activation energy (Figure 16.3). Thus many more molecules can cross the energy barrier (activation energy) in the presence of a catalyst than in its absence. Almost all the chemical reactions in our bodies are catalyzed by specific catalysts called enzymes.

OBJECTIVES

1. To investigate the relationship between reaction rate and the nature of reactants.
2. To measure the rate of reaction as a function of concentration.
3. To demonstrate the effect of temperature on the rate of reaction.
4. To investigate the effect of surface area and the effect of a catalyst on the rate of reaction.

PROCEDURE

1. **Nature of reactants.** Do all of the reactions in the five test tubes described below at the same time. Label five test tubes (75 × 10 mm) 1 through 5. Place 1 mL of acid into each test tube as follows: no. (1) 3 M H_2SO_4; no. (2) 6 M HCl; no. (3) 6 M HNO_3; no. (4) 2 M H_3PO_4; and no. (5) 6 M CH_3COOH. Now place into each test tube, as quickly as you can do it, one 1-cm polished strip of magnesium ribbon. The reaction will convert the magnesium ribbon to the corresponding salts with the liberation of hydrogen gas. You can assess the rate of reaction, qualitatively, by observing the speed with which the gas is liberated (bubbling) and/or by noticing the time it takes for the magnesium ribbon to disappear. Assess the rates of reaction; then list, in decreasing order, the rates of reaction of magnesium with the various acids on your Report Sheet (1).
2. Place 1 mL of 6 M HCl in each of three labeled test tubes. Add a 1-cm polished strip of magnesium to the first, zinc to the second, and copper to the third. Do all of the reactions in the three test tubes at the same

time; assess the rates of reaction of the three metals by the speed of evolution of H_2 gas; then list, in decreasing order, the rates of reaction of the metals with the acid on your Report Sheet (2).

3. **Concentration.** The *iodine clock reaction* is a convenient reaction for observing concentration effects. The reaction is between potassium iodate, KIO_3, and sodium bisulfite, $NaHSO_3$; the net ionic reaction is given by the following equation.

$$\mathbf{IO_3^-(aq) + 3HSO_3^-(aq) \rightleftharpoons I^-(aq) + 3SO_4^{2-}(aq) + 3H^+(aq)}$$

We can monitor the rate of reaction by the disappearance of the bisulfite. We do so by adding more IO_3^- than HSO_3^- at the start of the reaction. When we have used up all the bisulfite, there is still some iodate left. This will then react with the product iodide, I^-, and results in the formation of I_2.

$$\mathbf{IO_3^-(aq) + 5I^-(aq) + 6H^+(aq) \rightleftharpoons 3I_2(aq) + 3H_2O(l)}$$

We can detect the appearance of iodine with the aid of starch indicator; this reagent forms a blue complex with iodine. The time it takes for the blue color to suddenly appear indicates when all the bisulfite was used up in the first reaction. That's why the name: iodine clock. Thus you should measure the time (with a stopwatch, if available) elapsed between mixing the two solutions and the appearance of the blue color. Place the reactants in two separate 150-mL beakers according to the outline in Table 16.1. Use a graduated pipet to measure each reactant and a graduated cylinder to measure the water. Simultaneously pour the two reactants into a third beaker and time the appearance of the blue color. Repeat the experiment with the other two trial concentrations. Record your data on the Report Sheet (3).

Table 16.1 *Reactant Concentration and Rate of Reaction*

	Beaker A			**Beaker B**	
Trial	*0.1 M KIO_3*	*Starch*	*Water*	*0.01 M $NaHSO_3$*	*Water*
1	2.0 mL	2 mL	46 mL	5 mL	45 mL
2	4.0 mL	2 mL	44 mL	5 mL	45 mL
3	6.0 mL	2 mL	42 mL	5 mL	45 mL

4. **Surface area.** Using a large mortar and pestle, crush and pulverize about 0.5 g of marble chips. Place the crushed marble chips into one large test tube (150 × 16 mm) and 0.5 g of uncrushed marble chips into another. Add 2 mL of 6 M HCl to each test tube and note the speed of bubbling of the CO_2 gas. Record your data on the Report Sheet (4).

5. **Temperature.** Add 5 mL of 6 M HCl to three clean test tubes (150 × 16 mm). Place the first test tube in an ice bath, the second in a beaker containing warm water (50°C), and the third in a beaker with tap water (20°C). Wait 5 min. To each test tube add a piece of zinc ribbon (1 cm × 0.5 cm × 0.5 mm). Note the time you added the zinc metal. Finally, note the time when the bubbling of gas stops in each test tube and the

zinc metal disappears. Record the time of reaction (time of the disappearance of the zinc − the time of the start of the reaction) on your Report Sheet (5).

6. **Catalyst.** Add 2 mL of 3% H_2O_2 solution to two clean test tubes (150 × 16 mm). The evolution of oxygen bubbles will indicate if hydrogen peroxide decomposed. Note if anything happens. Add a few grains of MnO_2 to one of the test tubes. Note the evolution of oxygen, if any. Record your data on the Report Sheet (6).

CHEMICALS AND EQUIPMENT

1. Mortar and pestle
2. 10-mL graduated pipet
3. 5-mL volumetric pipet
4. Magnesium ribbon
5. Zinc ribbon
6. Copper ribbon
7. 3 M H_2SO_4
8. 6 M HCl
9. 6 M HNO_3
10. 2 M H_3PO_4
11. 6 M CH_3COOH
12. 0.1 M KIO_3
13. 0.01 M $NaHSO_3$
14. Starch indicator
15. Marble chips
16. 3% hydrogen peroxide
17. Manganese dioxide, MnO_2

name

section

date

partner

grade

16 EXPERIMENT 16

Pre-Lab Questions

1. You are suffering from heartburn and wish to get relief as quickly as possible. One antacid comes in the form of a compressed pill; another is in the form of a loose powder. Which form would give faster relief for your heartburn, considering that each contained the same ingredients and were taken in equal amounts? Why?

2. You increase the temperature of a reaction you are doing from 30°C to 40°C. What do you expect to happen?

3. Which reaction will be faster? Explain your choice.

   a. $Ag^+(aq) + Cl^-(aq) \rightarrow AgCl(s)$

   b. $H_2(g) + Cl_2(g) \rightarrow 2HCl(g)$

4. In the reaction,

   $2H_2O_2(l) \rightarrow 2H_2O(l) + O_2(g)$

   the following data was obtained for the accumulation of oxygen:

Time, min.	*O_2 in mole/L*
0	0
80	0.016
160	0.032
240	0.048

   What is the rate of reaction? How could the rate of reaction be increased?

name ______ section ______ date ______

partner ______ grade ______

16 EXPERIMENT 16

Report Sheet

1. Nature of reactants **Name of the acid**

Fastest reaction ______

______

______

______

Slowest reaction ______

2. Nature of reactants **Name of the metal**

Fastest reaction ______

______

Slowest reaction ______

3. Effect of concentration

Trial no.	Time
1	______
2	______
3	______

4. Surface area

Fast reaction ______

Slow reaction ______

5. Effect of temperature

Trial at	4°C	20°C	50°C
Reaction time	______	______	______

6. Catalyst **Observation**

No catalyst ______

MnO_2 ______

Post-Lab Questions

1. In general, we expect that as the temperature of a reaction increases, the rate will also increase. Why do we expect this effect?

2. In the **PRE-LAB** questions, you were asked to speculate on the effectiveness of a powdered antacid versus a pressed pill to remedy heartburn. Is there anything in this experiment that would suggest that a powder would act faster than a pill?

3. What happened when the manganese dioxide, MnO_2, was added to the peroxide solution? Why did this occur?

4. Suppose, in the reaction between magnesium and HCl, you used a solution of acid that was 3 M HCl rather than 6 M HCl. What change would you observe in the reaction rate? Why?

5. You did reactions of zinc with 6 M HCl at 20°C and at 50°C. Should you expect the reaction to be faster at the higher temperature? How much faster? What did you see?

The Law of Chemical Equilibrium and Le Chatelier's Principle

BACKGROUND

Two important questions are asked about every chemical reaction: (1) How much product is produced and (2) How fast is it produced? The first question involves chemical equilibrium and the second question belongs to the domain of chemical kinetics. (We dealt with kinetics in Experiment 16.) Some reactions are irreversible and they go to completion (100% yield). When you ignite methane gas in your gas burner in the presence of air (oxygen), methane burns completely and forms carbon dioxide and water.

$$CH_4(g) + 2O_2(g) \rightarrow CO_2(g) + 2H_2O(g)$$

Other reactions do not go to completion. They are reversible. In such cases, the reaction can go in either direction: forward or backward. For example, the reaction

$$Fe^{3+}(aq) + SCN^-(aq) \rightleftharpoons FeSCN^{2+}(aq)$$

is often used to illustrate reversible reactions. This is so because it is easy to observe the progress of the reaction visually. The yellow Fe^{3+} ion reacts with thiocyanate ion to form a deep red complex ion, $FeSCN^{2+}$. This is the forward reaction. At the same time, the complex red ion also decomposes and forms the yellow iron(III) ion and thiocyanate ion. This is the backward (reverse) reaction. At the beginning when we mix iron(III) ion and thiocyanate ion, the rate of the forward reaction is at a maximum. As time goes on, this rate decreases because we have less and less iron(III) and thiocyanate to react. On the other hand, the rate of the reverse reaction (which began at zero) gradually increases. Eventually the two rates become equal. When this point is reached, we call the process a *dynamic equilibrium*, or just *equilibrium*. When in equilibrium at a particular temperature, a reaction mixture obeys the *Law of Chemical Equilibrium*. This law imposes a condition on the concentration of reactants and products expressed in the equilibrium constant (K). For the above reaction between iron(III) and thiocyanate ions, the equilibrium constant can be written as

$$K = [FeSCN^{2+}]/[Fe^{3+}][SCN^-]$$

or in general

$$K = [\text{products}]/[\text{reactants}]$$

The brackets, [ ], indicate concentration, in moles/L, at equilibrium. As the name implies, the *equilibrium constant* is a constant at a set temperature for a particular reaction. Its magnitude tells if a reaction goes to completion or if it is far from completion (reversible reaction). A number much smaller than 1 for K indicates that at equilibrium only a few molecules of products are formed, meaning the mixture consists mainly of reactants. We say that the equilibrium lies far to the left. On the other hand, a completion of a reaction (100% yield) would have a very large number (infinite?) for the equilibrium constant. In this case, obviously the equilibrium lies far to the right. The above reaction between iron(III) and thiocyanate has an equilibrium constant of 207, indicating that the equilibrium lies to the right but does not go to completion. Thus at equilibrium, both reactants and product are present, albeit the products far outnumber the reactants.

The Law of Chemical Equilibrium is based on the constancy of the equilibrium constant. This means that if one disturbs the equilibrium, for example by adding more reactant molecules, there will be an increase in the number of product molecules in order to maintain the product/reactant ratio unchanged and thus preserve the numerical value of the equilibrium constant. The *Le Chatelier Principle* expresses this as follows: *If an external stress is applied to a system in equilibrium, the system reacts in such a way as to partially relieve the stress.* In our present experiment, we demonstrate the Le Chatelier Principle in two manners: (1) disturbing the equilibrium by changing the concentration of a product or reactant; and (2) changing the temperature.

Concentration

1. In the first experiment, we add ammonia to a pale blue copper(II) sulfate solution. The ionic reaction is

$$\underset{\text{pale blue}}{Cu(H_2O)_4^{2+}(aq)} + \underset{\text{colorless}}{4NH_3(aq)} \rightleftharpoons \underset{\text{(color?)}}{Cu(NH_3)_4^{2+}(aq)} + 4H_2O(l)$$

A change in the color indicates the copper–ammonia complex formation. Adding a strong acid, HCl, to this equilibrium causes the ammonia, NH_3, to react with the acid:

$$NH_3(aq) + H^+(aq) \rightleftharpoons NH_4^+(aq)$$

Thus we removed some reactant molecules from the equilibrium mixture. As a result we expect the equilibrium to shift to the left, reforming hydrated copper(II) ions with the reappearance of pale blue color.

2. In the second reaction, we demonstrate the common ion effect. When we have a mixture of $H_2PO_4^-/HPO_4^{2-}$ solution, the following equilibrium exists:

$$H_2PO_4^-(aq) + H_2O(l) \rightleftharpoons H_3O^+(aq) + HPO_4^{2-}(aq)$$

If we add a few drops of aqueous HCl to the solution, we will have added a common ion, H_3O^+, that already was present in the equilibrium mixture. We expect, on the basis of the Le Chatelier Principle, that the equilibrium will shift to the left. Thus the solution will not become acidic.

3. In the iron(III)–thiocyanate reaction

$$\underset{\text{yellow}}{\mathbf{Fe^{3+}(aq)}} + \mathbf{3Cl^-(aq) + K^+(aq)} + \underset{\text{colorless}}{\mathbf{SCN^-(aq)}} \rightleftharpoons$$
$$\underset{\text{red}}{\mathbf{Fe(SCN)^{2+}(aq)}} + \underset{\text{colorless}}{\mathbf{3Cl^-(aq) + K^+(aq)}}$$

the chloride and potassium ions are spectator ions. Nevertheless, their concentration may also influence the equilibrium. For example, when the chloride ions are in excess, the yellow color of the Fe^{3+} will disappear with the formation of a colorless $FeCl_4^-$ complex

$$\underset{\text{yellow}}{\mathbf{Fe^{3+}(aq)}} + \mathbf{4Cl^-(aq)} \rightleftharpoons \underset{\text{colorless}}{\mathbf{FeCl_4^-(aq)}}$$

Temperature

1. Most reactions are accompanied by some energy changes. Frequently, the energy is in the form of heat. We talk of endothermic reactions if heat is consumed during the reaction. In endothermic reactions, we can consider heat as one of the reactants. Conversely, heat is evolved in an exothermic reaction, and we can consider heat as one of the products. Therefore, if we heat an equilibrium mixture of an endothermic reaction, it will behave as if we added one of its reactants (heat) and the equilibrium will shift to the right. Heating the equilibrium mixture of an exothermic reaction, the equilibrium will shift to the left. We will demonstrate the effect of temperature on the reaction:

$$\underset{\text{pale rose}}{\mathbf{Co(H_2O)_6^{2+}(aq)}} + \mathbf{4Cl^-(aq)} \rightleftharpoons \underset{\text{deeper strawberry}}{\mathbf{CoCl_4^{2-}(aq)}} + \mathbf{6H_2O(l)}$$

You will observe a change in the color depending on whether the equilibrium was established at room temperature or at 100°C (in boiling water). From the color change, you should be able to tell whether the reaction was endothermic or exothermic.

OBJECTIVES

1. To study chemical equilibria.
2. To investigate the effects of (1) changing concentrations and (2) changing temperature in equilibrium reactions.

PROCEDURE

Concentration Effects

1. Place 20 drops (about 1 mL) of 0.1 M $CuSO_4$ solution into a clean and dry test tube (100 × 13 mm). Add (dropwise) 1 M NH_3 solution, mixing the contents after each drop. Continue to add until the color changes. Note the new color and the number of drops of 1 M ammonia added and record it on your Report Sheet (1). To the equilibrium mixture thus obtained, add (dropwise, counting the number of drops added) 1 M HCl solution until the color changes back to pale blue. Report your observations on your Report Sheet (2).

2. Place 2 mL of $H_2PO_4^-/HPO_4^{2-}$ solution into a clean and dry test tube (100 × 13 mm). Use red and blue litmus papers and test to see whether the solution is acidic or basic. Record your findings on your Report Sheet (3). Add a drop of 1 M HCl to new pieces of red and blue litmus paper. Record your observation on the Report Sheet (4). Add one drop of 1 M HCl solution to the test tube. Mix it and test it with red and blue litmus paper. Record your observation on the Report Sheet (5).
3. Prepare a stock solution by adding 1 mL of 0.1 M iron(III) chloride, $FeCl_3$, and 1 mL of 0.1 M potassium thiocyanate, KSCN, to 50 mL distilled water in a 100-mL beaker. Set up four clean and dry test tubes (100 × 13 mm) and label them nos. 1, 2, 3, and 4. To each test tube add about 2 mL of the stock equilibrium mixture you just prepared. Use the solution in test tube no. 1 as the standard to which you can compare the color of the other solutions. To test tube no. 2, add 10 drops of 0.1 M iron(III) chloride solution; to test tube no. 3, add 10 drops of 0.1 M KSCN solution. To test tube no. 4, add five drops of saturated NaCl solution. Observe the color in each test tube and record your observations on the Report Sheet (6) and (7).

Temperature Effects

1. Set up two clean and dry test tubes (100 × 13 mm). Label them nos. 1 and 2. Prepare a boiling water bath by heating a 400-mL beaker containing about 200 mL water to a boil.

CAUTION

Concentrated HCl is toxic and can cause skin burns. Wear gloves when dispensing. Do not allow skin contact. If you do come into contact with the acid, immediately wash the exposed area with plenty of water for at least 15 min. Do not inhale the HCl vapors. Dispense in the hood.

2. Place 5 drops of 1 M $CoCl_2$ solution in test tube no. 1. Add concentrated HCl dropwise until a color change occurs. Record your observation on the Report Sheet (8).
3. Place 1 mL $CoCl_2$ solution in test tube no. 2. Note the color. Immerse the test tube into the boiling water bath. Report your observations on the Report Sheet (9) and (10).

CHEMICALS AND EQUIPMENT

1. 0.1 M $CuSO_4$
2. 1 M NH_3
3. 1 M HCl
4. Saturated NaCl
5. Concentrated HCl
6. 0.1 M KSCN
7. 0.1 M $FeCl_3$
8. 1 M $CoCl_2$
9. $H_2PO_4^-/HPO_4^{2-}$ solution
10. Litmus paper

name | section | date

partner | grade

17 EXPERIMENT 17

Pre-Lab Questions

1. For the reaction at 20°C,

   $NH_3(aq) + H^+(aq) \rightleftharpoons NH_4^+(aq)$

   the equilibrium constant is calculated to be $K = 4.5 \times 10^8$.

   a. What is favored in this reaction: starting material or products?

   b. If the reaction is exothermic and the temperature is increased to 40°C, how would the equilibrium concentration of NH_3 change? Explain your answer.

2. If the reaction between iron(III) ion and thiocyanate ion yielded an equilibrium concentration of 0.15 M for each of these ions, what is the equilibrium concentration of the red iron(III)–thiocyanate complex? (*Hint:* The equilibrium constant for the reaction is in the **Background** section.) Show your work.

3. In the reaction below,

   $NH_3(g) + H_2O(l) \rightleftharpoons NH_2^- + H_3O^+$

   the equilibrium constant is 10^{-34}. Would this reaction be practical to run? Explain your answer.

name | section | date

partner | grade

17 EXPERIMENT 17

Report Sheet

1. What is the color of the copper–ammonia complex? ____________

   How many drops of 1 M ammonia did you add to cause a change in color? ____________

2. How many drops of 1 M HCl did you add to cause a change in color back to pale blue? ____________

3. Testing the phosphate solution, what was the color of the red litmus paper? ____________

   What was the color of the blue litmus paper? ____________

4. Testing the 1 M HCl solution, what was the color of the red litmus paper? ____________

   What was the color of the blue litmus paper? ____________

5. After adding one drop of 1 M HCl to the phosphate solution and testing it with litmus paper, what was the color of the red litmus paper? ____________

   What was the color of the blue litmus paper? ____________

   Was your phosphate solution acidic, basic, or neutral

   a. before the addition of HCl? ____________

   b. after the addition of HCl? ____________

   Was your solution after adding HCl acidic, basic, or neutral? ____________

6. Compare the colors in each of the test tubes containing the iron(III) chloride–thiocyanate mixtures:

   no. 1 ____________________

   no. 2 ____________________

   no. 3 ____________________

   no. 4 ____________________

7. In which direction did the equilibrium shift in test tube

   no. 2 ______________________

   no. 3 ______________________

   no. 4 ______________________

8. What is the color of the $CoCl_2$ solution

   a. before the addition of HCl? ____________

   b. after the addition of HCl? ____________

9. What is the color of the $CoCl_2$ solution

   a. at room temperature? ____________

   b. at boiling water temperature? ____________

10. In which direction did the equilibrium shift upon heating? ____________

11. From the above shift, determine if the reaction was exothermic or endothermic. ____________

Post-Lab Questions

1. Consider the reaction:

$Fe^{3+}(aq) + 4Cl^{-}(aq) \rightleftharpoons FeCl_4^{-}(aq)$
yellow colorless

Explain what would happen to the color of a dilute solution containing $FeCl_4^{-}$ if

a. you added a solution containing silver ion, Ag^{+}. (Silver ion reacts with chloride ion in solution to form the precipitate AgCl.)

b. you added concentrated HCl.

c. you added concentrated H_2SO_4.

2. Adding HCl or H_2SO_4 to a mixture of $H_2PO_4^{-}/HPO_4^{2-}$ would cause a common ion effect. What is the common ion, and what would be the overall observation?

3. The Haber process is an important reaction for the fixation of nitrogen; nitrogen is converted into ammonia, an important component in the production of fertilizers.

$N_2(g) + 3H_2(g) \rightleftharpoons 2NH_3(g) + 22{,}000\ \text{cal.}$

Consider the reaction is at equilibrium. Explain how the equilibrium is shifted when

a. more nitrogen is added?

b. more hydrogen is added?

c. ammonia is removed?

d. the temperature is increased?

e. the temperature is decreased?

At 25°C the equilibrium constant is 7.4×10^2. What is favored: starting material or product?

pH and Buffer Solutions

BACKGROUND

We frequently encounter acids and bases in our daily life. Fruits, such as oranges, apples, etc., contain acids. Household ammonia, a cleaning agent, and Liquid Plumber are bases. *Acids* are compounds that can donate a proton (hydrogen ion). *Bases* are compounds that can accept a proton. This classification system was proposed simultaneously by Johannes Brønsted and Thomas Lowry in 1923, and it is known as the Brønsted-Lowry theory. Thus any proton donor is an acid, and a proton acceptor is a base.

When HCl reacts with water

$$HCl + H_2O \rightleftharpoons H_3O^+ + Cl^-$$

HCl is an **acid** and H_2O is a **base** because HCl **donated a proton,** thereby becoming Cl^-, and water **accepted a proton,** thereby becoming H_3O^+.

In the reverse reaction (from right to left) the H_3O^+ is an acid and Cl^- is a base. As the arrow indicates, the equilibrium in this reaction lies far to the right. That is, out of every 1000 HCl molecules dissolved in water, 990 are converted to Cl^- and only 10 remain in the form of HCl at equilibrium. But H_3O^+ (hydronium ion) is also an acid and can donate a proton to the base, Cl^-. Why do hydronium ions not give up protons to Cl^- with equal ease and form more HCl? This is because different acids and bases have different strengths. HCl is a stronger acid than hydronium ion, and water is a stronger base than Cl^-.

In the Brønsted-Lowry theory, every acid–base reaction creates its *conjugate acid–base pair*. In the above reaction HCl is an acid, which, after giving up a proton, becomes a conjugate base, Cl^-. Similarly, water is a base, which, after accepting a proton, becomes a conjugate acid, the hydronium ion.

conjugate base–acid pair (H_2O / H_3O^+)

$$HCl + H_2O \rightleftharpoons H_3O^+ + Cl^-$$

conjugate acid–base pair (HCl / Cl^-)

Some acids can give up only one proton. These are *monoprotic* acids. Examples are (H)Cl, (H)NO_3, HCOO(H), and CH_3COO(H). The hydrogens

circled are the ones donated. Other acids yield two or three protons. These are called *diprotic* or *triprotic* acids. Examples are H_2SO_4, H_2CO_3, and H_3PO_4. However, in the Brønsted-Lowry theory, each acid is considered monoprotic, and a diprotic acid (such as carbonic acid) donates its protons in two distinct steps:

1. $H_2CO_3 + H_2O \rightleftharpoons H_3O^+ + HCO_3^-$
2. $HCO_3^- + H_2O \rightleftharpoons H_3O^+ + CO_3^{2-}$

Thus the compound HCO_3^- is a conjugate base in the first reaction and an acid in the second reaction. A compound that can act either as an acid or a base is called *amphiprotic.*

In the self-ionization reaction

$$\mathbf{H_2O + H_2O \rightleftharpoons H_3O^+ + OH^-}$$

one water acts as an acid (proton donor) and the other as a base (proton acceptor). In pure water, the equilibrium lies far to the left, that is, only very few hydronium and hydroxyl ions are formed. In fact, only 1×10^{-7} moles of hydronium ion and 1×10^{-7} moles of hydroxide ion are found in one liter of water. The dissociation constant for the self-ionization of water is

$$\mathbf{K_d = \frac{[H_3O^+][OH^-]}{[H_2O]^2}}$$

This can be rewritten as

$$\mathbf{K_w = K_d[H_2O]^2 = [H_3O^+][OH^-]}$$

K_w, the **ion product of water,** is still a constant because very few water molecules reacted to yield hydronium and hydroxide ions; hence the concentration of water essentially remained constant. At room temperature, the K_w has the value of

$$\mathbf{K_w = 1 \times 10^{-14} = [1 \times 10^{-7}] \times [1 \times 10^{-7}]}$$

This value of the ion product of water applies not only to pure water but to any aqueous (water) solution. This is very convenient because if we know the concentration of the hydronium ion, we automatically know the concentration of the hydroxide ion and vice versa. For example, if in a 0.01 M HCl solution HCl dissociates completely, the hydronium ion concentration is $[H_3O^+] = 1 \times 10^{-2}$ M. This means that the $[OH^-]$ is

$$\mathbf{[OH^-] = K_w/[H_3O^+] = 1 \times 10^{-14}/1 \times 10^{-2} = 1 \times 10^{-12}\ M}$$

To measure the strength of an aqueous acidic or basic solution, P. L. Sorensen introduced the pH scale.

$$\mathbf{pH = -log[H_3O^+]}$$

In pure water, we have seen that the hydronium ion concentration is 1×10^{-7} M. The logarithm of this is −7 and, thus, the pH of pure water is 7. Because water is an amphiprotic compound, pH 7 means a neutral solution. On the other hand, in a 0.01 M HCl solution (dissociating completely), we have $[H_3O^+] = 1 \times 10^{-2}$ M. Thus its pH is 2. The pH scale shows that acidic solutions have a pH less than 7 and basic solutions have a pH greater than 7.

pH	0	1	2	3	4	5	6	7	8	9	10	11	12	13	14
				acidic				neutral			basic				

The pH of a solution can be measured conveniently by special instruments called pH meters. All that must be done is to insert the electrodes of the pH meter into the solution to be measured and read the pH from a scale. pH of a solution can also be obtained, although less precisely, by using pH indicator paper. The paper is impregnated with organic compounds that change their color at different pH values. The color shown by the paper is then compared with a color chart provided by the manufacturer.

There are certain solutions that resist a change in the pH even when we add to them acids or bases. Such systems are called *buffers.* A mixture of a weak acid and its conjugate base usually forms a good buffer system. An example is carbonic acid, which is the most important buffer in our blood and maintains it close to pH 7.4. Buffers resist large changes in pH because of the Le Chatelier Principle governing equilibrium conditions. In the carbonic acid–bicarbonate (weak acid–conjugate base) buffer system,

$$\mathbf{H_2CO_3 + H_2O \rightleftharpoons HCO_3^- + H_3O^+}$$

any addition of an acid, H_3O^+, will shift the equilibrium to the left. Thus this reduces the hydronium ion concentration, returning it to the initial value so that it stays constant; hence the change in pH is small. If a base, OH^-, is added to such a buffer system, it will react with the H_3O^+ of the buffer. But the equilibrium then shifts to the right, replacing the reacted hydronium ions, hence again, the change in pH is small.

Buffers stabilize a solution at a certain pH. This depends on the nature of the buffer and its concentration. For example, the carbonic acid–bicarbonate system has a pH of 6.37 when the two ingredients are at equimolar concentration. A change in the concentration of the carbonic acid relative to its conjugate base can shift the pH of the buffer. The Henderson-Hasselbalch equation below gives the relationship between pH and concentration.

$$\mathbf{pH = pK_a + \log\frac{[A^-]}{[HA]}}$$

In this equation the pK_a is the $-\log K_a$, where K_a is the dissociation constant of carbonic acid,

$$\mathbf{K_a = \frac{[HCO_3^-][H_3O^+]}{[H_2CO_3]}}$$

[HA] is the concentration of the acid, and $[A^-]$ is the concentration of the conjugate base. The pK_a of the carbonic acid–bicarbonate system is 6.37. When equimolar conditions exist, then $[HA] = [A^-]$. In this case, the second term in the Henderson-Hasselbalch equation is zero. This is so because $[A^-]/[HA] = 1$, and the $\log 1 = 0$. Thus at equimolar concentration of the acid–conjugate base, the pH of the buffer equals the pK_a; in the carbonic acid–bicarbonate system this is 6.37. If, however, we have ten times more bicarbonate than carbonic acid, $[A^-]/[HA] = 10$, then $\log 10 = 1$ and the pH of the buffer will be

$$pH = pK_a + \log [A^-]/[HA] = 6.37 + 1.0 = 7.37$$

This is what happens in our blood—the bicarbonate concentration is ten times that of the carbonic acid and this keeps our blood at a pH of 7.4. Any large change in the pH of our blood may be fatal (acidosis or alkalosis). Other buffer systems work the same way. For example, the second buffer system in our blood is

$$H_2PO_4^- + H_2O \rightleftharpoons HPO_4^{2-} + H_3O^+$$

The pK_a of this buffer system is 7.21. It requires a 1.6 to 1.0 molar ratio of HPO_4^{2-} to $H_2PO_4^-$ to maintain our blood at pH 7.4.

OBJECTIVES

1. To learn how to measure the pH of a solution.
2. To understand the operation of buffer systems.

PROCEDURE

Measurement of pH

1. Add one drop of 0.1 M HCl to the first depression of a spot plate. Dip a 2-cm long universal pH paper into the solution. Remove the excess liquid from the paper by touching the plate. Compare the color of the paper to the color chart provided (Figure 18.1). Record the pH on your Report Sheet (1).
2. Repeat the same procedure with 0.1 M acetic acid, 0.1 M sodium acetate, 0.1 M carbonic acid (or club soda or seltzer), 0.1 M sodium bicarbonate, 0.1 M ammonia, and 0.1 M NaOH. For each solution, use a different depression of the spot plate and a new piece of pH paper. Record your results on the Report Sheet (1).
3. Depending on the availability of the number of pH meters this may be a class exercise (demonstration), or 6–8 students may use one pH meter. Add 5 mL of 0.1 M acetic acid to a dry and clean 10-mL beaker. Wash the electrode over a 200-mL beaker with distilled or deionized water contained in a wash bottle. The 200-mL beaker serves to collect the wash water. Gently wipe the electrode with Kimwipes (or other soft tissues) to dryness. Insert the dry electrode into the acetic acid solution. Your pH meter has been calibrated by your instructor. Switch

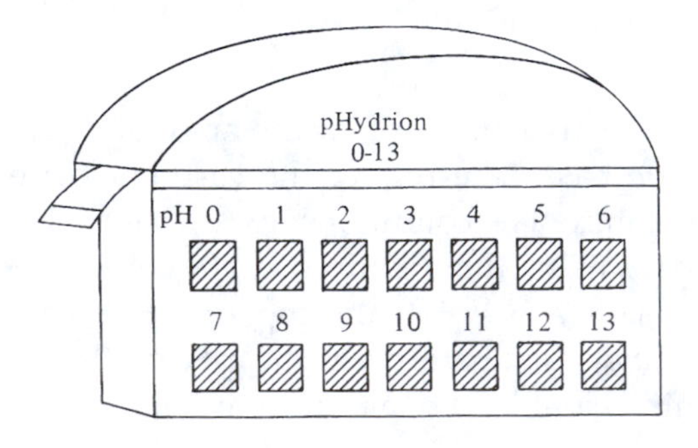

Figure 18.1
pH paper dispenser.

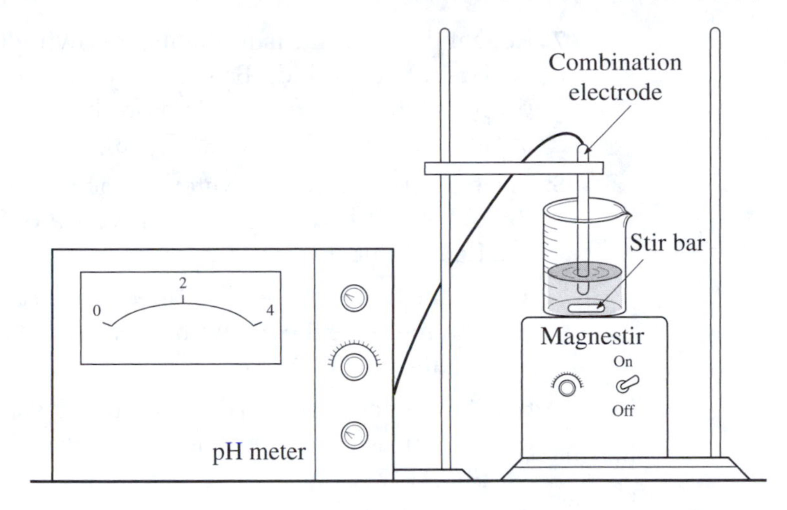

Figure 18.2
pH meter.

"on" the pH meter and read the pH from the position of the needle on your scale. Alternatively, if you have a digital pH meter, a number corresponding to the pH will appear (Figure 18.2).

CAUTION

Make sure the electrode is immersed into the solution but does not touch the walls or the bottom of the beaker. Electrodes are made of thin glass, and they break easily if you don't handle them gently.

4. Repeat the same procedures with 0.1 M sodium acetate, 0.1 M carbonic acid, 0.1 M sodium bicarbonate, and 0.1 M ammonia. Make certain that for each solution you use a dry and clean beaker, and before each measurement wash the electrode with distilled water and dry with Kimwipes. Record your data on the Report Sheet (2).

Buffer Systems

5. Prepare four buffer systems in four separate, labeled, dry, and clean 50-mL beakers, as follows:
   - **a.** 5 mL 0.1 M acetic acid + 5 mL 0.1 M sodium acetate
   - **b.** 1 mL 0.1 M acetic acid + 10 mL 0.1 M sodium acetate
   - **c.** 5 mL 0.1 M carbonic acid + 5 mL 0.1 M sodium bicarbonate
   - **d.** 1 mL 0.1 M carbonic acid + 10 mL 0.1 M sodium bicarbonate

   Measure the pH of each buffer system with the aid of universal pH paper. Record your data on the Report Sheet (3), (6), (9), and (12).
6. Divide each of the buffers you prepared (from above: a, b, c, d) into two halves (5 mL each) and place them into clean and dry 10-mL beakers.
   - **a.** To the first 5-mL sample of buffer (a), add 0.5 mL 0.1 M HCl. Mix and measure the pH with the aid of universal pH paper. Record your data on the Report Sheet (4).
   - **b.** To the second 5-mL sample of buffer (a), add 0.5 mL 0.1 M NaOH. Mix and measure the pH with universal pH paper. Record your data on the Report Sheet (5).

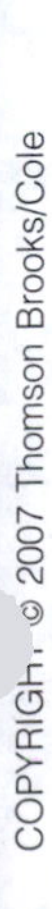

7. Repeat the same measurements, following the steps in **6a** and **6b**, using buffers (b), (c) and (d). Be sure to use clean, dry 10-mL beakers for each preparation. Record your data on the Report Sheet for the appropriate buffer system at the spaces (7), (8), (10), (11), (13), and (14).
8. Place 5 mL of distilled water in each of two 10-mL beakers. Measure the pH of distilled water with universal pH paper. Record the result on the Report Sheet (15).
   - **a.** To the first sample of distilled water, add 0.5 mL of 0.1 M HCl. Mix and measure the pH with universal pH paper. Record the result on the Report Sheet (16).
   - **b.** To the second sample of distilled water, add 0.1 M NaOH. Mix and measure the pH as before. Record the result on the Report Sheet (17).
9. Dispose of all solutions in properly labeled liquid waste containers.

CHEMICALS AND EQUIPMENT

1. pH meter
2. pH paper
3. Kimwipes
4. Wash bottle
5. 0.1 M HCl
6. 0.1 M acetic acid (CH_3COOH)
7. 0.1 M sodium acetate ($CH_3COO^-Na^+$)
8. 0.1 M carbonic acid or club soda or seltzer
9. 0.1 M $NaHCO_3$
10. 0.1 M $NH_3(aq)$ (aqueous ammonia)
11. 0.1 M NaOH
12. Spot plate
13. 10-mL beakers

name | section | date

partner | grade

18

EXPERIMENT 18

Pre-Lab Questions

1. Sulfuric acid, H_2SO_4, is a diprotic acid. What does this mean?

2. Write the formulas for the conjugate bases of the following acids:

a. HCl

b. HCOOH

c. H_2SO_4

3. The pH of blood is 7.4 and that of saliva is 6.4. Which of the two is more acidic? How much more acidic is it?

4. What do you need in order to make an effective buffer?

5. The pK_a of acetic acid, CH_3COOH, is 4.75. What is the pH of a solution in which the acetate/acetic acid ratio is (a) 10 and (b) 0.1?

name ______ section ______ date ______

partner ______ grade ______

18 EXPERIMENT 18

Report Sheet

pH of solutions	**1.** *by pH paper*	**2.** *by pH meter*
0.1 M HCl	______	not done
0.1 M acetic acid	______	______
0.1 M sodium acetate	______	______
0.1 M carbonic acid	______	______
0.1 M sodium bicarbonate	______	______
0.1 M ammonia	______	______
0.1 M NaOH	______	not done

Buffer system a

	pH
3. 5 mL 0.1 M CH_3COOH + 5 mL 0.1 M $CH_3COO^-Na^+$	______
4. after addition 0.5 mL 0.1 M HCl	______
5. after addition 0.5 mL 0.1 M NaOH	______

Buffer system b

6. 1 mL 0.1 M CH_3COOH + 10 mL 0.1 M $CH_3COO^-Na^+$	______
7. after addition 0.5 mL 0.1 M HCl	______
8. after addition 0.5 mL 0.1 M NaOH	______

Buffer system c

9. 5 mL 0.1 M H_2CO_3 + 5 mL 0.1 M $NaHCO_3$	______
10. after addition 0.5 mL 0.1 M HCl	______
11. after addition 0.5 mL 0.1 M NaOH	______

Buffer system d

12. 1 mL 0.1 M H_2CO_3 + 10 mL 0.1 M $NaHCO_3$ __________

13. after addition 0.5 mL 0.1 M HCl __________

14. after addition 0.5 mL 0.1 M NaOH __________

No buffer system

15. distilled water __________

16. after addition of 0.5 mL 0.1 M HCl __________

17. after addition of 0.5 mL 0.1 M NaOH __________

Post-Lab Questions

1. From your data on the Report Sheet, which buffer system was the best in resisting a change in pH?

2. Which gives a more accurate reading of the pH: pH paper or a pH meter?

3. From your data, does distilled water fit the definition of a buffer? Explain your answer.

4. Write equations to show how a buffer made up of equimolar amounts of acetic acid, CH_3COOH, and acetate, CH_3COO^-, would behave when (a) HCl was added and (b) NaOH was added.

5. Calculate the expected pH values of the buffer systems from the experiments (a, b, c, d), using the Henderson-Hasselbalch equation shown in the **Background** section. Use for pK_a values: carbonic acid = 6.37 and acetic acid = 4.75.

a.

b.

c.

d.

Are these calculated values in agreement with your measured pH values?

Analysis of Vinegar by Titration

BACKGROUND

In order to measure how much acid or base is present in a solution we often use a method called *titration*. If a solution is acidic, titration consists of adding base to it until all the acid is neutralized. To do this, we need two things: (1) a means of measuring how much base is added and (2) a means of telling just when the acid is completely neutralized.

How much base is added requires the knowledge of the number of moles of the base. The number of moles can be calculated if we know the volume (**V**) and the molarity (**M**) of the base.

$$\mathbf{moles = V \times M}$$

When we deal with monoprotic acids and bases that produce one OH^- per molecule, the titration is completed when the number of moles of acid equals the number of moles of base.

$$\mathbf{Moles_{acid} = Moles_{base}}$$

or

$$\mathbf{V_{acid}M_{acid} = V_{base}M_{base}}$$

This is called the *titration equation*.

We use an *indicator* to tell us when the titration is completed. Indicators are organic compounds that change color when there is a change in the pH of the solution. The *end point* of the titration is when a sudden change in the pH of the solution occurs. Therefore, we can tell the completion of the titration when we observe a change in the color of our solution to which a few drops of indicator have been added.

Commercial vinegar contains 5–6% acetic acid. Acetic acid, CH_3COOH, is a monoprotic acid. It is a weak acid and when titrated with a strong base such as NaOH, upon completion of the titration, there is a sudden change in the pH in the range from 6.0 to 9.0. The best way to monitor such a change is to use the indicator phenolphthalein, which changes from colorless to a pink hue at pH 8.0–9.0.

With the aid of the titration equation, we can calculate the concentration of acetic acid in the vinegar. To do so, we must know the volume of the acid (5 mL), the molarity of the base (0.2 M), and the volume of the base used to reach the end point of the titration. This will be read from the *buret,* which is filled with the 0.2 M NaOH solution at the beginning of the titration to its maximum capacity. The base is then slowly added (dropwise) from the buret to the vinegar in an Erlenmeyer flask. Continuous swirling ensures proper mixing. The titration is stopped when the indicator shows a permanent pink coloration. The buret is read again. The volume of the base added is the difference between the initial volume (25 mL) and the volume left in the buret at the end of titration.

OBJECTIVES

1. To learn the techniques of titration.
2. To determine the concentration of acetic acid in vinegar.

PROCEDURE

1. Rinse a 25-mL buret (or 50-mL buret) with about 5 mL of 0.2 M NaOH solution. (Be sure to record the exact concentration of the base.) After rinsing, fill the buret with 0.2 M NaOH solution about 2 mL above the 0.0-mL mark. Use a clean and dry funnel for filling. Tilting the buret at a 45-degree angle, slowly turn the stopcock to allow the solution to fill the tip. Collect the excess solution dripping from the tip into a beaker to be discarded later. The air bubbles must be completely removed from the tip. If you do not succeed the first time, repeat it until the liquid in the buret forms one continuous column from top to bottom. Clamp the buret onto a ring stand (Figure 19.1a). By slowly opening the stopcock, allow the bottom of the meniscus to drop to the 0.0-mL mark. Collect the excess solution dripping from the tip into a beaker to be discarded later. Read the meniscus carefully to the nearest 0.1 mL (Figure 19.1b).

CAUTION

Do not use your mouth when using the pipet. Use the Spectroline pipet filler; directions for its use are in Experiment 2. Review the section in Experiment 2 if you do not remember its proper use.

2. With the aid of a 5-mL volumetric pipet, add 5 mL vinegar to a 100-mL Erlenmeyer flask. Allow the vinegar to drain completely from the pipet by holding the pipet in such a manner that its tip touches the wall of the flask. Record the volume of the vinegar for trial 1 on your Report Sheet (1). Record also the molarity of the base (2) and the initial reading of the base in the buret on your Report Sheet (3). Add a few drops of phenolphthalein indicator to the flask and about 10 mL of distilled water. The distilled water is added to dilute the natural color that some commercial vinegars have. In this way, the natural color will not interfere with the color change of the indicator.

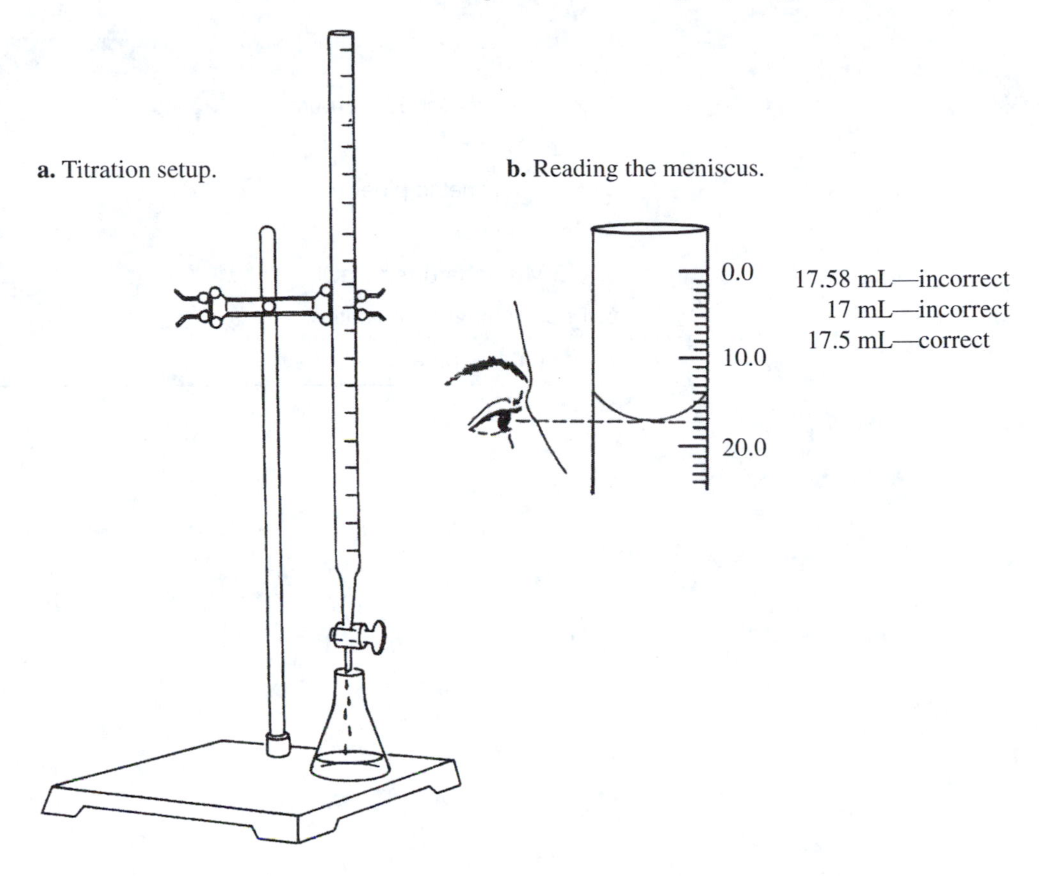

Figure 19.1
Titration setup.

3. While holding the neck of the Erlenmeyer flask in your left hand and swirling it, open the stopcock of the buret slightly with your right hand and allow the *dropwise* addition of the base to the flask. At the point where the base hits the vinegar solution the color may *temporarily* turn pink, but this color will disappear upon mixing the solution by swirling. Continue the titration until a faint *permanent* pink coloration appears. Stop the titration. Record the readings of the base in your buret on your Report Sheet (4). Read the meniscus to the nearest 0.1 mL (Figure 19.1b).

CAUTION

Be careful not to add too much base. It is an error called *overtitration.* If the indicator in your flask turns deep pink or purple, you have overtitrated and will need to repeat the entire titration with a new sample of vinegar.

4. Repeat the procedures in steps 1–3 with a new 5-mL vinegar sample for trial 2. Record these results on your Report Sheet.
5. With the aid of the titration equation, calculate the molarity of the vinegar (6) for trials 1 and 2.
6. Average the two molarities. Using the molar mass of 60 g/mole for acetic acid, calculate the percent concentration of acetic acid in vinegar.

CHEMICALS AND EQUIPMENT

1. 25-mL buret (or 50-mL buret)
2. Buret clamp
3. 5-mL volumetric pipet
4. Small funnel
5. 0.2 M standardized NaOH
6. Phenolphthalein indicator
7. Spectroline pipet filler

name

section

date

partner

grade

19

EXPERIMENT 19

Pre-Lab Questions

1. For the reaction taking place in this experiment,

   a. write the complete balanced equation

   b. write the net ionic equation

2. Explain the function of the phenolphthalein in the experiment.

3. Write the titration equation.

4. What is the significance of the "end point"?

5. The formula of acetic acid found in vinegar is CH_3COOH. Only one hydrogen in the molecule acts as an acid hydrogen. Which one is donated? Write the formula of the conjugate base after the hydrogen is lost.

6. What is the molarity of an unknown acid if 35.5 mL of the acid can be titrated to an end point by 15.5 mL of 0.2501 M NaOH? Show your work and mind significant figures.

______________ *name*	______ *section*	______ *date*
______________ *partner*	______ *grade*	

19 EXPERIMENT 19

Report Sheet

Titration	*Trial 1*	*Trial 2*
1. Volume of vinegar sample	______	______
2. Molarity of NaOH solution	______	______
3. Initial reading of NaOH in buret	______ mL	______ mL
4. Final reading of NaOH in buret	______ mL	______ mL
5. Volume of NaOH used in titration: (4) – (3)	______ mL	______ mL
6. Molarity of acetic acid in vinegar: (2) × [(5)/(1)]	______	______
7. Average molarity of acetic acid		______
8. Percent (w/v) of acetic acid in vinegar: % = (7) × 60 × 0.10		______ %

$[\% \text{ w/v} = \text{g}/100 \text{ mL} = (7) \cancel{\text{mol}}/1000 \cancel{\text{mL}} \times 60 \text{ g}/\cancel{\text{mol}} \times 0.10 \times 1000 \cancel{\text{mL}}/100 \text{ mL}]$

Post-Lab Questions

1. Why is it necessary to exclude any air bubbles in the buret, particularly in the tip, when it is filled with NaOH solution?

2. Suppose you have the following solutions: (a) 100 mL of 0.1 M HCl; (b) 100 mL of 0.1 M CH_3COOH.

   a. Which solution is more acidic? (*Hint:* Which gives a more acidic pH?) Why?

   b. In a titration with NaOH, would either of the solutions require more base in order to reach an end point? Explain your answer.

3. What was the purpose of adding water to the sample of vinegar used for the titration? The added volume was not considered in the calculations. Why was it ignored?

4. A normal bottle of vinegar is about 750 mL. Based on your results, how many grams of acetic acid are in the bottle?

5. You have a 5% formic acid (HCOOH) solution and a 5% vinegar (acetic acid, CH_3COOH) solution. You determine each solution's acid concentration by titrating 5 mL of each with 0.2000 M NaOH. Which solution contains more acid? Show your calculations.

Analysis of Antacid Tablets

BACKGROUND

The natural environment of our stomach is quite acidic. Gastric juice, which is mostly hydrochloric acid, has a pH of 1.0. Such a strong acidic environment denatures proteins and helps with their digestion by enzymes such as pepsin. Not only is the denatured protein more easily digested by enzymes than the native protein, but the acidic environment helps to activate pepsin. The inactive form of pepsin, pepsinogen, is converted to the active form, pepsin, by removing a chunk of its chain, 42 amino acid units. This can only occur in an acidic environment, and pepsin molecules catalyze this reaction (autocatalysis). But too much acid in the stomach is not good either. In the absence of food, the strong acid, HCl, denatures the proteins in the stomach wall itself. If this goes on unchecked, it may cause stomach or duodenal ulcers.

We feel the excess acidity in our stomach. Such sensations are called "heartburn" or "sour stomach." To relieve "heartburn," we take antacids in tablet or liquid form. Antacid is a medical term. It implies a substance that neutralizes acid. Drugstore antacids contain a number of different active ingredients. Almost all of them are weak bases (hydroxides and/or carbonates). Table 20.1 lists the active ingredients of some commercial antacids.

HCl in the gastric juice is neutralized by these active ingredients as the following reactions show:

$$NaHCO_3 + HCl \longrightarrow NaCl + H_2O + CO_2$$
$$CaCO_3 + 2HCl \longrightarrow CaCl_2 + H_2O + CO_2$$
$$Al(OH)_3 + 3HCl \longrightarrow AlCl_3 + 3H_2O$$
$$Mg(OH)_2 + 2HCl \longrightarrow MgCl_2 + 2H_2O$$
$$AlNa(OH)_2CO_3 + 4HCl \longrightarrow AlCl_3 + NaCl + 3H_2O + CO_2$$

Besides the active ingredients, antacid tablets also contain inactive ingredients, such as starch, which act as a binder or filler. The efficacy of an antacid tablet is its ability to neutralize HCl. The more HCl that is neutralized, the more effective the antacid pill. (Perhaps you have heard the competing advertisement claims of different commercial antacids: "Tums neutralizes one-third more stomach acid than Rolaids.")

Table 20.1 *Active Ingredients of Some Drugstore Antacids*

Alka-Seltzer: sodium bicarbonate and citrate
Bromo-Seltzer: sodium bicarbonate and citrate
Chooz, Tums: calcium carbonate
Di-gel, Gelusil, Maalox: aluminum hydroxide and magnesium hydroxide
Gaviscon, Remegel: aluminum hydroxide and magnesium carbonate
Rolaids: aluminum sodium dihydroxy carbonate

Antacids are not completely harmless. The HCl production in the stomach is regulated by the stomach pH. If too much antacid is taken, the pH becomes too high; the result will be the so-called "acid rebound." This means that, ultimately, more HCl will be produced than was present before taking the antacid pill.

In the present experiment, you will determine the amount of HCl neutralized by two different commercial antacid tablets. To do so we use a technique called *back-titration.* We add an **excess** amount of 0.2 M HCl to the antacid tablet. The excess acid (more than is needed for neutralization) helps to dissolve the tablet. Then the active ingredients in the antacid tablet will neutralize part of the added acid. The remaining HCl is determined by titration with NaOH. A standardized NaOH solution of known concentration (0.2 M) is used and added slowly until all the HCl is neutralized. We observe this end point of the titration when the added indicator, thymol blue, changes its color from red to yellow. Because HCl is monoprotic and NaOH yields one OH^- for each molecule, we can use the titration equation to obtain the volume of the excess 0.2 M HCl (the volume not neutralized by the antacid).

$$\mathbf{V_{acid} \times M_{acid} = V_{base} \times M_{base}}$$
$$\mathbf{V_{acid} = (V_{base} \times M_{base})/M_{acid}}$$

Once this is known, the amount of HCl neutralized by the antacid pill is obtained as the difference between the initially added volume and the back-titrated volume:

$$\mathbf{V_{HCl\ neutralized\ by\ the\ pill} = V_{HCl\ initially\ added} - V_{HCl\ back\text{-}titrated}}$$

In this way we can compare the effectiveness of different drugstore antacids.

OBJECTIVES

1. To learn the technique of back-titration.
2. To compare the efficacies of drugstore antacid tablets.

PROCEDURE

1. Rinse a 25-mL buret (or 50-mL buret) with about 5 mL 0.20 M NaOH. After rinsing, fill the buret with 0.20 M NaOH solution about 2 mL above the top mark. Use a clean and dry funnel for filling. Tilting the filled buret at a 45-degree angle, turn the stopcock open to allow the

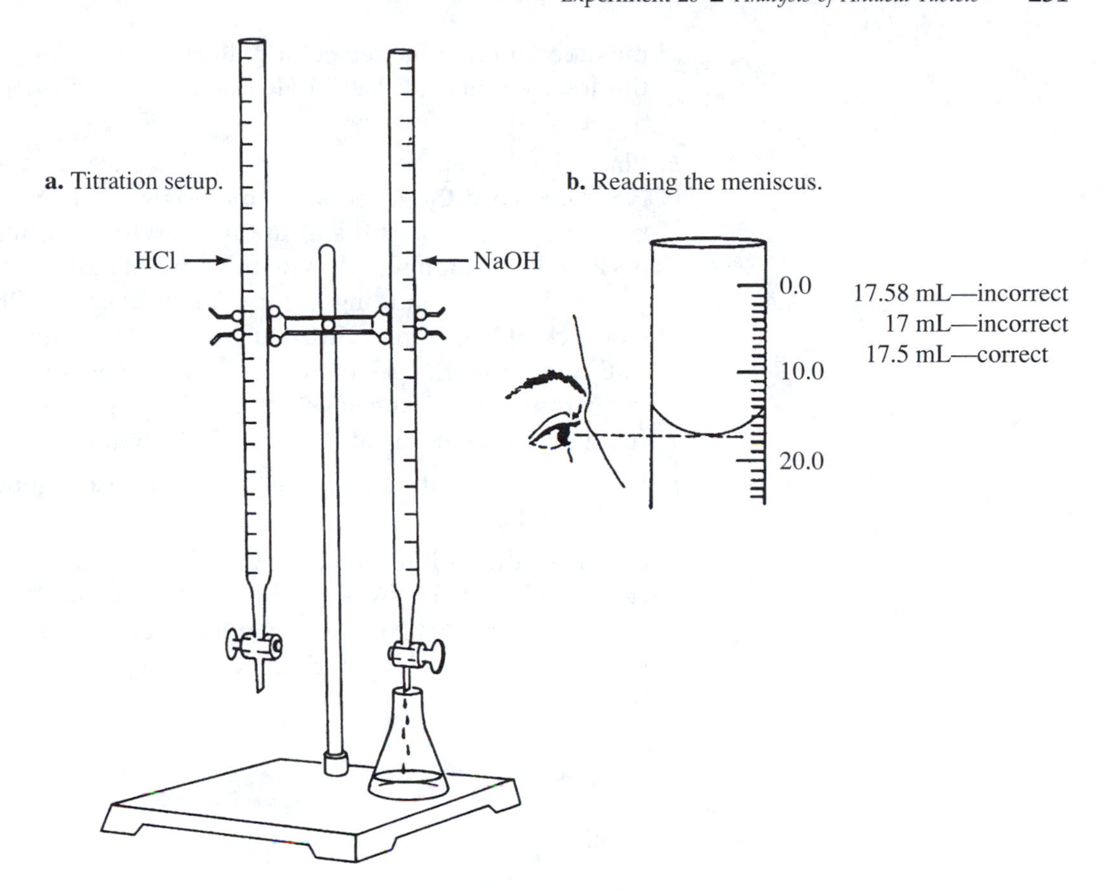

Figure 20.1
Titration setup.

solution to fill the tip of the buret. Collect the excess solution dripping from the tip into a beaker to be discarded later. The air bubbles should be completely removed from the tip by this maneuver. If you do not succeed the first time, repeat it until the liquid in the buret forms one continuous column from top to bottom. Clamp the buret onto a ring stand (Figure 20.1a). By slowly opening the stopcock, allow the bottom of the meniscus to drop to the 0.0-mL mark. Collect the excess solution dripping from the tip into a beaker to be discarded later. (Carefully read the meniscus here, and in all other readings, to the nearest 0.1 mL; see Figure 20.1b.)

2. Repeat the above procedure with a 100-mL buret and fill it to the 0.0-mL mark with 0.20 M HCl. Clamp this, too, onto a ring stand.
3. Obtain two different antacid tablets from your instructor. Note the name of each tablet on your Report Sheet (1). Weigh each tablet on a balance to the nearest 0.001 g. Report the weight on your Report Sheet (2). Place each tablet in separate 250-mL Erlenmeyer flasks. Label the flasks. Add about 10 mL water to each flask. With the help of stirring rods (one for each flask), break up the tablets.
4. Add exactly 50 mL 0.20 M HCl to each Erlenmeyer flask from the buret. Also, add a few drops of thymol blue indicator. Gently stir with the stirring rods to disperse the tablets. (Some of the inactive ingredients may not go into solution and will settle as a fine powder on the bottom of the flask.) At this point the solution should be red (the color of thymol blue at acidic pH). If either of your solutions does not have red coloration, add 10 mL 0.20 M HCl from the refilled buret and

make certain that the red color will persist for more than 30 sec. Record the total volume of 0.20 M HCl added to each flask on your Report Sheet (3).

5. Place the Erlenmeyer flask under the buret containing the 0.20 M NaOH. Record the level of the meniscus of the NaOH solution in the buret before you start the titration (4). While holding and swirling the neck of the Erlenmeyer flask with your left hand, titrate the contents of your solution by adding (dropwise) 0.20 M NaOH by opening the stopcock of the buret with your right hand. Continue to add NaOH until the color changes to yellow and stays yellow for 30 sec. after the last drop. Record the level of the NaOH solution in the buret by reading the meniscus at the end of titration (5).
6. Refill the buret with 0.20 M NaOH as before and repeat the titration for the second antacid.
7. Calculate the volume of the acid obtained in the back-titration and record it on your Report Sheet (6). Calculate the volume of the 0.20 M HCl neutralized by the antacid tablets (7). Calculate the grams of HCl neutralized by 1 g antacid tablet. Record it on your Report Sheet (8).

CHEMICALS AND EQUIPMENT

1. 25-mL buret (or 50-mL buret)
2. 100-mL buret
3. Buret clamp
4. Ring stand
5. Balance
6. Antacid tablets
7. 0.20 M NaOH
8. 0.20 M HCl
9. Thymol blue indicator

name | section | date

partner | grade

20

EXPERIMENT 20

Pre-Lab Questions

1. Below are some common foods and their pH values. Which ones are most likely to cause "heartburn"? Explain why you made those choices.

   soda, 2.0–4.0; tomatoes, 4.0–4.4; black coffee, 5.0–5.1;
   milk, 6.3–6.6; pure water, 7.0

2. If a patient had an elevated blood pressure level and had to restrict his or her sodium intake, which antacid(s) would you recommend for such a patient (see Table 20.1)?

3. How does a *back-titration* work?

4. Some antacids will generate carbon dioxide, CO_2, when they neutralize the acid in gastric juice. Write the net ionic equation for the reaction of the antacid containing the anion carbonate, CO_3^{2-}, or the anion bicarbonate, HCO_3^-, with acid (H^+).

5. Write balanced reactions for the neutralization of the bases, magnesium hydroxide, $Mg(OH)_2$, and aluminum hydroxide, $Al(OH)_3$, found in Maalox, for example.

name ______ section ______ date ______

partner ______ grade ______

20 EXPERIMENT 20

Report Sheet

	(a)	(b)
1. Name of the antacid tablet	______	______
2. Weight of the antacid tablet	______ g	______ g
3. Total volume of 0.20 M HCl added to the antacid before titration	______ mL	______ mL
4. Reading of 0.20 M NaOH in the buret before titration	______ mL	______ mL
5. Reading of 0.20 M NaOH in the buret after titration	______ mL	______ mL
6. Volume of 0.20 M HCl obtained in back-titration: (5) − (4)	______ mL	______ mL
7. Volume of 0.20 M HCl neutralized by one antacid tablet: (3) − (6)	______ mL	______ mL
8. Grams of HCl neutralized by 1 g antacid tablet: [(7)/(2)] × 0.20 × 36.5	______	______

Post-Lab Questions

1. Gelusil is an example of an antacid that contains magnesium hydroxide, $Mg(OH)_2$. Consider the reaction between this basic material and HCl. How many moles of HCl do you need to neutralize (a) 0.20, (b) 0.425, and (c) 2.3 moles of $Mg(OH)_2$?

2. From your results, which antacid would neutralize more stomach acid per tablet? What was the active ingredient?

3. How many grams of magnesium hydroxide, $Mg(OH)_2$, do you need in an antacid tablet to neutralize 20 mL of stomach acid? (Stomach acid is 0.1 M HCl. There are 3.65 g of HCl in 1 L of stomach acid or 1 g of HCl in 274 g of stomach acid.) Show your work.

4. The indicator that you used in the titration was thymol blue. This indicator changes color at the end point from red (acidic) to yellow at pH 2.0. Suppose you chose a different indicator, for example, phenolphthalein, which changes color at the end point from colorless (acidic) to pink at pH 8.5. Would you need the same, more, or less volume of NaOH in the back-titration? Explain your answer.

5. You are a manufacturer of antacids. You could use either $Mg(OH)_2$ or $Al(OH)_3$ in your formulation. You can buy either of these for $2.00 per pound. Which of these will provide a better product (i.e., give more relief) for the same amount of money? Explain your answer.

Structure in Organic Compounds: Use of Molecular Models. I

BACKGROUND

The study of organic chemistry usually involves molecules that contain carbon. Thus a convenient definition of *organic chemistry* is the chemistry of carbon compounds.

There are several characteristics of organic compounds that make their study interesting:

1. Carbon forms strong bonds to itself as well as to other elements; the most common elements found in organic compounds, other than carbon, are hydrogen, oxygen, and nitrogen.
2. Carbon atoms are generally tetravalent. This means that carbon atoms in most organic compounds are bound by four covalent bonds to adjacent atoms.
3. Organic molecules are three-dimensional and occupy space. The covalent bonds that carbon makes to adjacent atoms are at discrete angles to each other. Depending on the type of organic compound, the angle may be 180, 120, or 109.5 degrees. These angles correspond to compounds that have triple bonds (1), double bonds (2), and single bonds (3), respectively.

$$-C\equiv C- \qquad (1)$$

$$>C=C< \qquad (2)$$

$$-\overset{|}{\underset{|}{C}}-\overset{|}{\underset{|}{C}}- \qquad (3)$$

4. Organic compounds can have a limitless variety in composition, shape, and structure.

Thus, while a molecular formula tells the number and type of atoms present in a compound, it tells nothing about the structure. The structural formula is a two-dimensional representation of a molecule and shows the sequence in which the atoms are connected and the bond type. For example, the molecular formula, C_4H_{10}, can be represented by two different structures: butane (4) and 2-methylpropane (isobutane) (5).

```
                                    H   H   H
                                    |   |   |
    H   H   H   H                 H—C — C — C—H
    |   |   |   |                   |   |   |
  H—C — C — C — C—H                 H   |   H
    |   |   |   |                     H—C—H
    H   H   H   H                       |
                                        H
```

Butane (4)

2-Methylpropane (5)
(Isobutane)

Consider also the molecular formula, C_2H_6O. Two structures correspond to this formula: dimethyl ether (6) and ethanol (ethyl alcohol) (7).

```
    H       H                H   H
    |       |                |   |
  H—C — O — C—H            H—C — C — O—H
    |       |                |   |
    H       H                H   H
```

Dimethyl ether (6)

Ethanol (7)
(Ethyl alcohol)

In the pairs above, each structural formula represents a different compound. Each compound has its own unique set of physical and chemical properties. Compounds with the same molecular formula but with different structural formulas are called *isomers*.

The three-dimensional character of molecules is expressed by its stereochemistry. By looking at the *stereochemistry* of a molecule, the spatial relationships between atoms on one carbon and the atoms on an adjacent carbon can be examined. Because rotation can occur around carbon–carbon single bonds in open chain molecules, the atoms on adjacent carbons can assume different spatial relationships with respect to each other. The different arrangements that atoms can assume as a result of a rotation about a single bond are called *conformations*. A specific conformation is called a *conformer*. Whereas individual isomers can be isolated, conformers cannot because interconversion, by rotation, is too rapid.

Conformers may be represented by projections through the use of two conventions, as shown in Figure 21.1. These projections attempt to show on a flat surface how three-dimensional objects, in this case organic molecules, might look in three-dimensional space.

The *sawhorse projection* views the carbon–carbon bond at an angle and, by showing all the bonds and atoms, shows their spatial arrangements. The *Newman projection* provides a view along a carbon–carbon bond by sighting directly along the carbon–carbon bond. The near carbon is represented by a circle, and bonds attached to it are represented by lines going to the center of the circle. The carbon behind is not visible (it is blocked by the near carbon), but the bonds attached to it are partially visible and are

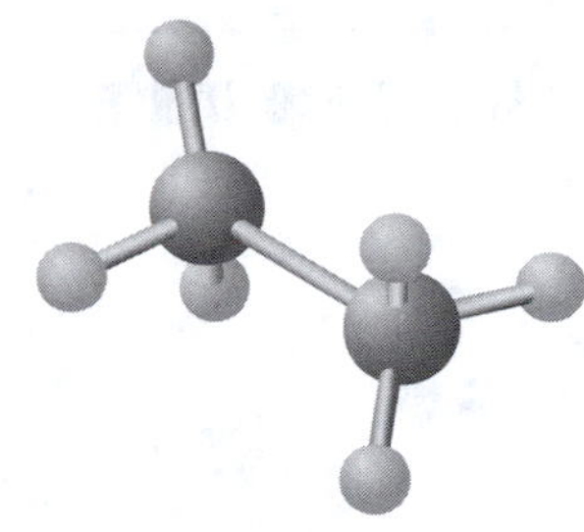

Figure 21.1
Molecular representations.

a. Sawhorse projection of ethane

b. Newman projection of ethane

c. Ball-and-stick model of ethane

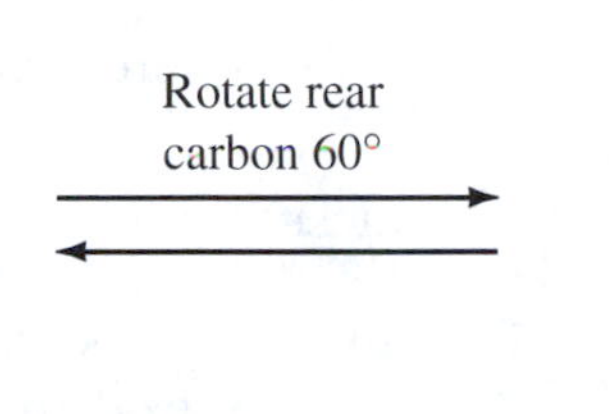

Rotate rear carbon 60°

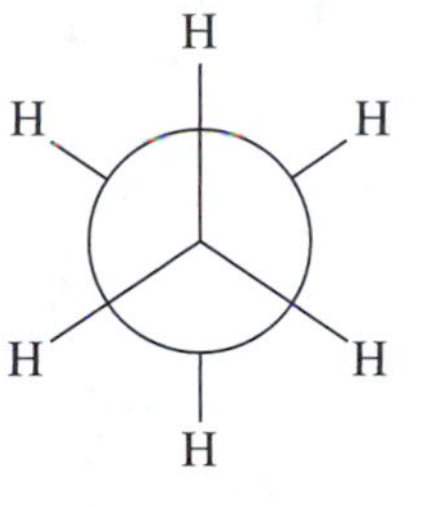

Figure 21.2
Two conformers of ethane.

a. Eclipsed conformation of ethane

b. Staggered conformation of ethane

represented by lines going to the edge of the circle. With Newman projections, rotations show the spatial relationships of atoms on adjacent carbons easily. Two conformers that represent extremes are shown in Figure 21.2.

The *eclipsed* conformation has the bonds (and the atoms) on the adjacent carbons as close as possible. The *staggered* conformation has the bonds (and the atoms) on adjacent carbons as far as possible. One conformation can interconvert into the other by rotation around the carbon–carbon bond axis.

The three-dimensional character of molecular structure is shown through molecular model building. With molecular models, the number and types of bonds between atoms and the spatial arrangements of the atoms can be visualized for the molecules. This allows comparison of isomers and of conformers for a given set of compounds. The models also will let you see what is meant by *chemical equivalence*. Here *equivalence* relates to those positions or to those hydrogens on carbon(s) in an organic molecule that are equal in terms of chemical reactivity. In the case of hydrogen, replacement of any one of the equivalent hydrogens in a molecule by a substituent (any atom or group of atoms, for example, Cl or OH, respectively) leads to the identical substituted molecule.

OBJECTIVES

1. To use models to visualize structure in organic molecules.
2. To build and compare isomers having a given molecular formula.
3. To explore the three-dimensional character of organic molecules.
4. To demonstrate equivalence of hydrogens in organic molecules.

PROCEDURE

Obtain a set of ball-and-stick molecular models from the laboratory instructor. The set contains the following parts (other colored spheres may be substituted as available):

- 2 Black spheres representing *Carbon;* this tetracovalent element has four holes
- 6 Yellow spheres representing *Hydrogen;* this monovalent element has one hole
- 2 Colored spheres representing the *halogen Chlorine;* this monovalent element has one hole
- 1 Blue sphere representing *Oxygen;* this divalent element has two holes
- 8 Sticks to represent bonds

1. With your models, construct the molecule methane. Methane is a simple hydrocarbon consisting of one carbon and four hydrogens. After you put the model together, answer the questions below in the appropriate space on the Report Sheet.

a. With the model resting so that three hydrogens are on the desk, examine the structure. Move the structure so that a different set of three hydrogens are on the desk each time. Is there any difference between the way that the two structures look (1a)?

b. Does the term *equivalent* adequately describe the four hydrogens of methane (1b)?

c. Tilt the model so that only two hydrogens are in contact with the desk and imagine pressing the model flat onto the desktop. Draw the way in which the methane molecule would look in two-dimensional space (1c). This is the usual way that three-dimensional structures are written.

d. Using a protractor, measure the angle H—C—H on the model (1d).

2. Replace one of the hydrogens of the methane model with a colored sphere, which represents the halogen chlorine. The new model is chloromethane (methyl chloride), CH_3Cl. Position the model so that the three hydrogens are on the desk.

a. Grasp the atom representing chlorine and tilt it to the right, keeping two hydrogens on the desk. Write the structure of the projection on the Report Sheet (2a).

b. Return the model to its original position and then tilt as before, but this time to the left. Write this projection on the Report Sheet (2b).

c. While the projection of the molecule changes, does the structure of chloromethane change (2c)?

3. Now replace a second hydrogen with another chlorine sphere. The new molecule is dichloromethane, CH_2Cl_2.

a. Examine the model as you twist and turn it in space. Are the projections given below isomers of the molecule CH_2Cl_2 or representations of the same structure only seen differently in three dimensions (3a)?

```
     H              Cl              H              Cl
     |              |               |              |
Cl — C — H     H — C — Cl     Cl — C — Cl     H — C — H
     |              |               |              |
     Cl             H               H              Cl
```

4. Construct the molecule ethane, C_2H_6. Note that you can make ethane from the methane model by removing a hydrogen and replacing the hydrogen with a methyl group, —CH_3.
   a. Write the structural formula for ethane (4a).
   b. Are all the hydrogens attached to the carbon atoms equivalent (4b)?
   c. Draw a sawhorse representation of ethane. Draw a staggered and an eclipsed Newman projection of ethane (4c).
   d. Replace any hydrogen in your model with chlorine. Write the structure of the molecule chloroethane (ethyl chloride), C_2H_5Cl (4d).
   e. Twist and turn your model. Draw two Newman projections of the chloroethane molecule (4e).
   f. Do the projections that you drew represent different isomers or conformers of the same compound (4f)?
5. Dichloroethane, $C_2H_4Cl_2$
   a. In your molecule of chloroethane, if you choose to remove another hydrogen note that you now have a choice among the hydrogens. You can either remove a hydrogen from the carbon to which the chlorine is attached, or you can remove a hydrogen from the carbon that has only hydrogens attached. First, remove the hydrogen from the carbon that has the chlorine attached and replace it with a second chlorine. Write its structure on the Report Sheet (5a).
   b. Compare this structure to the model that would result from removal of a hydrogen from the other carbon and its replacement by chlorine. Write its structure (5b) and compare it to the previous example. One isomer is 1,1-dichloroethane; the other is 1,2-dichloroethane. Label the structures drawn on the Report Sheet with the correct name.
   c. From what you did in **a** and **b**, above, you can make some conclusions about the hydrogens that are equivalent to each other in chloroethane. Draw the structure of chloroethane and label the hydrogens that are equivalent to each other (for example, as H_a). Are all the hydrogens of chloroethane equivalent? Are some of the hydrogens equivalent? How many sets of equivalent hydrogens are there?
6. Butane
   a. Butane has the formula C_4H_{10}. With help from a partner, construct a model of butane by connecting the four carbons in a series (C—C—C—C) and then adding the needed hydrogens. First,

orient the model in such a way that the carbons appear as a straight line. Next, tilt the model so that the carbons appear as a zig-zag line. Then, twist around any of the C—C bonds so that a part of the chain is at an angle to the remainder. Draw each of these structures in the space on the Report Sheet (6a). Note that the structures you draw are for the same molecule but represent a different orientation and projection.

b. Sight along the carbon–carbon bond of C_2 and C_3 on the butane chain: $\overset{1}{C}H_3—\overset{2}{C}H_2—\overset{3}{C}H_2—\overset{4}{C}H_3$. Draw a staggered Newman projection. Rotate the C_2 carbon clockwise by 60 degrees; draw the eclipsed Newman projection. Again, rotate the C_2 carbon clockwise by 60 degrees; draw the Newman projection. Is the last projection staggered or eclipsed (6b)? Continue rotation of the C_2 carbon clockwise by 60-degree increments and observe the changes that take place.

c. Examine the structure of butane for equivalent hydrogens. In the space on the Report Sheet (6c), redraw the structure of butane and label the hydrogens that are equivalent to each other. On the basis of this examination, predict how many monochlorobutane isomers (C_4H_9Cl) could be obtained from the structure you drew in 6c (6d). Test your prediction by replacement of hydrogen by chlorine on the models. Draw the structures of these isomers (6e).

d. Reconstruct the butane system. First, form a three-carbon chain, then connect the fourth carbon to the center carbon of the three-carbon chain. Add the necessary hydrogens. Draw the structure of 2-methylpropane (isobutane) (6f). Can any manipulation of the model, by twisting or turning of the model or by rotation of any of the bonds, give you the butane system? If these two, butane and 2-methylpropane (isobutane), are *isomers*, then how may we recognize that any two structures are isomers (6g)?

e. Examine the structure of 2-methylpropane for equivalent hydrogens. In the space on the Report Sheet (6h), redraw the structure of 2-methylpropane and label the equivalent hydrogens. Predict how many monochloroisomers of 2-methylpropane could be formed (6i) and test your prediction by replacement of hydrogen by chlorine on the model. Draw the structures of these isomers (6j).

7. C_2H_6O

a. There are two isomers with the molecular formula C_2H_6O: ethanol (ethyl alcohol) and dimethyl ether. With your partner, construct both of these isomers. Draw these isomers on the Report Sheet (7a) and name each one.

b. Manipulate each model. Can either be turned into the other by a simple twist or turn (7b)?

c. For each compound, label the hydrogens that are equivalent. How many sets of equivalent hydrogens are there in ethanol (ethyl alcohol) and dimethyl ether (7c)?

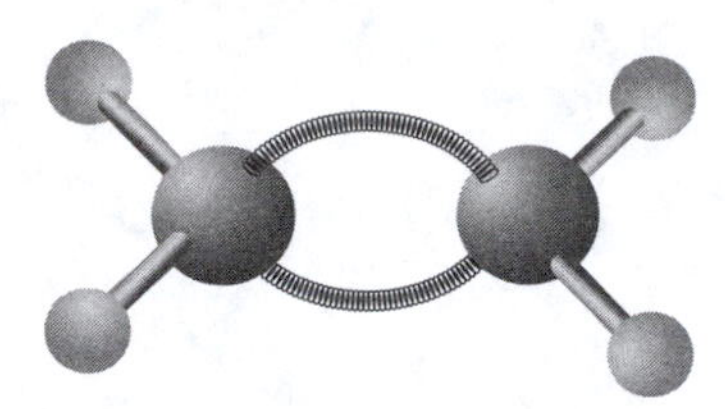

Figure 21.3
Using springs to construct a double bond.

8. Optional: Butenes
   a. If springs are available for the construction of double bonds (Figure 21.3), construct 2-butene, $CH_3—CH═CH—CH_3$. There are two isomers for compounds of this formulation: the isomer with the two $—CH_3$ groups on the same side of the double bond, *cis*-2-butene; and the isomer with the two $—CH_3$ groups on opposite sides of the double bond, *trans*-2-butene. Draw these two structures on the Report Sheet (8a).
   b. Can you twist, turn, or rotate one model into the other? Explain (8b).
   c. How many bonds are connected to any single carbon of these structures (8c)?
   d. With the protractor, measure the C—C═C angle (8d).
   e. Construct methylpropene, $CH_3—\underset{\displaystyle CH_3}{\underset{|}{C}}═CH_2$.

      Can you have a *cis*- or a *trans*- isomer in this system (8e)?
9. Optional: Butynes
   a. If springs are available for the construction of triple bonds, construct 2-butyne, $CH_3—C≡C—CH_3$. Can you have a *cis*- or a *trans*- isomer in this system (9a)?
   b. With the protractor, measure the C—C═C angle (9b).
   c. Construct a second butyne with your molecular models and springs. How does this isomer differ from the one in (a) above (9c)?

CHEMICALS AND EQUIPMENT

1. Molecular models (you may substitute other available colors for the spheres):
   2 Black spheres
   6 Yellow spheres
   2 Colored spheres (e.g., green)
   1 Blue sphere
   8 Sticks
2. Protractor
3. Optional: 3 springs

name

section

date

partner

grade

21

EXPERIMENT 21

Pre-Lab Questions

1. How do you define *organic chemistry?*

2. What are the four most common elements found in organic compounds?

3. Match the bond angle on carbon with the bond type shown below: (a) 109.5 degrees; (b) 120 degrees; (c) 180 degrees.

$-C\equiv C-$ $\qquad$ $-\overset{|}{\underset{|}{C}}-\overset{|}{\underset{|}{C}}-$ $\qquad$ $>C=C<$

4. Define *isomers.* Draw structures of two isomers that demonstrate this definition.

5. Use Newman projections to show ethane in the eclipsed conformation and in the staggered conformation. Correctly label each. How can you convert one into the other?

name | section | date

partner | grade

21 EXPERIMENT 21

Report Sheet

1. Methane

a.

b.

c.

d.

2. Chloromethane (methyl chloride)

a.

b.

c.

3. Dichloromethane

a.

4. Ethane and chloroethane (ethyl chloride)

a.

b.

c.

d.

e.

f.

5. Dichloroethane

a.

b.

c.

6. Butane

a.

b.

c.

d.

e.

f.

g.

h.

i.

j.

7. C_2H_6O

a.

b.

c. Ethanol (ethyl alcohol) has _____________ set(s) of equivalent hydrogens.

Dimethyl ether has _____________ set(s) of equivalent hydrogens.

8. Butenes

a.

b.

c.

d. C—C═C angle

e.

9. Butynes

a.

b.

c.

Post-Lab Questions

1. Write the structural formulas for the two (2) isomers of chloropropane, C_3H_7Cl.

2. For each of the isomers in the example above, identify equivalent hydrogens. Identify the equivalent sets by letters, e.g., H_a, H_b, etc.

3. Draw a staggered and an eclipsed conformer for 1,2-dichloroethane, $ClCH_2CH_2Cl$.

4. Why can you not convert *cis*-2-butene into *trans*-2-butene simply by undergoing a rotation?

Stereochemistry: Use of Molecular Models. II

BACKGROUND

In Experiment 21, we looked at some molecular variations that acyclic organic molecules can take:

1. *Constitutional isomerism.* Molecules can have the same molecular formula but different arrangements of atoms.
   a. *skeletal isomerism:* structural isomers where differences are in the order in which atoms that make up the skeleton are connected; e.g., C_4H_{10}

$$CH_3CH_2CH_2CH_3 \qquad CH_3-\underset{}{\overset{CH_3}{\overset{|}{C}H}}-CH_3$$

Butane — 2-Methylpropane

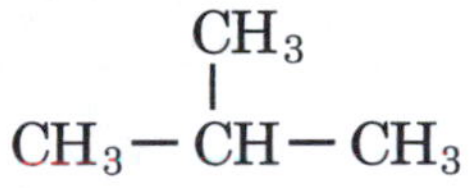

   b. *positional isomerism:* structural isomers where differences are in the location of a functional group; e.g., C_3H_7Cl

$$CH_3CH_2CH_2-Cl \qquad CH_3-\overset{Cl}{\overset{|}{C}H}-CH_3$$

1-Chloropropane — 2-Chloropropane

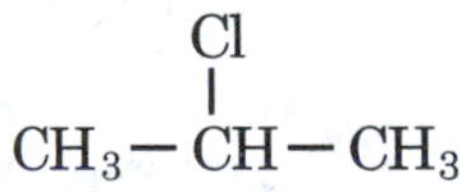

2. *Stereoisomerism.* Molecules have the same order of attachment of atoms but differ in the arrangement of the atoms in three-dimensional space.
   a. *cis-/trans- isomerism:* molecules that differ due to the geometry of substitution around a double bond; e.g., C_4H_8

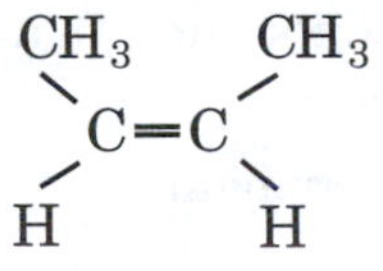

$$\begin{matrix} CH_3 & & CH_3 \\ & C=C & \\ H & & H \end{matrix} \qquad \begin{matrix} CH_3 & & H \\ & C=C & \\ H & & CH_3 \end{matrix}$$

cis-2-Butene 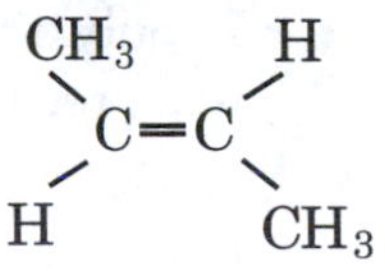 — *trans*-2-Butene

b. *conformational isomerism:* variation in acyclic molecules as a result of a rotation about a single bond; e.g., ethane, $CH_3—CH_3$

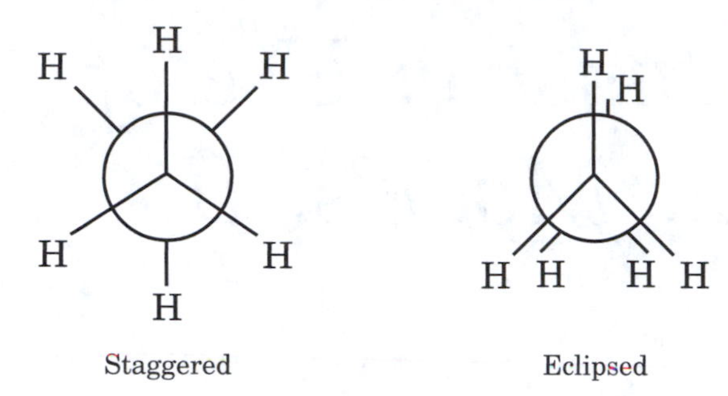

In this experiment, we will further investigate stereoisomerism by examining a cyclic system, cyclohexane, and several acyclic tetrahedral carbon systems. The latter possess more subtle characteristics as a result of the spatial arrangement of the component atoms. We will do this by building models of representative organic molecules, then studying their properties.

OBJECTIVES

1. To use models to study the conformations of cyclohexane.
2. To use models to distinguish between chiral and achiral systems.
3. To define and illustrate enantiomers, diastereomers, and meso forms.
4. To learn how to represent these systems in two-dimensional space.

PROCEDURE

You will build models and then you will be asked questions about the models. You will provide answers to these questions in the appropriate places on the Report Sheet. In doing this laboratory, it will be convenient if you tear out the Report Sheet and keep it by the **Procedure** as you work through the exercises. In this way, you can answer the questions without unnecessarily turning pages back and forth.

Cyclohexane

Obtain a model set of ''atoms'' that contain the following:

- 8 Carbon components—model atoms with 4 holes at the tetrahedral angle (e.g., black)
- 2 Substituent components (halogens)—model atoms with 1 hole (e.g., red)
- 18 Hydrogen components—model atoms with 1 hole (optional) (e.g., white)
- 24 Connecting links—bonds

1. Construct a model of cyclohexane by connecting 6 carbon atoms in a ring; then into each remaining hole insert a connecting link (bond) and, if available, add a hydrogen to each.
   a. Is the ring rigid or flexible, that is, can the ring of atoms move and take various arrangements in space, or is the ring of atoms locked into only one configuration (1a)?

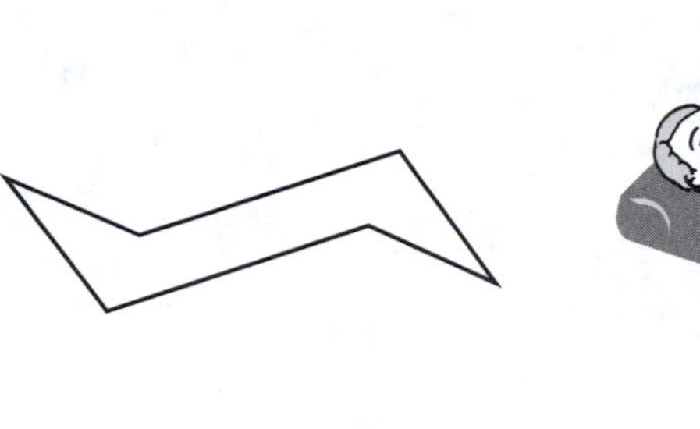

a. The chair conformation

b. A lounge chair

Figure 22.1
The chair conformation for a 6-carbon ring.

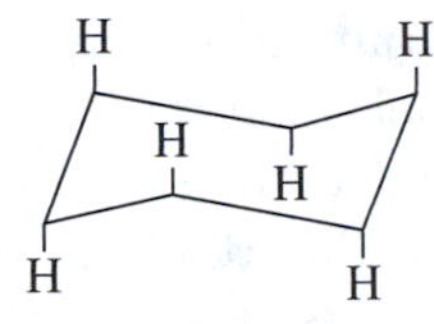

a. Axial position

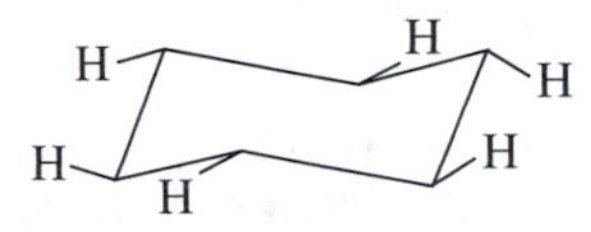

b. Equatorial position

Figure 22.2
Axial and equatorial hydrogens in the chair conformation.

b. Of the many configurations, which appears best for the ring—a planar or a puckered arrangement (1b)?

c. Arrange the ring atoms into a *chair* conformation (Figure 22.1a) and compare it to the picture of the lounge chair (Figure 22.1b). (Does the term fit the picture?)

2. With the model in the chair conformation, rest it on the tabletop.

a. How many hydrogens are in contact with the tabletop (2a)?

b. How many hydrogens point in a direction 180 degrees opposite to these (2b)?

c. Take your pencil and place it into the center of the ring perpendicular to the table. Now, rotate the ring around the pencil; we'll call this an *axis of rotation*. How many hydrogens are on bonds parallel to this axis (2c)? These hydrogens are called the *axial* hydrogens, and the bonds are called the *axial* bonds.

d. If you look at the perimeter of the cyclohexane system, the remaining hydrogens lie roughly in a ring perpendicular to the axis through the center of the molecule. How many hydrogens are on bonds lying in this ring (2d)? These hydrogens are called *equatorial* hydrogens, and the bonds are called the *equatorial* bonds.

e. Compare your model to the diagrams in Figure 22.2 and be sure you are able to recognize and distinguish between axial and equatorial positions.

In the space provided on the Report Sheet (2e), draw the structure of cyclohexane in the chair conformation with all 12 hydrogens attached. Label all the axial hydrogens H_a, and all the equatorial hydrogens H_e. How many hydrogens are labeled H_a (2f)? How many hydrogens are labeled H_e (2g)?

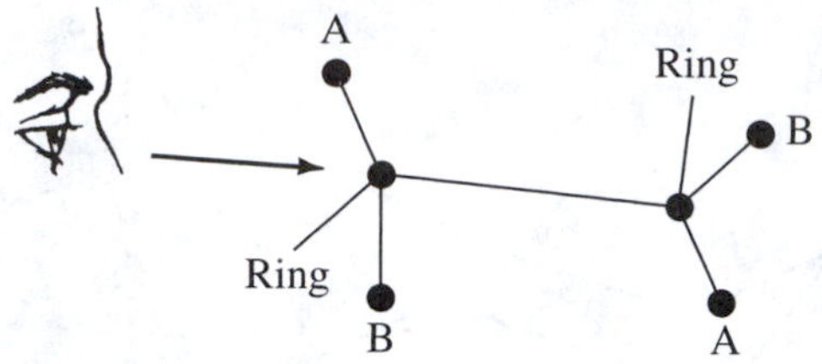

Figure 22.3
Cyclohexane ring viewed on edge.

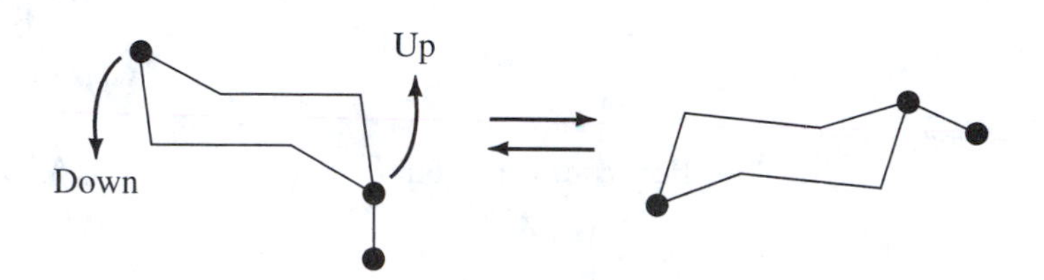

Figure 22.4
A "ring flip."

3. Look along any bond connecting any two carbon atoms in the ring. (Rotate the ring and look along a new pair of carbon atoms.) How are the bonds connected to these two carbons arranged? Are they staggered or are they eclipsed (3a)? In the space provided on the Report Sheet (3b), draw the Newman projection for the view (see Experiment 21 for an explanation of this projection); for the bond connecting a ring carbon, label that group "ring."
4. Pick up the cyclohexane model and view it from the side of the chair. Visualize the "ring" around the perimeter of the system perpendicular to the axis through the center. Of the 12 hydrogens, how many are pointed "up" relative to the plane (4a)? How many are pointed "down" (4b)?
5. Orient your model so that you look at an edge of the ring and it conforms to Figure 22.3. Are the two axial positions labeled *A cis* or *trans* to each other (5a)? Are the two equatorial positions labeled *B cis* or *trans* to each other (5b)? Are the axial and equatorial positions *A* and *B cis* or *trans* to each other (5c)? Rotate the ring and view new pairs of carbons in the same way. See whether the relationships of positions vary from the above. Position your eye as in Figure 22.3 and view along the carbon–carbon bond. In the space provided on the Report Sheet (5d), draw the Newman projection. Using this projection, review your answers to 5a, 5b, and 5c.
6. Replace one of the axial hydrogens with a colored component atom. Do a "ring flip" by moving one of the carbons *up* and moving the carbon farthest away from it *down* (Figure 22.4). In what position is the colored component after the ring flip (6a)—axial or equatorial? Do another ring flip. In what position is the colored component now (6b)? Observe all the axial positions and follow them through a ring flip.
7. Refer to Figure 22.3 and replace both positions labeled *A* by colored component atoms. Are they *cis* or *trans* (7a)? Do a ring flip. Are the two colored components *cis* or *trans* (7b)? Does the geometry change for the two components as the ring undergoes a ring flip (7c)? Repeat the exercise, replacing atoms in positions labeled *A* and *B* and answer the same three questions for this model.
8. Replace one of the colored components with a methyl, —CH_3, group. Manipulate the model so that the —CH_3 group is in an axial position; examine the model. Do a ring flip placing the —CH_3 in an equatorial

position; examine the model. Which of the chair conformations, —CH_3 axial or —CH_3 equatorial, is more crowded (8a)? What would account for one of the conformations being more crowded than the other (8b)? Which would be of higher energy and thus less stable (8c)? In the space provided on the Report Sheet (8d), draw the two conformations and connect with equilibrium arrows. Given your answers to 8a, 8b, and 8c, toward which conformation will the equilibrium lie (indicate by drawing one arrow bigger and thicker than the other)?

9. *A substituent group in the equatorial position of a chair conformation is more stable than the same substituent group in the axial position.* Do you agree or disagree? Explain your answer (9).

For the exercises in 10–15, although we will *not* be asking you to draw each and every conformation, we encourage you to practice drawing them in order to gain experience and facility in creating drawings on paper. Your instructor may make these exercises optional.

10. Construct *trans*-1,2-dimethylcyclohexane. By means of ring flips, examine the model with the two —CH_3 groups axial and the two —CH_3 groups equatorial. Which is the more stable conformation? Explain your answer (10).
11. Construct *cis*-1,2-dimethylcyclohexane by placing one —CH_3 group axial and the other equatorial. Do ring flips and examine the two chair conformations. Which is the more stable conformation? Explain your answer (11a). Given the two isomers, *trans*-1,2-dimethylcyclohexane and *cis*-1,2-dimethylcyclohexane, which is the more stable isomer? Explain your answer (11b).
12. Construct *cis*-1,3-dimethylcyclohexane by placing both —CH_3 groups in the axial positions. Do ring flips and examine the two chair conformations. Which is the more stable conformation? Explain your answer (12).
13. Construct *trans*-1,3-dimethylcyclohexane by placing one —CH_3 group axial and the other equatorial. Do ring flips and examine the two chair conformations. Which is the more stable conformation? Explain your answer (13a). Given the two isomers, *trans*-1,3-dimethylcyclohexane and *cis*-1,3-dimethylcyclohexane, which is the more stable isomer? Explain your answer (13b).
14. Construct *trans*-1,4-dimethylcyclohexane by placing both —CH_3 groups axial. Do ring flips and examine the two chair conformations. Which is the more stable conformation? Explain your answer (14).
15. Construct *cis*-1,4-dimethylcyclohexane by placing one —CH_3 group axial and the other equatorial. Do ring flips and examine the two chair conformations. Which is the more stable conformation? Explain your answer (15a). Given the two isomers, *trans*-1,4-dimethylcyclohexane and *cis*-1,4-dimethylcyclohexane, which is the more stable isomer? Explain your answer (15b).
16. Before we leave the cyclohexane ring system, there are some additional ring conformations we can examine. As we move from one cyclohexane chair conformation to another, the *boat* is a transitional conformation between them (Figure 22.5). Examine a model of the boat conformation by viewing along a carbon–carbon bond, as shown by

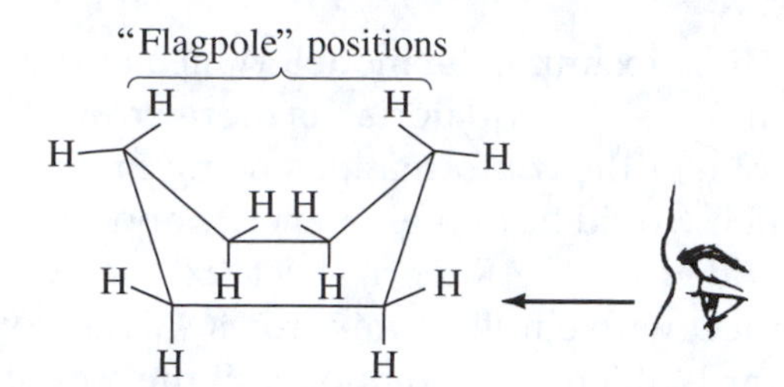

Figure 22.5
The boat conformation.

Figure 22.5. In the space provided on the Report Sheet (16a), draw the Newman projection for this view and compare with the Newman projection of 5d. By examining the models and comparing the Newman projections, explain which conformation, the chair or the boat, is more stable (16b). Replace the "flagpole" hydrogens by —CH_3 groups. What happens when this is done (16c)? The steric strain can be relieved by twisting the ring and separating the two bulky groups. What results is a *twist boat*.

17. Review the conformations the cyclohexane ring can assume as it moves from one chair conformation to another:

$$\text{chair} \rightleftharpoons \text{twist boat} \rightleftharpoons \text{boat} \rightleftharpoons \text{twist boat} \rightleftharpoons \text{chair}$$

Chiral Molecules

For this exercise, obtain a small hand mirror and a model set of "atoms" containing the following:

- 8 Carbon components—model atoms with four holes at the tetrahedral angle (e.g., black)
- 32 Substituent components—model atoms with one hole in four colors (e.g., 8 red; 8 white; 8 blue; 8 green; or any other colors your set may have)
- 28 Connecting links—bonds

Enantiomers

1. Construct a model consisting of a tetrahedral carbon center with four different component atoms attached: red, white, blue, green; each color represents a *different* group or atom attached to carbon. Does this model have a *plane of symmetry* (1a)? A plane of symmetry can be described as a cutting plane—a plane that when passed through a model or object *divides it into two equivalent halves*; the elements on one side of the plane are the exact reflection of the elements on the other side. If you are using a pencil to answer these questions, examine the pencil. Does it have a plane of symmetry (1b)?
2. Molecules without a plane of symmetry are *chiral*. In the model you constructed in no. 1, the tetrahedral carbon is the stereocenter; the molecule is chiral. A simple test for a stereocenter in a molecule is to look for a stereocenter with four different atoms or groups attached to it; this molecule will have no plane of symmetry. On the Report Sheet (2) are three structures; label the stereocenter in each structure with an asterisk (*).
3. Now take the model you constructed in no. 1 and place it in front of a mirror. Construct the model of the image projected in the mirror. You now have two models. If one is the object, what is the other (3a)? Do

either have a plane of symmetry (3b)? Are both chiral (3c)? Now try to superimpose one model onto the other, that is, to place one model on top of the other in such a way that all five elements (i.e., the colored atoms) fall exactly one on top of the other. Can you superimpose one model onto the other (3d)? *Enantiomers* are two molecules that are related to each other such that they are *nonsuperimposable mirror images of each other*. Are the two models you have a pair of enantiomers (3e)?

4. Molecules with a plane of symmetry are *achiral*. Replace the blue substituent with a second green one. The model should now have three different substituents attached to the carbon. Does the model now have a plane of symmetry (4a)? Passing the cutting plane through the model, what colored elements does it cut in half (4b)? What is on the left half and right half of the cutting plane (4c)? Place this model in front of the mirror. Construct the model of the image projected in the mirror. You now have two models—an object and its mirror image. Are these two models superimposable on each other (4d)? Are the two models representative of different molecules or identical molecules (4e)?

Each stereoisomer in a pair of enantiomers has the property of being able to rotate monochromatic plane-polarized light. The instrument chemists use to demonstrate this property is called a *polarimeter* (see your text for a further description of the instrument). A pure solution of a single one of the enantiomers (referred to as an *optical isomer*) can rotate the light in either a clockwise (dextrorotatory, +) or a counterclockwise (levorotatory, –) direction. Thus those molecules that are optically active possess a "handedness" or chirality. Achiral molecules are optically inactive and do not rotate the light.

Meso Forms and Diastereomers

5. With your models, construct a pair of enantiomers. From each of the models, remove the same common element (e.g., the white component) and the connecting links (bonds). Reconnect the two central carbons by a bond. What you have constructed is the *meso* form of a molecule, such as *meso*-tartaric acid. How many chiral carbons are there in this compound (5a)?

$$\begin{array}{c} HOOC-C_aH-C_bH-COOH \\ \qquad\quad | \qquad\quad | \qquad\qquad\qquad \\ \qquad\; OH \quad\; OH \qquad\qquad\quad \end{array}$$

Tartaric acid

Is there a plane of symmetry (5b)? Is the molecule chiral or achiral (5c)?

6. In the space provided on the Report Sheet (6), use circles to indicate the four different groups for carbon C_a and squares to indicate the four different groups for carbon C_b.

7. Project the model into a mirror and construct a model of the mirror image. Are these two models superimposable or nonsuperimposable (7a)? Are the models identical or different (7b)?

8. Now take one of the models you constructed in no. 7, and on one of the carbon centers exchange any two colored component groups. Does the new model have a plane of symmetry (8a)? Is it chiral or achiral (8b)? How many stereocenters are present (8c)? Take this model and one of the models you constructed in no. 7 and see whether they are

superimposable. Are the two models superimposable (8d)? Are the two models identical or different (8e)? Are the two models mirror images of each other (8f)? Here we have a pair of molecular models, each with two stereocenters, that are not mirror images of each other. These two examples represent *diastereomers*, stereoisomers that are not related as mirror images.

9. Take the new model you constructed in no. 8 and project it into a mirror. Construct a model of the image in the mirror. Are the two models superimposable (9a)? What term describes the relationship of the two models (9b)?

Thus if we let these three models represent different isomers of tartaric acid, we find that there are three stereoisomers for tartaric acid—a *meso* form and a pair of enantiomers. A *meso* form with any one of the enantiomers of tartaric acid represents a pair of diastereomers. Although it may not be true for this compound because of the *meso* form, in general, if you have **n** stereocenters, there are 2^n stereoisomers possible.

Drawing Stereoisomers

This section will deal with conventions for representing these three-dimensional systems in two-dimensional space.

10. Construct models of a pair of enantiomers; use tetrahedral carbon and four differently colored components for the four different groups: red, green, blue, white. Hold one of the models in the following way:
    a. Grasp the blue group with your fingers and rotate the model until the green and red groups are pointing toward you (Figure 22.6a). (Use the model that has the green group on the left and the red group on the right.)
    b. Holding the model in this way, the blue and white groups point away from you.
    c. If we use a drawing that describes a bond pointing toward you as a wedge and a bond pointing away from you as a dashed line, the model can be drawn as shown in Figure 22.6b.

    If this model were compressed into two-dimensional space, we would get the projection shown in Figure 22.6c. This is termed a *Fischer projection* and is named after a pioneer in stereochemistry, Emile Fischer. The Fischer projection has the following requirements:

    (1) The center of the cross represents the chiral carbon and is in the plane of the paper.

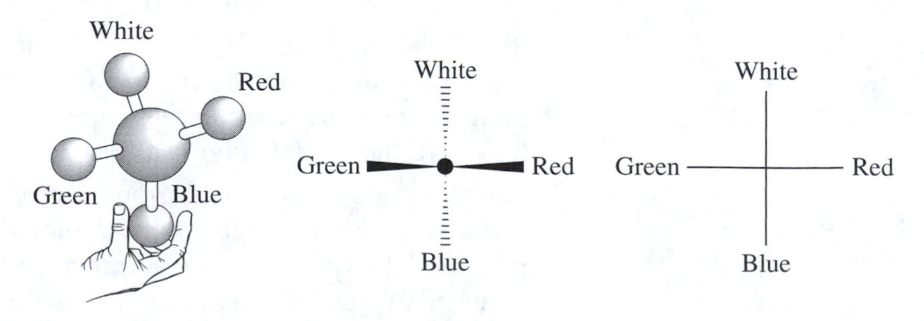

a. Holding the model **b.** Dashed-line-wedge **c.** Fischer projection

Figure 22.6
Projections in two-dimensional space.

(2) The horizontal line of the cross represents those bonds projecting *out front* from the plane of the paper.

(3) The vertical line of the cross represents bonds projecting *behind* the plane of the paper.

d. In the space provided on the Report Sheet (10), use the enantiomer of the model in Figure 22.6a and draw both the dashed-line-wedge and Fischer projection.

11. Take the model shown in Figure 22.6a and rotate it by 180 degrees (turn upside down). Draw the Fischer projection (11a). Does this keep the requirements of the Fischer projection (11b)? Is the projection representative of the same system or of a different system (i.e., the enantiomer) (11c)?

In general, if you have a Fischer projection and rotate it in the plane of the paper by 180 degrees, the resulting projection is of the *same* system. Test this assumption by taking the Fischer projection in Figure 22.6c, rotating it in the plane of the paper by 180 degrees, and comparing it to the drawing you did for no. 11a.

12. Again, take the model shown in Figure 22.6a. Exchange the red and the green components. Does this exchange give you the enantiomer (12a)? Now exchange the blue and the white components. Does this exchange return you to the original model (12b)?

In general, for a given stereocenter, whether we use the dashed-line wedge or the Fischer projection, an odd-numbered exchange of groups leads to the mirror image of that center; an even-numbered exchange of groups leads back to the original system.

13. Test the above by starting with the Fischer projection given below and carrying out the operations directed in a, b, and c; use the space provided on the Report Sheet (13) for the answers.

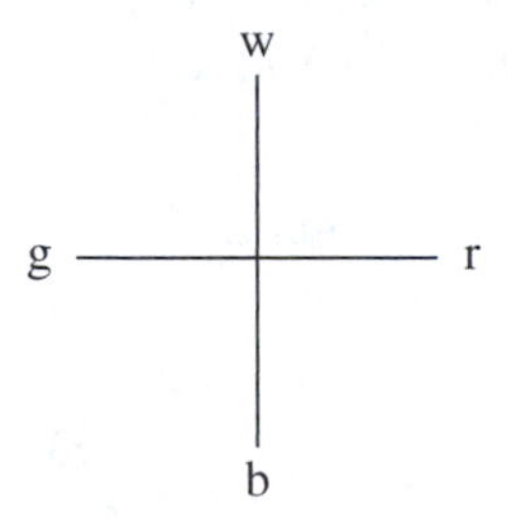

a. Exchange *r* and *g*; draw the Fischer projection you obtain; label this new projection as either the same as the starting model or the enantiomer.

b. Using the new Fischer projection from above, exchange *b* and *w*; draw the Fischer projection you now have.

c. Now rotate the last Fischer projection you obtained by 180 degrees; draw the Fischer projection you now have; label this as either the same as the starting model or the enantiomer.

14. Let us examine models with two stereocenters by using tartaric acid as the example, HOOC—CH(OH)—CH(OH)—COOH; use your colored components to represent the various groups. Hold your models so that

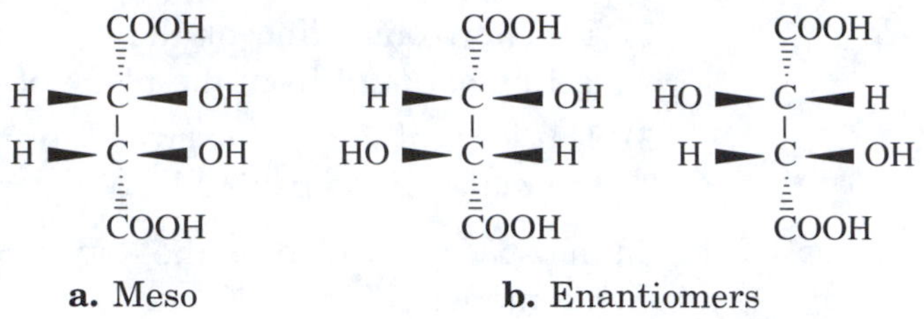

Figure 22.7
The stereoisomers of tartaric acid.

a. Meso b. Enantiomers

each stereoisomer is oriented as in Figure 22.7. In the space provided on the Report Sheet (14), draw each of the corresponding Fischer projections.

Circle the Fischer projection that shows a plane of symmetry. Underline all the Fischer projections that would be optically active.

15. Use the Fischer projection of *meso*-tartaric acid and carry out even and odd exchanges of the groups; follow these exchanges with a model. Does an odd exchange lead to an enantiomer, a diastereomer, or to a system identical to the *meso* form (15a)? Does an even exchange lead to an enantiomer, a diastereomer, or to a system identical to the *meso* form (15b)?

R/S Convention for Chiral Centers

A system of assigning configuration to a chiral carbon was devised by the chemists R. S. Cahn, C. K. Ingold, and V. Prelog. The system uses the letters *R* and *S* to designate the configuration at the chiral carbon: *R*, from the Latin *rectus*, or right; *S*, from the Latin *sinister*, or left. The designation arises from the priority order assigned to the four groups attached to the chiral carbon. Priority order is based on the atomic number of the atoms directly attached to the chiral carbon: the higher the atomic number, the higher the priority. In the case where two atoms have the same atomic number, you must move along the bonds to the next atoms out from the chiral center until an atom of different atomic number is reached. In order to apply the Cahn-Ingold-Prelog system, hold a model of the molecule so that the atom with *lowest* priority is pointed directly away from you. Then examine the remaining three attachments in terms of the order of their priorities (Figure 22.8):

a. If movement of your eye is from the highest to the lowest priority in a clockwise direction, the configuration of the chiral carbon is defined as *R*.

b. If movement of your eye is from the highest to the lowest priority in a counter-clockwise direction, the configuration of the chiral carbon is defined as *S*.

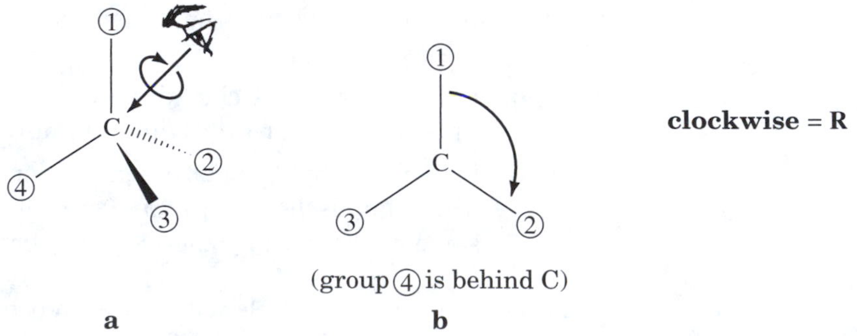

clockwise = R

(group ④ is behind C)

a b

Figure 22.8
An example of assignment using *R*/*S* priority rules: 1 > 2 > 3 > 4.

Now let us work with a model to see how the rules apply.

16. Construct a model of 2-bromobutane, $CH_3CHBrCH_2CH_3$. (You should have enough carbon components for the carbon skeleton and colored components for H and Br.) Answer the following questions in the appropriate spaces on the Report Sheet (16).

a. What are the four groups attached to the chiral carbon (16a)? Assign to these groups priorities: 1 = highest to 4 = lowest.

b. View the model as shown in Figure 22.8a and draw the arrangement you see. Is it *R* or *S* (16b)?

c. Exchange the H and the Br. Again view the model and draw the arrangement. Is it *R* or *S* (16c)?

d. Are the configurations opposite one another? Are they enantiomers (16d)?

CHEMICALS AND EQUIPMENT

Model kits vary in size and color of components. Use what is available; other colors may be substituted.

1. Cyclohexane model kit: 8 carbons (black, 4 holes); 18 hydrogens (white, 1 hole); 2 substituents (red, 1 hole); 24 bonds.
2. Chiral model kit: 8 carbons (black, 4 holes); 32 substituents (8 red, 1 hole; 8 white, 1 hole; 8 blue, 1 hole; 8 green, 1 hole); 28 bonds.
3. Hand mirror

name　　　　section　　　　date

partner　　　　grade

22

EXPERIMENT 22

Pre-Lab Questions

1. Is a planar hexagon the most stable conformation for the cyclohexane ring? If not, draw the most stable conformation for the ring.

2. Define enantiomers.

3. How does a chiral molecule differ from an achiral molecule?

4. If someone refers to a molecule as being "optically active," what does that person mean?

5. Give priority order to the members of the halogen family following the Cahn-Ingold-Prelog rules.

name section date

partner grade

22

EXPERIMENT 22

Report Sheet

Cyclohexane

1. a.

   b.

2. a.

   b.

   c.

   d.

   e.

   f.

   g.

3. a.

   b.

4. a.

   b.

5. a.

   b.

   c.

   d.

6. a.

b.

7. *Trial 1* *Trial 2*

a.

b.

c.

8. a.

b.

c.

d.

9.

10. e,e or a,a

11. a. a,e or e,a

b.

12. a,a or e,e

13. a. a,e or e,a

b.

14. a,a or e,e

15. a. a,e or e,a

b.

16. a.

b.

c.

Enantiomers

1. a.

b.

2.

$CH_3-CH(OH)-CH_2CH_3$ $CH_3-CH(OH)-COOH$ $ClCH_2-CH(Br)-CH_3$

3. a.

b.

c.

d.

e.

4. a.

b.

c.

d.

e.

Meso forms and diastereomers

5. a.

b.

c.

6.

$$\begin{array}{ccccccc} & \mathrm{H} & & \mathrm{H} & \\ & | & & | & \\ \mathrm{HOOC}- & \mathrm{C_a} & - & \mathrm{C_b} & -\mathrm{COOH} \\ & | & & | & \\ & \mathrm{HO} & & \mathrm{OH} & \end{array} \qquad \begin{array}{ccccc} & \mathrm{H} & & \mathrm{H} & \\ & | & & | & \\ \mathrm{HOOC}- & \mathrm{C_a} & - & \mathrm{C_b} & -\mathrm{COOH} \\ & | & & | & \\ & \mathrm{HO} & & \mathrm{OH} & \end{array}$$

7. a.

   b.

8. a.

   b.

   c.

   d.

   e.

   f.

9. a.

   b.

Drawing stereoisomers

10.

11. a.

   b.

   c.

12. a.

b.

13. a.

b.

c.

14.

15. a.

b.

R/S convention for chiral centers

16. a.

b.

c.

d.

Post-Lab Questions

1. Why is methyl cyclohexane at lower energy when the methyl group is in an equatorial position?

2. *meso*-Tartaric acid has two chiral centers yet is optically inactive. Explain this observation.

3. Draw the Fischer projections for the pair of enantiomers of lactic acid, CH_3—CH(OH)—COOH. Determine the configuration for each chiral carbon: R; S.

4. Look at the following pairs of structures. Determine whether they are identical, enantiomers, or diastereomers.

a.

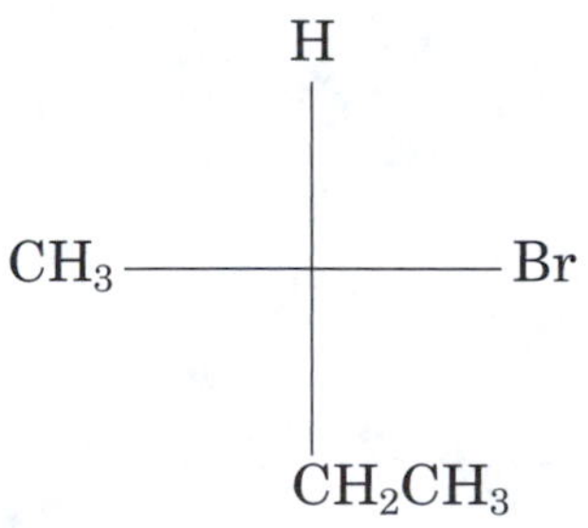

H

CH_3 — Br

CH_2CH_3

and

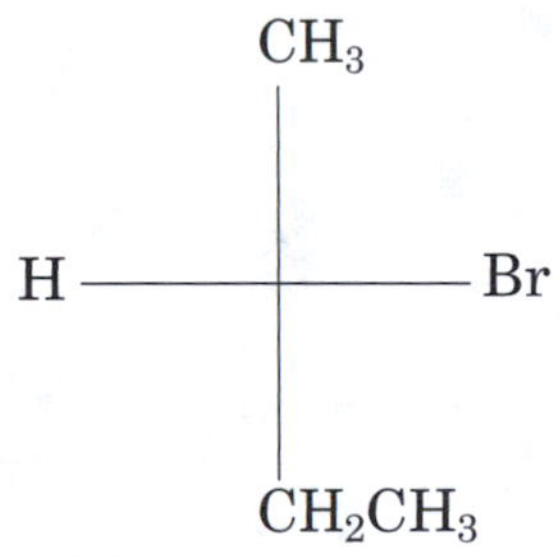

CH_3

H — Br

CH_2CH_3

b.

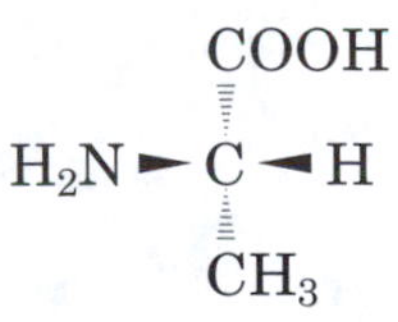

COOH

H_2N ► C ◄ H

CH_3

and

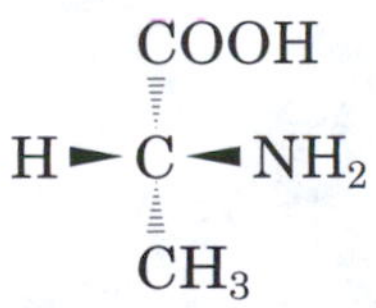

COOH

H ► C ◄ NH_2

CH_3

Column and Paper Chromatography: Separation of Plant Pigments

BACKGROUND

Chromatography is a widely used experimental technique for the separation of a mixture of compounds into its individual components. Two kinds of chromatographic techniques will be explored: column chromatography and paper chromatography.

In column chromatography, a mixture of components dissolved in a solvent is poured over a column of solid adsorbent and is eluted with the same or a different solvent. This is a solid–liquid system; the stationary phase (the adsorbent) is solid and the mobile phase (the eluent) is liquid. In paper chromatography, the paper adsorbs water from the atmosphere of the developing chromatogram. (The water is present in the air as vapor, and it may be supplied as one component in the eluting solution.) The water is the stationary phase. The (other) component of the eluting solvent is the mobile phase and carries with it the components of the mixture. This is a liquid–liquid system.

Column chromatography generally is used for preparative purposes, when one deals with a relatively large amount of the mixture, and the components need to be isolated in milligram or gram quantities.

Paper chromatography, on the other hand, is an analytical technique. Microgram or even picogram quantities can be separated by this technique, and they can be characterized by their R_f value. This value is an index of how far a certain spot moved on the paper.

$$R_f = \frac{\textbf{Distance of the center of the sample spot from the origin}}{\textbf{Distance of the solvent front from the origin}}$$

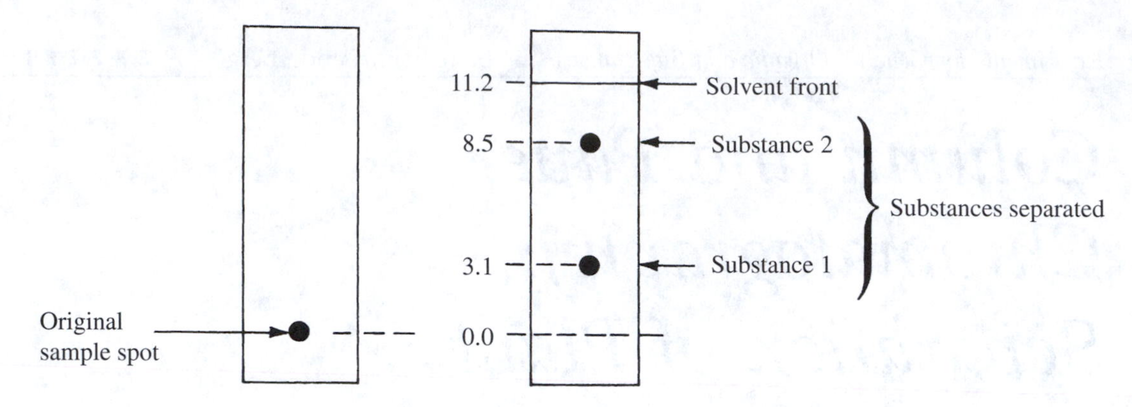

Figure 23.1
Illustration of chromatograms before and after elution.

For example, in Figure 23.1 the R_f values for two substances are as follows:

R_f (substance 1) = 3.1 cm/11.2 cm = 0.28

R_f (substance 2) = 8.5 cm/11.2 cm = 0.76

One is able to identify the components by comparison of the R_f values found in the mixture to individual components whose R_f values are known. In tomato paste, there are two main pigment components: β-carotene (yellow-orange) and lycopene (red) pigments. These compounds are classed as alkenes; their structures are given below:

Lycopene (red)

β-Carotene (yellow-orange)

The colors of these pigments are due to the numerous double bond functional groups in their structure. In this experiment, we will use color and R_f values for the spots to identify the components in the tomato paste.

There is another observation that may be made with compounds having carbon-to-carbon double bonds. When bromine reacts with the double bonds, it adds to the bonds, and there are color changes that result; for example, the red bromine solution becomes colorless, because it adds to the double bond and there is no longer free bromine present.

$$-\overset{|}{C}=\overset{|}{C}- + Br_2 \longrightarrow -\overset{|}{\underset{Br}{\underset{|}{C}}}-\overset{|}{\underset{Br}{\underset{|}{C}}}-$$

Red → Colorless

In the tomato juice "rainbow" experiment, we stir bromine water into the tomato juice. The slow stirring allows the bromine water to penetrate deeper and deeper into the cylinder in which the tomato juice was placed. As the bromine penetrates, more and more double bonds will react. Therefore, you may be able to observe a continuous change, a "rainbow" of colors, starting with the reddish tomato color at the bottom of the cylinder, where no reaction occurred (because the bromine did not reach the bottom), to lighter colors on the top of the cylinder, where most of the double bonds have reacted with the bromine. The colored pigments in the tomato juice react with the bromine and lose their color as the reaction progresses.

OBJECTIVES

1. To compare separation of components of a mixture by two different techniques.
2. To demonstrate the effect of bromination on plant pigments of tomato juice.

PROCEDURE

A. Paper Chromatography

1. Obtain a sheet of Whatman no. 1 filter paper, cut to size, 10 × 20 cm.
2. Plan the **spotting** of the samples as illustrated in Figure 23.2. Five spots will be applied. The first and fifth spots will be β-carotene solutions supplied by your instructor. The second, third, and fourth spots will have your tomato paste extracts in different concentrations. Use a pencil to mark lightly the spots according to Figure 23.2.
3. Pigments of tomato paste will be **extracted** in two steps.
   a. Weigh about 10 g of tomato paste in a 50-mL beaker. Add 15 mL of 95% ethanol. Stir the mixture vigorously with a spatula until the paste will not stick to the stirrer. Place a small amount of glass wool (the size of a pea) in a small funnel, blocking the funnel exit. Place the funnel into a 50-mL Erlenmeyer flask and pour the tomato paste–ethanol mixture into the funnel. When the filtration is completed, squeeze the glass wool lightly with your spatula. In this step, we removed the water from the tomato paste and the

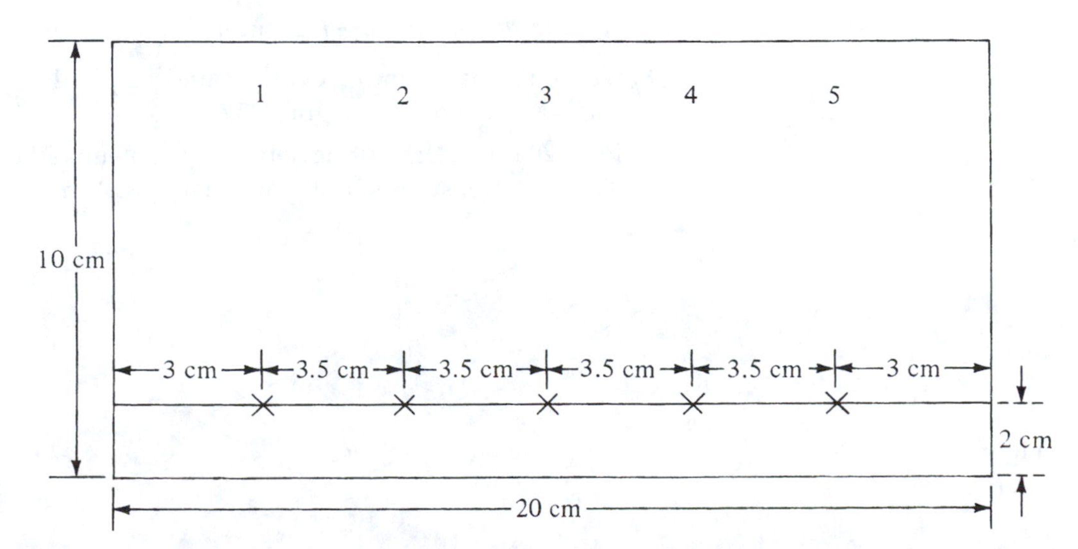

Figure 23.2
Preparation of chromatographic paper for spotting.

aqueous components are in the filtrate, which we discard. The residue in the glass wool will be used to extract the pigments.

b. Place the residue in the glass wool in a 50-mL beaker. Add 10 mL petroleum ether and stir the mixture for about 2 min. to extract the pigments. Filter the extract as before, through a new funnel with glass wool blocking the exit, into a new and clean 50-mL beaker. Place the beaker under the hood on a hot plate (or water bath). Evaporate the solvent to about 1 mL volume. Use low heat and take care not to evaporate all the solvent. After evaporation cover the beaker with aluminum foil.

CAUTION

No open flame, such as from a Bunsen burner, is to be used. The solvents are flammable.

Spotting

4. Place your chromatographic paper on a clean area (another filter paper) in order not to contaminate it. Use separate capillaries, one for your tomato paste extract and one for the β-carotene solution. First apply your capillary to the extracted pigment by dipping it into the solution as illustrated in Figure 23.3. Apply the capillary lightly to the chromatographic paper by touching, sequentially, the spots marked 2, 3, and 4. Make sure, when touching the paper, that you make only small spots, not larger than 2-mm diameter, by quickly withdrawing the capillary from the paper each time you touch it. (See Figure 23.4.)
5. While these spots dry, use your second capillary to apply β-carotene to positions marked 1 and 5.
6. Return to the first capillary and apply more extract on top of the spots at positions marked 3 and 4 as you did before. Let them dry (Figure 23.5).
7. Finally, apply extract once more on top of the spot at position 4. Let all the spots dry. The unused extract in your beaker should be covered with aluminum foil. Place it in your drawer in the dark to save it for the second part of this experiment.

Figure 23.3
Withdrawing samples by a capillary tube.

Developing the paper chromatogram

8. Curve the paper into a cylinder and staple the edges above the 2-cm line as it is shown in Figure 23.6.
9. Pour 20 mL of the eluting solvent (petroleum ether : toluene : acetone in 45 : 1 : 4 ratio, supplied by your instructor) into a 600-mL beaker.

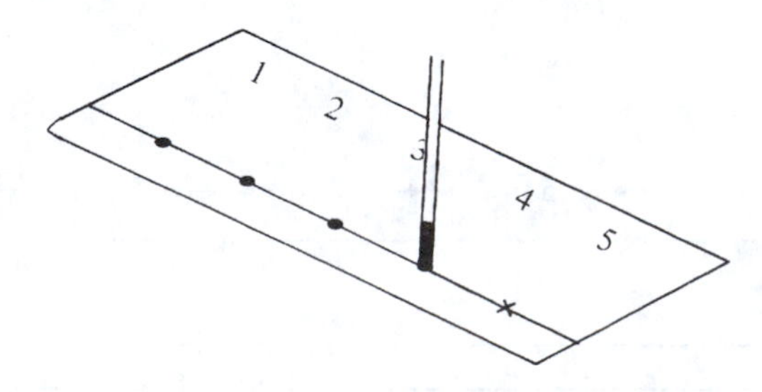

Figure 23.4
Spotting.

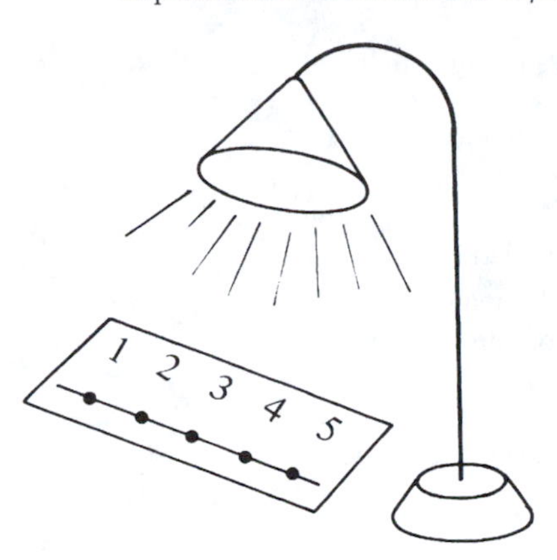

Figure 23.5
Drying chromatographic spots.

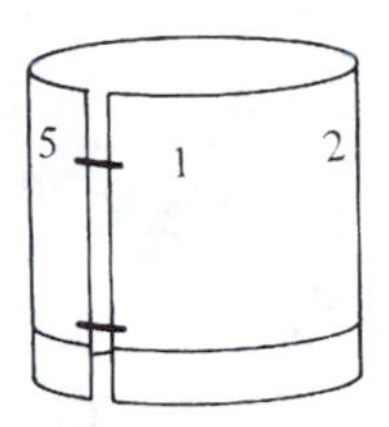

Figure 23.6
Stapling.

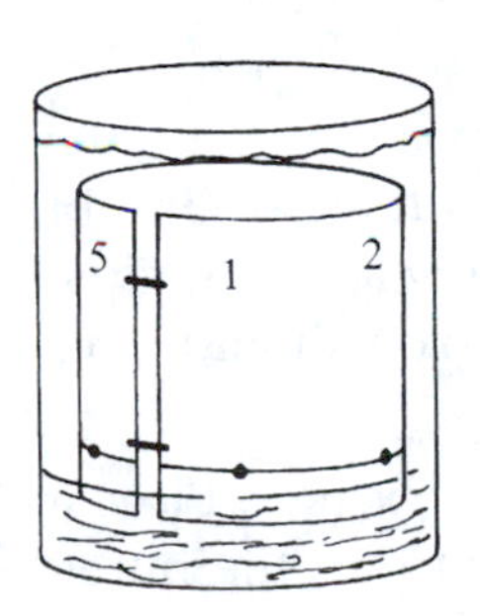

Figure 23.7
Developing the chromatogram.

10. Place the stapled chromatogram into the 600-mL beaker, the spots being at the bottom near the solvent surface but **not covered by it.** Cover the beaker with aluminum foil (Figure 23.7). Allow the solvent front to migrate up to 0.5–1 cm below the edge of the paper. This may take from 15 min. to 1 hr. Make certain by frequent inspection that the **solvent front does not run over the edge of the paper.**

    [*While you are waiting for the paper chromatography to develop, you can do the column chromatography experiment. B.*]

11. Remove the chromatogram from the beaker when the solvent front reaches 0.5–1 cm from the edge. You must remove the filter paper from the 600-mL beaker before the solvent front reaches the edges of the paper. Mark the position of the solvent front with a pencil. Put the paper standing on its edges under the hood and let it dry.

    [*While waiting for the paper to dry, you can do the short tomato juice "rainbow" experiment. C.*]

12. Remove the staples from the dried chromatogram. Mark the spots of the pigments by circling with a pencil. Note the colors of the spots. Measure the distance of the center of each spot from its origin. Calculate the R_f values.

CAUTION

If you use the iodine, heat in the hood. Iodine vapor is toxic. Do not breathe the vapor. Also do not touch the crystals; they will stain your fingers.

13. If the spots on the chromatogram are faded, we can visualize them by exposing the chromatogram to iodine vapor. Place your chromatogram into a wide-mouth jar containing a few iodine crystals. Cap the jar and warm it slightly on a hot plate to enhance the sublimation of iodine. The iodine vapor will interact with the faded pigment spots and make them visible. After a few minutes of exposure to iodine vapor, remove the chromatogram and mark the spots **immediately** with a pencil. The spots will fade again with exposure to air. Measure the distance of the center of the spots from the origin and calculate the R_f values.

14. Record the results of the paper chromatography on the Report Sheet. (Is there more than one spot visible for the tomato paste?)

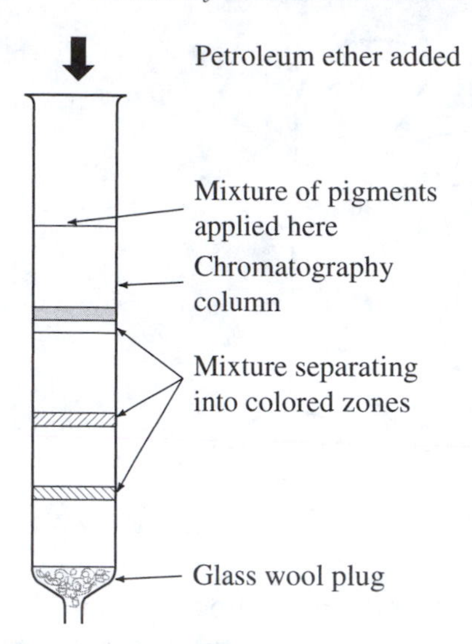

Figure 23.8
Chromatographic column.

B. Column Chromatography

1. While you are waiting for the chromatogram to develop (step 10, above), you can perform the column chromatography experiment.
2. Use a 25-mL buret. (You may use a chromatograhic column, if available, of 1.6 cm diameter and about 13 cm long; see Figure 23.8. If you use the column instead of the buret, all quantities below should be doubled.)
3. Add a small piece of glass wool (about the size of a pea) and with the aid of a glass rod, push it down near the stopcock. Add 15–16 mL of petroleum ether to the buret. Open the stopcock slowly and allow the solvent to fill the tip of the buret. Close the stopcock. You should have 12–13 mL of solvent above the glass wool.
4. Weigh 20 g of aluminum oxide (alumina) in a 100-mL beaker. Place a small funnel on top of your buret. Pour the alumina into the buret. Allow the alumina to settle to form a 20-cm column. *Drain* the solvent but *do not allow the column to run dry. Always have at least 0.5 mL of clear solvent above the alumina in the column.* If alumina adheres to the walls of the buret, wash it down with more solvent.
5. Transfer by pipet 0.5–1 mL of the extract you stored in your drawer onto the column. The pipet containing the extract should be placed near the surface of the solvent on top of the column. Touching the walls of the buret with the tip of the pipet, allow the extract to drain slowly on top of the column. Open the stopcock slightly. Allow the sample to enter the column, *but make sure there is a small amount of solvent above the alumina in the column. (The column should never run dry.)* Add 10 or more mL of petroleum ether and wash the sample into the column by opening the stopcock and collecting the eluted solvent in a beaker.
6. As the solvent elutes the sample, you observe the migration of the pigments and their separation into at least two bands. When the fastest-moving pigment band reaches near the bottom of the column, close the stopcock and observe the color of the pigment bands and how far they migrated from the top of the column. Record your observation

on the Report Sheet. This concludes the column chromatographic part of the experiment. Discard your solvent in a bottle supplied by your instructor for a later distillation. Clean your buret as directed by your instructor.

C. Tomato Juice "Rainbow"

While waiting for the paper to dry you can perform the following short experiment. Weigh about 15 g of tomato paste in a beaker. Add about 30 mL of water and stir. Transfer the tomato juice into a 50-mL graduated cylinder and, with the aid of a pipet, add 5 mL of saturated bromine water (dropwise). With a glass rod, stir the solution very gently. Observe the colors and their positions in the cylinder. Record your observations on the Report Sheet.

CHEMICALS AND EQUIPMENT

1. Melting-point capillaries open at both ends
2. 25-mL buret (or chromatography column)
3. Glass wool
4. Whatman no. 1 filter paper, 10 × 20 cm, cut to size
5. Heat lamp (optional)
6. Stapler
7. Hot plate (with or without water bath)
8. Tomato paste
9. Aluminum oxide (alumina)
10. Petroleum ether (b.p. 30–60°C)
11. 95% ethanol
12. Acetone
13. Toluene
14. 0.5% β-carotene in petroleum ether
15. Saturated bromine water
16. Iodine crystals
17. Ruler
18. Wide-mouth jar

name | section | date

partner | grade

23 EXPERIMENT 23

Pre-Lab Questions

1. The structures of the two main pigments, lycopene and β-carotene, are shown in the **Background** section.

   a. These pigments belong to what class of organic compounds?

   b. What functional groups are present in these pigments?

   c. What solvents would be good for these pigments: polar or nonpolar? Why?

2. What marks are placed on your chromatography paper in order to calculate the R_f value of a spot?

3. What happens to the double bonds in the pigments when bromine is added?

4. In the column chromatography, (a) what is the "stationary phase" and (b) what is the "mobile phase"?

name | section | date

partner | grade

23 EXPERIMENT 23

Report Sheet

Paper Chromatography

Sample *Spots*	*Distance from origin to solvent front (cm) (a)*	*Distance from origin to center of spot (cm) (b)*	R_f *(b)/(a)*	*Color*
β-carotene				
lane 1				
lane 5				
Tomato Extract				
lane 2 (a)				
(b)				
(c)				
(d)				
lane 3 (a)				
(b)				
(c)				
(d)				
lane 4 (a)				
(b)				
(c)				
(d)				

Column Chromatography

Number of bands	*Distance migrated from top of the column (cm)*	*Color*
1		
2		
3		

"Rainbow"

Describe the colors observed in the tomato juice "rainbow" experiment, starting from the bottom of the cylinder.

1. red

2.

3.

4.

5.

6.

Post-Lab Questions

1. What is the evidence that your sample of tomato paste contains lycopene?

2. Tomato juice is red. How does this suggest that the juice contains the pigment lycopene rather than β-carotene?

3. What is the effect of the amount of sample applied to the paper on the separation of the tomato pigments? Compare the results on lanes 2, 3, and 4 of the paper chromatogram.

4. If a molecule of lycopene reacts with bromine, how many bromine molecules would be necessary to completely saturate it?

5. Compare the structures of lycopene and β-carotene. What is the basic difference between the two?

Identification of Hydrocarbons

BACKGROUND

The number of known organic compounds totals into the millions. Of these compounds, the simplest types are those that contain only hydrogen and carbon atoms. These are known as *hydrocarbons*. Because of the number and variety of hydrocarbons that can exist, some means of classification is necessary.

One means of classification depends on the way in which carbon atoms are connected. *Chain* aliphatic hydrocarbons are compounds consisting of carbons linked either in a single chain or in a branched chain. *Cyclic* hydrocarbons are aliphatic compounds that have carbon atoms linked in a closed polygon (also referred to as a *ring*). For example, hexane (single) and 2-methylpentane (branched) are chain aliphatic molecules, while cyclohexane is a cyclic aliphatic compound.

$CH_3CH_2CH_2CH_2CH_2CH_3$

Hexane

$CH_3CH(CH_3)CH_2CH_2CH_3$

2-Methylpentane

$(CH_2)_6$ (ring of six CH_2 groups)

Cyclohexane

Another means of classification depends on the type of bonding that exists between carbons. Hydrocarbons that contain only carbon-to-carbon single bonds are called *alkanes*. These are also referred to as *saturated* molecules. Hydrocarbons containing at least one carbon-to-carbon double bond are called *alkenes*, and compounds with at least one carbon-to-carbon triple bond are called *alkynes*. These compounds are referred to as *unsaturated* molecules. Finally, a class of cyclic hydrocarbons that contain a closed loop (sextet) of electrons is called *aromatic*. Table 24.1 distinguishes between the families of hydrocarbons.

Table 24.1 *Types of Hydrocarbons*

Class	*Characteristic Bond Type*		*Example*	
I. Aliphatic				
1. Alkane*	$-\overset{\vert}{\underset{\vert}{C}}-\overset{\vert}{\underset{\vert}{C}}-$	single	$CH_3CH_2CH_2CH_2CH_2CH_3$	hexane
2. Alkene†	$>C=C<$	double	$CH_3CH_2CH_2CH_2CH=CH_2$	1-hexene
3. Alkyne†	$-C\equiv C-$	triple	$CH_3CH_2CH_2CH_2C\equiv CH$	1-hexyne
II. Cyclic				
1. Cycloalkane*	$-\overset{\vert}{\underset{\vert}{C}}-\overset{\vert}{\underset{\vert}{C}}-$	single		cyclohexane
2. Cycloalkene†	$>C=C<$	double		cyclohexene
3. Aromatic				benzene
			CH_3	toluene

*Saturated †Unsaturated

With so many compounds possible, identification of the bond type is an important step in establishing the molecular structure. Quick, simple tests on small samples can establish the physical and chemical properties of the compounds by class.

Some of the observed physical properties of hydrocarbons result from the nonpolar character of the compounds. In general, hydrocarbons do not mix with polar solvents such as water or ethanol (ethyl alcohol). On the other hand, hydrocarbons mix with relatively nonpolar solvents such as ligroin (a mixture of alkanes), carbon tetrachloride (CCl_4), or dichloromethane (CH_2Cl_2). Because the density of most hydrocarbons is less than that of water, they will float. Crude oil and crude oil products (home heating oil and gasoline) are mixtures of hydrocarbons; these substances, when spilled on water, spread quickly along the surface because they are insoluble in water.

The chemical reactivity of hydrocarbons is determined by the type of bond in the compound. Although saturated hydrocarbons (alkanes) will burn (undergo *combustion*), they are generally unreactive to most reagents. (Alkanes do undergo a substitution reaction with halogens but require ultraviolet light.) Unsaturated hydrocarbons, alkenes and alkynes, not only burn, but also react by *addition* of reagents to the double or triple bonds. The addition products become saturated, with fragments of the reagent becoming attached to the carbons of the multiple bond. Aromatic compounds, with a higher carbon-to-hydrogen ratio than nonaromatic compounds, burn with a sooty flame as a result of unburned carbon particles being present. These compounds undergo *substitution* in the presence of catalysts rather than an addition reaction.

1. *Combustion.* The major component in "natural gas" is the hydrocarbon methane (CH_4). Other hydrocarbons used for heating or cooking purposes are propane (C_3H_8) and butane (C_4H_{10}). The products from combustion are carbon dioxide and water (heat is evolved, also).

$$CH_4 + 2O_2 \longrightarrow CO_2 + 2H_2O + \text{Heat}$$

$$CH_3CH_2CH_3 + 5O_2 \longrightarrow 3CO_2 + 4H_2O + \text{Heat}$$

2. *Reaction with bromine.* Unsaturated hydrocarbons react rapidly with bromine in a solution of carbon tetrachloride or cyclohexane. The reaction is the addition of the elements of bromine to the carbons of the multiple bonds.

$$\underset{}{CH_3CH{=}CHCH_3} + \underset{\text{Red}}{Br_2} \longrightarrow \underset{\text{Colorless}}{CH_3\overset{\overset{\displaystyle Br}{|}}{C}H{-}\overset{\overset{\displaystyle Br}{|}}{C}HCH_3}$$

$$CH_3C{\equiv}CCH_3 + \underset{\text{Red}}{2Br_2} \longrightarrow \underset{\text{Colorless}}{CH_3\overset{\overset{\displaystyle Br}{|}}{\underset{\underset{\displaystyle Br}{|}}{C}}{-}\overset{\overset{\displaystyle Br}{|}}{\underset{\underset{\displaystyle Br}{|}}{C}}CH_3}$$

The bromine solution is red; the product that has the bromine atoms attached to carbon is colorless. Thus a reaction has taken place when there is a loss of color from the bromine solution and a colorless solution remains. Because alkanes have only single C—C bonds present, no reaction with bromine is observed; the red color of the reagent would persist when added. Aromatic compounds resist addition reactions because of their "aromaticity": *the possession of a closed loop (sextet) of electrons*. These compounds react with bromine in the presence of a catalyst such as iron filings or aluminum chloride.

$$C_6H_5\text{–}H + Br_2 \xrightarrow[\text{or } AlCl_3]{Fe} C_6H_5\text{–}Br + HBr$$

Note that a substitution reaction has taken place and the gas HBr is produced.

3. *Reaction with concentrated sulfuric acid.* Alkenes react with cold concentrated sulfuric acid by addition. Alkyl sulfonic acids form as products and are soluble in H_2SO_4.

$$CH_3-CH=CH-CH_3 + \underset{(H_2SO_4)}{HOSO_2OH} \longrightarrow CH_3-\underset{\displaystyle H}{\underset{|}{CH}}-\underset{\displaystyle OSO_2OH}{\underset{|}{CH}}-CH_3$$

Saturated hydrocarbons are unreactive (additions are not possible); alkynes react slowly and require a catalyst ($HgSO_4$); aromatic compounds also are unreactive because addition reactions are difficult.

4. *Reaction with potassium permanganate.* Dilute or alkaline solutions of $KMnO_4$ oxidize unsaturated compounds. Alkanes and aromatic compounds are generally unreactive. Evidence that a reaction has occurred is observed by the loss of the purple color of $KMnO_4$ and the formation of the brown precipitate manganese dioxide, MnO_2.

$$3CH_3-CH=CH-CH_3 + \underset{\text{Purple}}{2KMnO_4} + 4H_2O \longrightarrow$$

$$3CH_3-\underset{\displaystyle OH}{\underset{|}{CH}}-\underset{\displaystyle OH}{\underset{|}{CH}}-CH_3 + \underset{\text{Brown}}{2MnO_2} + 2KOH$$

Note that the product formed from an alkene is a glycol.

OBJECTIVES

1. To investigate the physical properties, solubility, and density of some hydrocarbons.
2. To compare the chemical reactivity of an alkane, an alkene, and an aromatic compound.
3. To use physical and chemical properties to identify an unknown.

PROCEDURE

CAUTION

Assume the organic compounds are highly flammable. Use only small quantities. Keep away from open flames. Assume the organic compounds are toxic and can be absorbed through the skin. Avoid contact; wash if any chemical spills on your person. Handle concentrated sulfuric acid carefully. Flush with water if any spills on your person. Potassium permanganate and bromine are toxic; bromine solutions are also corrosive. Although the solutions are diluted, they may cause burns to the skin. Wear gloves when working with these chemicals.

General Instructions

1. The hydrocarbons hexane, cyclohexene, and toluene (alkane, alkene, and aromatic) are available in dropper bottles.
2. The reagents 1% Br_2 in cyclohexane, 1% aqueous $KMnO_4$, and concentrated H_2SO_4 are available in dropper bottles.

3. Unknowns are in dropper bottles labeled A, B, and C. They may include an alkane, an alkene, or an aromatic compound.
4. Test tubes of 100 × 13 mm will be suitable for all the tests. When mixing the components, grip the test tube between thumb and fore-finger; it should be held firmly enough to keep from slipping but loosely enough so that when the third and fourth fingers tap it, the contents will be agitated enough to mix.
5. Record all data and observations in the appropriate places on the Report Sheet.
6. Dispose of all organic wastes as directed by the instructor. *Do not pour them into the sink!*

Physical Properties of Hydrocarbons

1. *Water solubility of hydrocarbons.* Label six test tubes with the name of the substance to be tested. Place into each test tube 5 drops of the appropriate hydrocarbon: hexane, cyclohexene, toluene, unknown A, unknown B, unknown C. Add about 5 drops of water dropwise into each test tube. Water is a polar solvent. Is there any separation of components? Which component is on the bottom; which component is on the top? Mix the contents as described above. What happens when the contents are allowed to settle? What do you conclude about the density of the hydrocarbon? Is the hydrocarbon *more* dense than water or *less* dense than water? Record your observations. Save these solutions for comparison with the next part.
2. *Solubility of hydrocarbons in ligroin.* Label six test tubes with the name of the substance to be tested. Place into each test tube 5 drops of the appropriate hydrocarbon: hexane, cyclohexene, toluene, unknown A, unknown B, unknown C. Add about 5 drops of ligroin dropwise into each test tube. Ligroin is a nonpolar solvent. Is there a separation of components? Is there a bottom layer and a top layer? Mix the contents as described above. Is there any change in the appearance of the contents before and after mixing? Compare these test tubes with those from the previous part. Record your observations. Can you make any conclusion about the density of the hydrocarbons from what you actually see?

Chemical Properties of Hydrocarbons

1. *Combustion.* The instructor will demonstrate this test in the fume hood. Place 5 drops of each hydrocarbon and unknown on separate watch glasses. Carefully ignite each sample with a match. Observe the flame and color of the smoke for each of the samples. Record your observations on the Report Sheet.
2. *Reaction with bromine.* Label six clean, dry test tubes with the name of the substance to be tested. Place into each test tube 5 drops of the appropriate hydrocarbon: hexane, cyclohexene, toluene, unknown A, unknown B, unknown C. Carefully add (dropwise and with shaking) 1% Br_2 in cyclohexane. Keep count of the number of drops needed to have the color persist; do not add more than 10 drops. Record your observations. To any sample that gives a negative test after adding 10 drops of bromine solution (i.e., the red color persists), add 5 more drops of 1% Br_2 solution and a small quantity of aluminum chloride, the amount on the tip of a micro-spatula; shake the misture. Hold, with forceps, a piece of moistened blue litmus paper and lower it into the test tube until it is just above the surface of the liquid. Be careful and

try not to touch the sides of the test tube or the liquid with the litmus paper. Record any change in the color of the solution and the litmus paper.

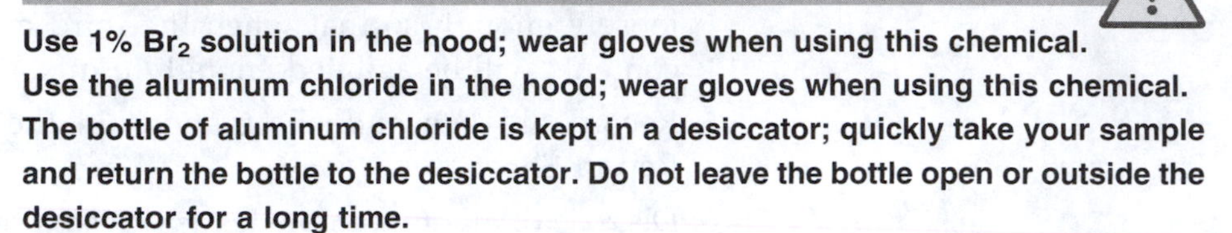

CAUTION

Use 1% Br_2 solution in the hood; wear gloves when using this chemical.
Use the aluminum chloride in the hood; wear gloves when using this chemical.
The bottle of aluminum chloride is kept in a desiccator; quickly take your sample and return the bottle to the desiccator. Do not leave the bottle open or outside the desiccator for a long time.

3. *Reaction with $KMnO_4$.* Label six clean, dry test tubes with the name of the substance to be tested. Place into each test tube 5 drops of the appropriate hydrocarbon: hexane, cyclohexene, toluene, unknown A, unknown B, unknown C. Carefully add (dropwise) 1% aqueous $KMnO_4$ solution; after each drop, shake to mix the solutions. Keep count of the number of drops needed to have the color of the permanganate solution persist; do not add more than 10 drops. Record your observations.
4. *Reaction with concentrated H_2SO_4.* Label six clean, dry test tubes with the name of the substance to be tested. Place into each test tube 5 drops of the appropriate hydrocarbon: hexane, cyclohexene, toluene, unknown A, unknown B, unknown C. Place all of the test tubes in an ice bath. *Wear gloves and carefully* add (with shaking) 3 drops of cold, concentrated sulfuric acid to each test tube. Note whether heat is evolved by feeling the test tube. Note whether the solution has become homogeneous or whether a color is produced. (The evolution of heat or the formation of a homogeneous solution or the appearance of a color is evidence that a reaction has occurred.) Record your observations.
5. *Unknowns.* By comparing the observations you made for your unknowns with that of the known hydrocarbons, you can identify unknowns A, B, and C. Record their identities on your Report Sheet.

CHEMICALS AND EQUIPMENT

1. 1% aqueous $KMnO_4$
2. 1% Br_2 in cyclohexane
3. Blue litmus paper
4. Concentrated H_2SO_4
5. Cyclohexene
6. Hexane
7. Aluminum chloride
8. Test tubes
9. Ligroin
10. Toluene
11. Unknowns A, B, and C
12. Watch glasses
13. Ice

name | section | date

partner | grade

EXPERIMENT 24

Pre-Lab Questions

1. Hexane is an example of a *saturated* hydrocarbon and cyclohexene is an example of an *unsaturated* hydrocarbon. Explain the difference between the two compounds.

2. Show the structural feature that distinguishes whether a hydrocarbon is an

   alkane

   alkene

   alkyne

   aromatic

3. When you make salad dressing, the oil and water do not mix but separate into layers. Why does this happen?

name | section | date

partner | grade

24 EXPERIMENT 24

Report Sheet

Physical properties of hydrocarbons

Solubility: Does the hydrocarbon mix with the solvent, *soluble,* or not mix with the solvent, *insoluble*? Use the observations you make for the solubility tests and determine whether the hydrocarbons are polar or nonpolar substances.

Density: For water, is the density *greater* than water (sinks) or *less* than water (floats)? For ligroin, can you tell anything about the relative densities?

	H_2O		Ligroin	
Hydrocarbon	*Solubility*	*Density*	*Solubility*	*Density*
Hexane				
Cyclohexene				
Toluene				
Unknown A				
Unknown B				
Unknown C				

Chemical properties of hydrocarbons

Hydrocarbon	*Combustion*	*Bromine Test*	*$KMnO_4$ Test*	*H_2SO_4 Test*
Hexane				
Cyclohexene				
Toluene				
Unknown A				
Unknown B				
Unknown C				

Unknown A is _______________.

Unknown B is _______________.

Unknown C is _______________.

Post-Lab Questions

1. Below are four organic compounds. The reagent shown is added to the compound. What should you see when the two materials are mixed together?

   a. $CH_3-CH=CH_2$ + Br_2

   b. + $KMnO_4$

   c. $CH_3-CH=CH-CH_3$ + H_2SO_4

   d. + $KMnO_4$

2. A student has two compounds in two separate bottles but with no labels on either one. One is an unbranched alkane, octane, C_8H_{18}; the other is 1-hexene, an unbranched alkene, C_6H_{12}. Based on your observations in this experiment, tell what you should see in the following tests:

	Octane	*1-Hexene*
a. Water solubility		
b. Ligroin solubility		
c. Combustion		
d. Density versus water		
e. Bromine test		
f. Permanganate test		

3. An unknown compound, believed to be a hydrocarbon, showed the following behavior: it burned with a yellow sooty flame; no heat or color appeared when sulfuric acid was added; permanganate solution remained purple; the red color of bromine solution was lost only after a catalyst was added. From the compounds below, circle the one that fits the observations.

   $CH_3CH_2CH_2CH_2CH_3$ 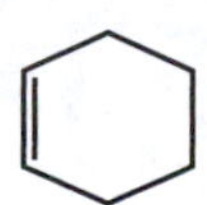 $H-C\equiv C-CH_2CH_2CH_3$

Identification of Alcohols and Phenols

BACKGROUND

Specific groups of atoms in an organic molecule can determine its physical and chemical properties. These groups are referred to as *functional groups.* Organic compounds that contain the functional group –OH, the hydroxyl group, are called *alcohols.*

Alcohols are important commercially and include uses as solvents, drugs, and disinfectants. The most widely used alcohols are methanol or methyl alcohol, CH_3OH, ethanol or ethyl alcohol, CH_3CH_2OH, and 2-propanol or isopropyl alcohol, $(CH_3)_2CHOH$. Methanol is found in automotive products such as antifreeze and "dry gas." Ethanol is used as a solvent for drugs and chemicals, but is more popularly known for its effects as an alcoholic beverage. 2-Propanol, also known as "rubbing alcohol," is an antiseptic.

Alcohols may be classified as either primary, secondary, or tertiary:

$R-CH_2-OH$	$R-\underset{\displaystyle OH}{\underset{\vert}{CH}}-R'$	$R-\overset{\displaystyle R'}{\overset{\vert}{\underset{\displaystyle R''}{\underset{\vert}{C}}}}-OH$
Primary alcohol	Secondary alcohol	Tertiary alcohol

Note that the classification depends on the number of carbon-containing groups, R (alkyl or aromatic), attached to the carbon bearing the hydroxyl group. Examples of each type are as follows:

CH_3CH_2-OH	$CH_3-\underset{\displaystyle CH_3}{\underset{\vert}{CH}}-OH$	$CH_3-\overset{\displaystyle CH_3}{\overset{\vert}{\underset{\displaystyle CH_3}{\underset{\vert}{C}}}}-OH$
Ethanol (Ethyl alcohol) a primary alcohol	2-Propanol (Isopropyl alcohol) a secondary alcohol	2-Methyl-2-propanol (*t*-Butyl alcohol) a tertiary alcohol

Table 25.1 *Selected Alcohols and Phenols*

Compound	*Name and Use*
CH_3OH	Methanol: solvent for paints, shellacs, and varnishes
CH_3CH_2OH	Ethanol: alcoholic beverages; solvent for medicines, perfumes, and varnishes
$CH_3-CH(OH)-CH_3$	2-Propanol (isopropyl alcohol): rubbing alcohol; astringent; solvent for cosmetics, perfumes, and skin creams
$CH_2(OH)-CH_2(OH)$	Ethylene glycol: antifreeze
$CH_2(OH)-CH(OH)-CH_2(OH)$	Glycerol (glycerin): sweetening agent; solvent for medicines; lubricant; moistening agent
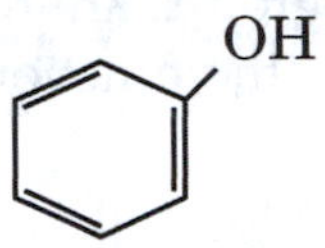 OH (benzene ring)	Phenol (carbolic acid): cleans surgical and medical instruments; topical antipruritic (relieves itching)
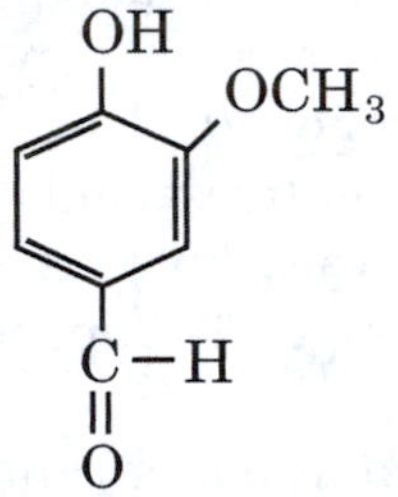 OH, OCH_3, C(=O)–H (benzene ring)	Vanillin: flavoring agent (vanilla flavor)
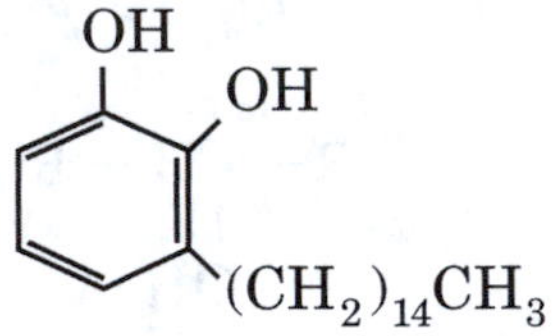 OH, OH, $(CH_2)_{14}CH_3$ (benzene ring)	Tetrahydrourushiol: irritant in poison ivy

Phenols bear a close resemblance to alcohols structurally because the hydroxyl group is present. However, since the –OH group is bonded directly to a carbon that is part of an aromatic ring, the chemistry is quite different from that of alcohols. Phenols are more acidic than alcohols; concentrated solutions of the compound phenol are quite toxic and can cause severe skin burns. Phenol derivatives are found in medicines; for example, thymol is used to kill fungi and hookworms. (Also see Table 25.1.)

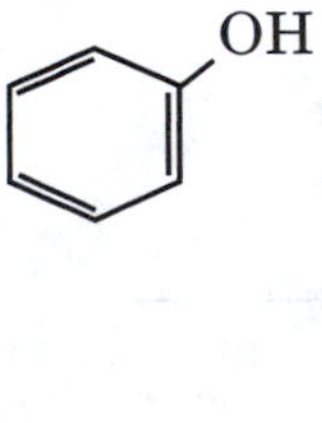

OH

Phenol

CH_3

OH

CH

CH_3 CH_3

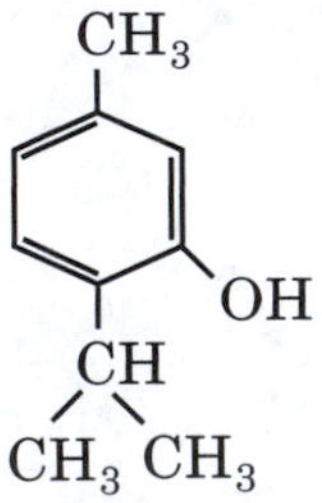

Thymol
(2-isopropyl-5-methylphenol)

In this experiment, you will examine physical and chemical properties of representative alcohols and phenols. You will be able to compare the differences in chemical behavior between these compounds and use this information to identify an unknown.

Physical Properties

Because the hydroxyl group is present in alcohols and phenols, these compounds are polar. The polarity of the hydroxyl group, coupled with its ability to form hydrogen bonds, enables many alcohols and phenols to mix with water. In general, for alcohols of comparable molecular weight, the greater the number of hydroxyl groups present in the molecule, the greater the number of hydrogen bonds that can form with water. As a result, the solubility in water increases. Because these compounds also contain nonpolar portions, they show additional solubility in many organic solvents, such as dichloromethane (CH_2Cl_2) and diethyl ether ($CH_3CH_2OCH_2CH_3$).

$$R-\overset{\delta-}{O}-\overset{\delta+}{H}\cdots\overset{\delta-}{O}(H)-\overset{\delta+}{H}$$

Hydrogen bonding of the hydroxyl group with water

Chemical Properties

The chemical behavior of the different classes of alcohols and phenols can be used as a means of identification. Quick, simple tests that can be carried out in test tubes will be performed.

1. *Lucas test.* This test is used to distinguish between water-soluble primary, secondary, and tertiary alcohols. Lucas reagent is a mixture of zinc chloride, $ZnCl_2$, in concentrated HCl. Upon addition of this reagent, a tertiary alcohol reacts rapidly and immediately gives an insoluble white layer. A secondary alcohol reacts slowly and, after heating slightly, gives the white layer within 10 min. A primary alcohol does not react. Any formation of a heterogeneous phase or appearance of an emulsion is a positive test.

$$\underset{\text{primary alcohol}}{CH_3CH_2—OH} + HCl + ZnCl_2 \rightarrow \text{no reaction}$$

$$\underset{\text{secondary alcohol}}{(CH_3)_2CH—OH} + HCl + ZnCl_2 \rightarrow \underset{\text{insoluble}}{(CH_3)_2CH—Cl} + H_2O \text{ (10 min. heat)}$$

$$\underset{\text{tertiary alcohol}}{(CH_3)_3C—OH} + HCl + ZnCl_2 \rightarrow \underset{\text{insoluble}}{(CH_3)_3C—Cl} + H_2O \text{ (<5 min.)}$$

2. *Chromic acid test.* This test is able to distinguish primary and secondary alcohols from tertiary alcohols. Using acidified dichromate solution, primary alcohols are oxidized to carboxylic acids; secondary alcohols are oxidized to ketones; tertiary alcohols are not oxidized. (Note that in alcohols that are oxidized, the carbon that has the hydroxyl group *loses a hydrogen.*) In the oxidation, the brown-red color of the chromic acid changes to a blue-green solution. Phenols are oxidized to nondescript

brown tarry masses. (Aldehydes are also oxidized under these conditions to carboxylic acids, but ketones remain intact; see Experiment 26 for further discussion.)

$$3CH_3CH_2-OH + 4H_2CrO_4 + 6H_2SO_4 \longrightarrow 3CH_3-\overset{\overset{\large O}{\|}}{C}-OH + 2Cr_2(SO_4)_3 + 13H_2O$$

primary alcohol — brown-red — carboxylic acid — blue-green

$$3CH_3-\overset{\overset{\large OH}{|}}{CH}-CH_3 + 2H_2CrO_4 + 3H_2SO_4 \longrightarrow 3CH_3-\overset{\overset{\large O}{\|}}{C}-CH_3 + Cr_2(SO_4)_3 + 8H_2O$$

secondary alcohol — brown-red — ketone — blue-green

$$(CH_3)_3C-OH + H_2CrO_4 + H_2SO_4 \longrightarrow \text{no reaction}$$

tertiary alcohol

3. *Iodoform test.* This test is more specific than the previous two tests. Only ethanol (ethyl alcohol) and alcohols with the part structure $CH_3CH(OH)$ react. These alcohols react with iodine in aqueous sodium hydroxide to give the yellow precipitate iodoform.

$$R\overset{\overset{\large OH}{|}}{C}HCH_3 + 4I_2 + 6NaOH \longrightarrow R\overset{\overset{\large O}{\|}}{C}-O^- Na^+ + 5NaI + 5H_2O + HCI_3(s)$$

iodoform
yellow

Phenols also react under these conditions. With phenol, the yellow precipitate triiodophenol forms.

$$C_6H_5OH + 3I_2 \longrightarrow C_6H_2I_3OH + 3HI$$

triiodophenol
yellow precipitate

4. *Acidity of phenol.* Phenol is also called carbolic acid. Phenol is an acid and will react with base; thus phenols readily dissolve in base solutions. In contrast, alcohols are not acidic.

$$C_6H_5-OH + HO^- \longrightarrow C_6H_5-O^- + H_2O$$

5. *Iron(III) chloride test.* Addition of aqueous iron(III) chloride to a phenol gives a colored solution. Depending on the structure of the phenol, the color can vary from green to purple.

$$C_6H_5-OH + FeCl_3 \longrightarrow C_6H_5-O-\underset{\underset{\large Cl}{|}}{Fe}-Cl + HCl$$

light yellow — violet color

OBJECTIVES

1. To learn characteristic chemical reactions of alcohols and phenols.
2. To use these chemical characteristics for identification of an organic compound.

PROCEDURE

CAUTION

Chromic acid is very corrosive. Any spill should be immediately flushed with water. Phenol is toxic. Also, contact with the solid will cause burns to skin; any contact should be thoroughly washed with large quantities of water. Solid phenol should be handled only with a spatula or forceps. Use gloves with these reagents. Dispose of reaction mixtures and excess reagents in proper containers as directed by your instructor. Do not flush them down the sink.

Physical Properties of Alcohols and Phenols

1. You will test the alcohols 1-butanol (a primary alcohol), 2-butanol (a secondary alcohol), 2-methyl-2-propanol (a tertiary alcohol), and phenol; you will also have as an unknown one of these compounds (labeled A, B, C, or D). As you run a test on a known, test the unknown at the same time for comparison. Note that the phenol will be provided as an aqueous solution.
2. Into separate test tubes (100 × 13 mm) labeled 1-butanol, 2-butanol, 2-methyl-2-propanol, and unknown, place 10 drops of each sample; dilute by mixing with 3 mL of distilled water. Into a separate test tube, place 2 mL of a prepared water solution of phenol. Are all the solutions homogeneous? Record your observations on the Report Sheet (1).
3. Test the pH of each of the aqueous solutions. Do the test by first dipping a clean glass rod into the solutions and then transferring a drop of liquid to pH paper. Use a broad pH indicator paper (e.g., pH range 1–12) and read the value of the pH by comparing the color to the chart on the dispenser. Record the results on the Report Sheet (2).

Chemical Properties of Alcohols and Phenols

1. *Iodoform test.* Prepare a water bath of 60°C in a 400-mL beaker. Place into separate clean, dry labeled test tubes (150 × 18 mm), (a) 1-butanol, (b) 2-butanol, (c) 2-methyl-2-propanol, (d) phenol, and (e) unknown, 5 drops of sample to be tested. Add to each test tube 2 mL of water. If the compound is not soluble, add 1,2-dimethoxy-ethane (dropwise) until the solution is homogeneous. Add to each test tube (dropwise) 2 mL of 10% NaOH; tap the test tube with your finger to mix. Warm the mixtures in the 60°C water bath. Add the prepared solution of I_2-KI test reagent, dropwise to each test tube, (with shaking) until each solution becomes brown. Cork each test tube and shake vigorously; remove the cork and return to the warm water bath. (If the color fades, add more I_2-KI test reagent until the dark color persists for 2 min. at 60°C.) Add 10% NaOH (dropwise) until each solution becomes colorless; cork and shake again. Return the test tubes to the warm water bath for 5 min. (*Remember to remove the corks when heating.*) Remove the test tubes from the water and dilute each solution with cold water; leave enough room so you can cork and shake one last time. Let cool and look for a light yellow precipitate. The formation of the yellow precipitate tends to be slow.

Put these test tubes to one side and make your observations when all the other tests are completed. Record your observations on the Report Sheet (3).

2. *Lucas test.* Place 5 drops of each sample into separate clean, dry test tubes (100 × 13 mm), labeled as before. Add 1 mL of Lucas reagent; mix well by stoppering each test tube with a cork, tapping the test tube sharply with your finger for a few seconds to mix; remove the cork after mixing and allow each test tube to stand for 5 min. Look carefully for any cloudiness that may develop during this time period. If there is no cloudiness after 10 min., warm the test tubes that are clear for 15 min. in a 60°C water bath. Record your observations on the Report Sheet (4).
3. *Chromic acid test.* Place into separate clean, dry test tubes (100 × 13 mm), labeled as before, 5 drops of sample to be tested. To each test tube add 10 drops of reagent grade acetone and 2 drops of chromic acid. Place the test tubes in a 60°C water bath for 5 min. Note the color of each solution. (Remember, the loss of the brown-red and the formation of a blue-green color is a positive test.) Record your observations on the Report Sheet (5).
4. *Iron(III) chloride test.* Place into separate clean, dry test tubes (100 × 13 mm), labeled as before, 5 drops of sample to be tested. Add 2 drops of iron(III) chloride solution to each. Note any color changes in each solution. (Remember, a purple color indicates the presence of a phenol.) Record your observations on the Report Sheet (6).
5. From your observations identify your unknown.

CHEMICALS AND EQUIPMENT

1. Aqueous phenol
2. Acetone (reagent grade)
3. 1-Butanol
4. 2-Butanol
5. 2-Methyl-2-propanol (*t*-butyl alcohol)
6. Chromic acid solution
7. 1,2-Dimethoxyethane
8. Iron(III) chloride solution
9. I_2-KI solution
10. 10% NaOH, sodium hydroxide
11. Lucas reagent
12. Corks
13. Hot plate
14. pH paper
15. Unknown

name | section | date

partner | grade

EXPERIMENT 25

Pre-Lab Questions

1. Define the term "functional group."

2. Draw the structure of a molecule that contains the hydroxyl functional group.

3. Draw structures of (a) a primary alcohol; (b) a secondary alcohol; and (c) a tertiary alcohol. In each example, circle the R group(s).

4. How does a phenol differ from a typical alcohol?

5. Circle the phenol group in each of the following examples if one is present:

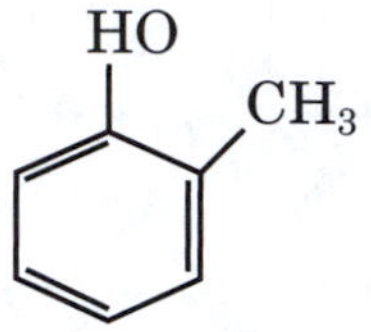

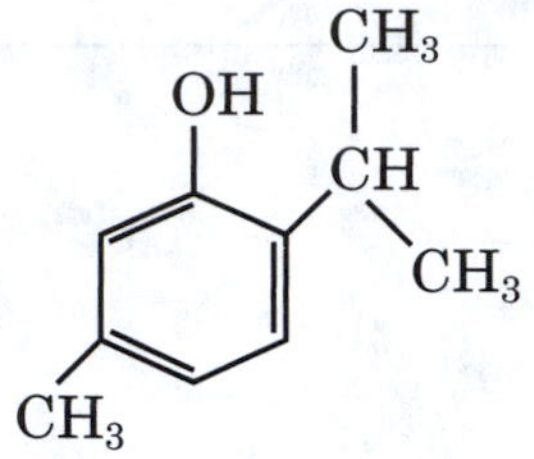

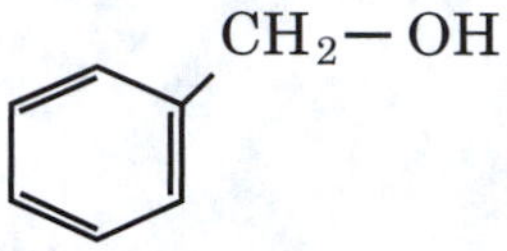

HO CH_3 (benzene ring)

OH, CH_3–CH–CH_3, CH_3 (benzene ring)

CH_2–OH (benzene ring)

6. Why is methanol, CH_3—OH (MW 32), soluble in water, but ethane, CH_3—CH_3 (MW 30), insoluble in water?

name _______________ section _______ date _______

partner _______________ grade _______

25 **EXPERIMENT 25**

Report Sheet

Test	*1-Butanol*	*2-Butanol*	*2-Methyl-2-propanol*	*Phenol*	*Unknown*
1. Water					
2. pH					
3. Iodoform					
4. Lucas					
5. Chromic acid					
6. Iron(III) chloride					

Identity of unknown:

Unknown no. _________. The unknown compound is _______________.

Post-Lab Questions

1. Based on the observations you made in this experiment, answer the following questions.

a. What happens to a phenol with iron(III) chloride reagent?

b. What happens to a tertiary alcohol with Lucas reagent?

c. Can you oxidize a secondary alcohol with chromic acid?

d. What should you see when 2-butanol is treated with NaI-I_2 reagent?

2. If you pour ethylene glycol, a liquid (see Table 25.1), into water, it mixes with the water in all proportions. On the other hand, if you bubble butane ($CH_3CH_2CH_2CH_3$), a gas of comparable molar mass, through water, it does not mix but just bubbles away. How do you explain these differences?

3. A student had two unknown alcohols. Unknown A gave a blue-green color with chromic acid, a precipitate in the iodoform test, and a precipitate with Lucas reagent, but only after heating for 15 min. Unknown B showed no color change with chromic acid, gave no precipitate in the iodoform test, but formed an immediate precipitate with Lucas reagent. To which classes do alcohols A and B belong? Give examples for each.

4. The compounds below give a purple color with iron(III) chloride solution. Circle the part in each structure responsible for the reaction that gives this test.

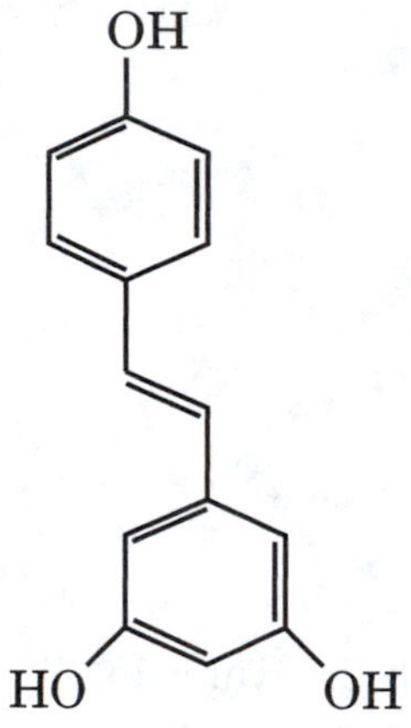

OH

HO OH

Resveratrol
(grapes)

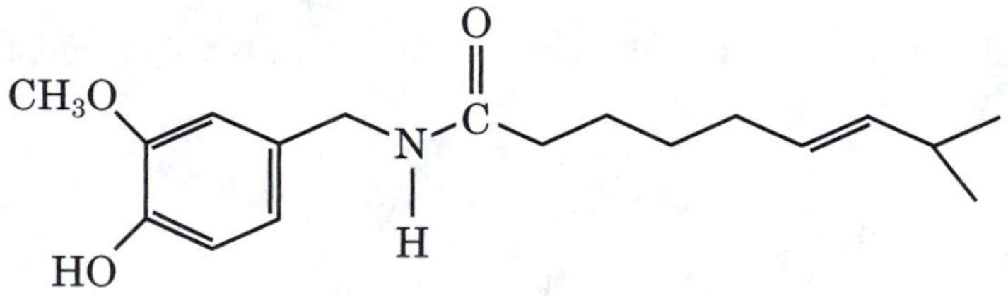

CH_3O

HO

O

C

N

H

Capsaicin
(hot pepper)

Identification of Aldehydes and Ketones

BACKGROUND

Aldehydes and ketones are representative of compounds that possess the carbonyl group:

$$\begin{array}{c} \mathrm{O} \\ \| \\ -\mathrm{C}- \end{array} \qquad \text{The carbonyl group}$$

Aldehydes have at least one hydrogen attached to the carbonyl carbon; in ketones, no hydrogens are directly attached to the carbonyl carbon, only carbon containing R groups:

$$\underset{\text{Aldehyde}}{\mathrm{R}-\overset{\overset{\displaystyle \mathrm{O}}{\|}}{\mathrm{C}}-\mathrm{H}} \qquad \underset{\text{Ketone}}{\mathrm{R}-\overset{\overset{\displaystyle \mathrm{O}}{\|}}{\mathrm{C}}-\mathrm{R}'} \qquad \text{(R and R}' \text{ can be alkyl or aromatic)}$$

Aldehydes and ketones of low molecular weight have commercial importance. Many others occur naturally. Table 26.1 has some representative examples.

In this experiment you will investigate the chemical properties of representative aldehydes and ketones.

Classification Tests

1. *Chromic acid test.* Aldehydes are oxidized to carboxylic acids by chromic acid; ketones are not oxidized. A positive test results in the formation of a blue-green solution from the brown-red color of chromic acid.

$$\underset{\text{aldehyde}}{3\mathrm{R}-\overset{\overset{\displaystyle \mathrm{O}}{\|}}{\mathrm{C}}-\mathrm{H}} + \underset{\text{brown-red}}{2\mathrm{H_2CrO_4}} + 3\mathrm{H_2SO_4} \longrightarrow 3\mathrm{R}-\overset{\overset{\displaystyle \mathrm{O}}{\|}}{\mathrm{C}}-\mathrm{OH} + \underset{\text{blue-green}}{\mathrm{Cr_4(SO_4)_3}} + 5\mathrm{H_2O}$$

$$\underset{\text{ketone}}{\mathrm{R}-\overset{\overset{\displaystyle \mathrm{O}}{\|}}{\mathrm{C}}-\mathrm{R}} \xrightarrow[\mathrm{H_2SO_4}]{\mathrm{H_2CrO_4}} \text{no reaction}$$

Table 26.1 *Representative Aldehydes and Ketones*

Compound		*Source and Use*
$HC(=O)H$	Formaldehyde	Oxidation of methanol; plastics; preservative
$CH_3C(=O)CH_3$	Acetone	Oxidation of 2-propanol; solvent
[structure]	Citral	Lemongrass oil; fragrance
[structure]	Jasmone	Oil of jasmine; fragrance

2. *Tollens' test.* Most aldehydes reduce Tollens' reagent (ammonia and silver nitrate) to give a precipitate of silver metal. The free silver forms a silver mirror on the sides of the test tube. (This test is sometimes referred to as the "silver mirror" test.) The aldehyde is oxidized to a carboxylic acid.

$$\underset{\text{aldehyde}}{R{-}\overset{\overset{\displaystyle O}{\|}}{C}{-}H} + 2Ag(NH_3)_2OH \longrightarrow \underset{\text{silver mirror}}{2Ag(s)} + R{-}\overset{\overset{\displaystyle O}{\|}}{C}{-}O^-NH_4^+ + H_2O + 3NH_3$$

3. *Iodoform test.* Methyl ketones give the yellow precipitate iodoform when reacted with iodine in aqueous sodium hydroxide.

$$\underset{\text{methyl ketone}}{R{-}\overset{\overset{\displaystyle O}{\|}}{C}{-}CH_3} + 3I_2 + 4NaOH \longrightarrow 3NaI + 3H_2O + R{-}\overset{\overset{\displaystyle O}{\|}}{C}{-}O^-Na^+ + \underset{\text{iodoform yellow}}{HCI_3(s)}$$

4. *2,4-Dinitrophenylhydrazine test.* All aldehydes and ketones give an immediate precipitate with 2,4-dinitrophenylhydrazine reagent. This reaction is general for both these functional groups. The color of the

precipitate varies from yellow to red. (Note that alcohols do not give this test.)

$$\text{R-C(=O)-H (aldehyde)} + \text{H}_2\text{N-NH-C}_6\text{H}_3(\text{NO}_2)_2 \longrightarrow \text{R-C(H)=N-NH-C}_6\text{H}_3(\text{NO}_2)_2 \text{ (yellow)}$$

$$\text{R-C(=O)-R (ketone)} + \text{H}_2\text{N-NH-C}_6\text{H}_3(\text{NO}_2)_2 \longrightarrow \text{R-C(R)=N-NH-C}_6\text{H}_3(\text{NO}_2)_2 \text{ (yellow)}$$

$$\text{ROH (alcohol)} + \text{H}_2\text{N-NH-C}_6\text{H}_3(\text{NO}_2)_2 \longrightarrow \text{no reaction}$$

Identification by forming a derivative

The classification tests (summarized in Table 26.2), when properly done, can distinguish between various types of aldehydes and ketones. However, these tests alone may not allow for the identification of a specific unknown aldehyde or ketone. A way to correctly identify an unknown compound is by using a known chemical reaction to convert it into another compound that is known. The new compound is referred to as a *derivative.* Then, by comparing the physical properties of the unknown and the derivative to the physical properties of known compounds listed in a table, an identification can be made.

The ideal derivative is a solid. A solid can be easily purified by crystallization and easily characterized by its melting point. Thus two similar aldehydes or two similar ketones usually have derivatives that have *different melting points.* The most frequently formed derivatives for aldehydes and ketones are 2,4-dinitrophenylhydrazone (2,4-DNP), oxime, and semicarbazone. Table 26.4 (p. 320) lists some aldehydes and ketones along with melting points of their derivatives. If, for example, we look at the properties of valeraldehyde and crotonaldehyde, though the boiling points are virtually the same, the melting points of the 2,4-DNP, oxime, and semicarbazone are different and provide a basis for identification.

Table 26.2 *Summary of Classification Tests*

Compound	*Reagent for Positive Test*
Aldehydes and ketones	2,4-Dinitrophenylhydrazine
Aldehydes	Chromic acid Tollens' reagent
Methyl ketones	Iodoform

1. *2,4-Dinitrophenylhydrazone.* 2,4-Dinitrophenylhydrazine reacts with aldehydes and ketones to form 2,4-dinitrophenylhydrazones (2,4-DNP).

$$R{-}\overset{\overset{\large O}{\|}}{C}{-}R' + H_2N{-}NH{-}C_6H_3(NO_2)_2 \xrightarrow{-H_2O} \begin{matrix} R \\ \quad\diagdown \\ \quad\; C{=}N{-}NH{-}C_6H_3(NO_2)_2 \\ \quad\diagup \\ R' \end{matrix}$$

2,4-dinitrophenylhydrazine — 2,4-dinitrophenylhydrazone (2,4-DNP)

(R′ = H or alkyl or aromatic)

The 2,4-DNP product is usually a colored solid (yellow to red) and is easily purified by recrystallization.

2. *Oxime.* Hydroxylamine reacts with aldehydes and ketones to form oximes.

$$R{-}\overset{\overset{\large O}{\|}}{C}{-}R' + NH_2OH \xrightarrow{-H_2O} \begin{matrix} R \\ \quad\diagdown \\ \quad\; C{=}NOH \\ \quad\diagup \\ R' \end{matrix}$$

hydroxylamine — oxime

(R′ = H or alkyl or aromatic)

These are usually derivatives with low melting points.

3. *Semicarbazone.* Semicarbazide, as its hydrochloride salt, reacts with aldehydes and ketones to form semicarbazones.

$$R{-}\overset{\overset{\large O}{\|}}{C}{-}R' + NH_2NH\overset{\overset{\large O}{\|}}{C}NH_2 \xrightarrow{-H_2O} \begin{matrix} R \\ \quad\diagdown \\ \quad\; C{=}NNH\overset{\overset{\large O}{\|}}{C}NH_2 \\ \quad\diagup \\ R' \end{matrix}$$

semicarbazide — semicarbazone

(R′ = H or alkyl or aromatic)

A pyridine base is used to neutralize the hydrochloride in order to free the semicarbazide so it may react with the carbonyl substrate.

OBJECTIVES

1. To learn the chemical characteristics of aldehydes and ketones.
2. To use these chemical characteristics in simple tests to distinguish between examples of aldehydes and ketones.
3. To identify aldehydes and ketones by formation of derivatives.

PROCEDURE

Classification Tests

1. Classification tests are to be carried out on four known compounds and one unknown. Any one test should be carried out on all five samples at the same time for comparison. Label test tubes as shown in Table 26.3.

Table 26.3 *Labeling Test Tubes*

Test Tube No.	*Compound*
1	Isovaleraldehyde (an aliphatic aldehyde)
2	Benzaldehyde (an aromatic aldehyde)
3	Cyclohexanone (a ketone)
4	Acetone (a methyl ketone)
5	Unknown

CAUTION

Chromic acid is toxic and corrosive. Handle with care and *promptly wash any spill.* Use gloves with this reagent.

2. *Chromic acid test.* Place 5 drops of each substance into separate, labeled test tubes (100 × 13 mm). Dissolve each compound in 20 drops of reagent-grade acetone (to serve as solvent); then add to each test tube 4 drops of chromic acid reagent, one drop at a time; after each drop, mix by sharply tapping the test tube with your finger. Let stand for 10 min. Aliphatic aldehydes should show a change within a minute; aromatic aldehydes take longer. Note the approximate time for any change in color or formation of a precipitate on the Report Sheet.
3. *Tollens' test.*

CAUTION

This reagent must be freshly prepared before it is to be used and any excess disposed of *immediately* after use. Organic residues should be discarded in appropriate waste containers. Unused Tollens' reagent should be collected from every student by the instructor. *Do not store Tollens' reagent; it is explosive when dry.* The instructor should dispose of the excess reagent by adding 1 M HNO_3 until acidic, warming on a hot plate. The solution can then be stored in a waste container for heavy metals.

Enough reagent for your use can be prepared in a 25-mL Erlenmeyer flask by mixing 5 mL of Tollens' solution A with 5 mL of Tollens' solution B. To the silver oxide precipitate that forms, add (dropwise, with shaking) 10% ammonia solution until the brown precipitate just dissolves. *Avoid an excess of ammonia.*

Place 5 drops of each substance into separately labeled clean, dry test tubes (100 × 13 mm). Dissolve the compound in *bis*(2-ethoxyethyl) ether by adding this solvent dropwise until a homogeneous solution is obtained. Then, add 2 mL (approx. 40 drops) of the prepared Tollens' reagent and mix by sharply tapping the test tube with your finger. Place the test tube in a 60°C water bath for 5 min. Remove the test tubes from the water and look for a silver mirror. If the tube is clean, a silver mirror will be formed; if not, a black precipitate of finely divided silver will appear. Record your results on the Report Sheet. (Clean your test tubes with 1 M HNO_3 and discard the solution in a waste container designated by your instructor.)

4. *Iodoform test.* Prepare a water bath of 60°C in a 400-mL beaker. Place into separately labeled clean, dry test tubes (150 × 18 mm), 5 drops of sample to be tested. Add to each test tube 2 mL of water. If the compound is not soluble, add 1,2-dimethoxyethane (dropwise) until the solution is homogeneous. Add to each test tube (dropwise) 2 mL of 10% NaOH; tap the test tube with your finger to mix. Warm the mixtures in the 60°C water bath. Add the prepared solution of I_2-KI test reagent dropwise to each test tube (with shaking) until each solution becomes brown. Cork each test tube and shake vigorously; return to the warm water bath. (If the color fades, add more I_2-KI test reagent until the dark color persists for 2 min. at 60°C.) Add 10% NaOH (dropwise) until each solution becomes colorless; cork and shake again. Return the test tubes to the warm water bath for 5 min. (*Remember to remove the corks when heating.*) Remove the test tubes from the water and dilute each solution with cold water; leave enough room so you can cork and shake one last time. Let cool and look for a light yellow precipitate. The formation of the yellow precipitate tends to be slow. Put these test tubes to one side and make your observations when all the other tests are completed. Record your observations on the Report Sheet.
5. *2,4-Dinitrophenylhydrazine test.* Place 5 drops of each substance into separately labeled clean, dry test tubes (100 × 13 mm) and add 20 drops of the 2,4-dinitrophenylhydrazine reagent to each. If no precipitate forms immediately, heat for 5 min. in a warm water bath (60°C); cool. Record your observations on the Report Sheet.

Formation of Derivatives

CAUTION

The chemicals used to prepare derivatives and some of the derivatives are potential carcinogens. Exercise care in using the reagents and in handling the derivatives. Avoid skin contact by wearing gloves.

1. This section is *optional.* Consult your instructor to determine whether this section is to be completed. Your instructor will indicate how many derivatives and which derivatives you should make.
2. *Waste.* Place all the waste solutions from these preparations in designated waste containers for disposal by your instructor.
3. *General procedure for recrystallization.* Heat a small volume (10–20 mL) of solvent to boiling on a steam bath (or carefully on a hot plate). Place crystals into a test tube (100 × 13 mm) and add the hot solvent (dropwise) until the crystals just dissolve (keep the solution hot, also). Allow the solution to cool to room temperature; then cool further in an ice bath. Collect the crystals on a Hirsch funnel by vacuum filtration. Use a trap between the Hirsch funnel setup and the aspirator (Figure 26.1).Wash the crystals with 10 drops of ice-cold solvent. Allow the crystals to dry by drawing air through the Hirsch funnel. Take a melting point (see Experiment 13 for a review of the technique).
4. *2,4-Dinitrophenylhydrazone* (2,4-DNP). Place 5 mL of the 2,4-dinitrophenylhydrazine reagent in a test tube (150 × 18 mm). Add 10 drops of the unknown compound; sharply tap the test tube with your finger to mix. If crystals do not form immediately, gently heat in a water bath

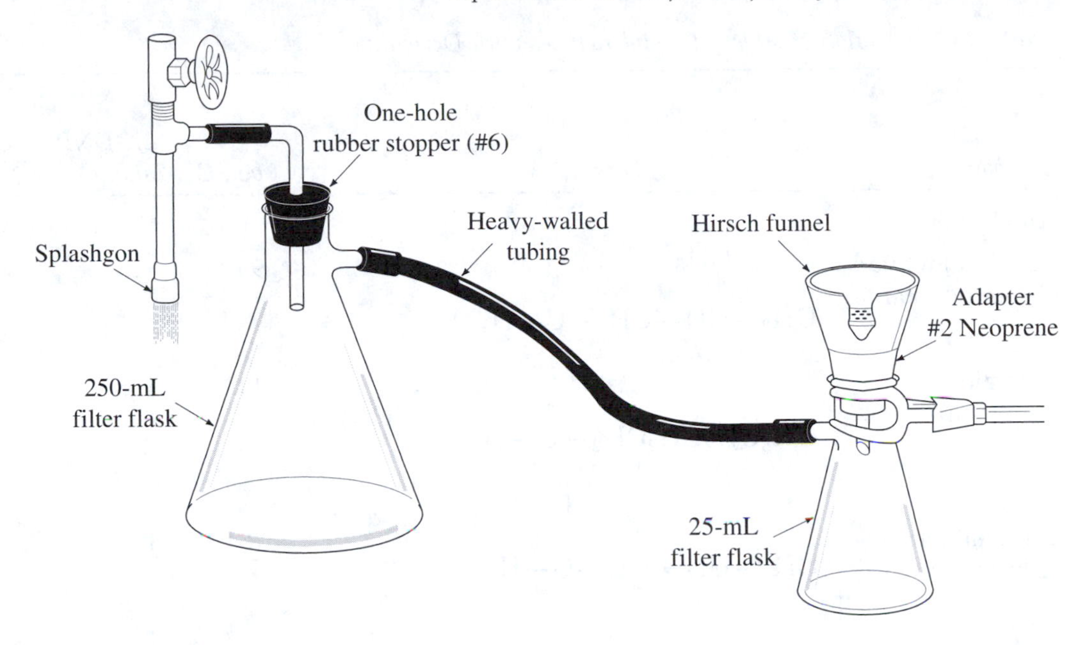

Figure 26.1
Vacuum filtration with a Hirsch funnel.

(60°C) for 5 min. Cool in an ice bath until crystals form. Collect the crystals by vacuum filtration using a Hirsch funnel (Figure 26.1). Allow the crystals to dry on the Hirsch funnel by drawing air through the crystals. Take a melting point and record on the Report Sheet. (The crystals are usually pure enough to give a good melting point. However, if the melting point range is too large, recrystallize from a minimum volume of ethanol.)

5. *Oxime.* Prepare fresh reagent by dissolving 1.0 g of hydroxylamine hydrochloride and 1.5 g of sodium acetate in 4 mL of distilled water in a test tube (150 × 18 mm). Add 20 drops of unknown and sharply tap the test tube with your finger to mix. Warm in a hot water bath (60°C) for 5 min. Cool in an ice bath until crystals form. (If no crystals form, scratch the inside of the test tube by rubbing with a glass rod.) Collect the crystals on a Hirsch funnel by vacuum filtration (Figure 26.1). Allow the crystals to air dry on the Hirsch funnel by drawing air through the crystals. Take a melting point and record on the Report Sheet. (Recrystallize, if necessary, from a minimum volume of ethanol.)
6. *Semicarbazone.* Place 2.0 mL of the semicarbazide reagent in a test tube (150 × 18 mm); add 10 drops of unknown. If the solution is not clear, add methanol (dropwise) until a clear solution results. Add 2.0 mL of pyridine and gently warm in a hot bath (60°C) for 5 min. Crystals should begin to form. (If there are no crystals, place the test tube in an ice bath and scratch the inside of the test tube by rubbing with a glass rod.) Collect the crystals on a Hirsch funnel by vacuum filtration (Figure 26.1). Allow the crystals to air dry on the Hirsch funnel by drawing air through the crystals. Take a melting point and record on the Report Sheet. (Recrystallize, if necessary, from a minimum volume of ethanol.)
7. Based on the observations you recorded on the Report Sheet, and by comparing the melting points of the derivatives for your unknown to the knowns listed in Table 26.4, identify your unknown.

Table 26.4 *Selection of Aldehydes and Ketones with Derivatives*

Compound	*Formula*	*b.p.°C*	*2,4-DNP m.p.°C*	*Oxime m.p.°C*	*Semi-carbazone m.p.°C*
Aldehydes					
Isovaleraldehyde (3-methylbutanal)	$CH_3-CH(CH_3)-CH_2-C(=O)-H$	93	123	49	107
Valeraldehyde (pentanal)	$CH_3CH_2CH_2CH_2-C(=O)-H$	103	106	52	—
Crotonaldehyde (2-butenal)	$CH_3-CH=CH-C(=O)-H$	104	190	119	199
Caprylaldehyde (octanal)	$CH_3CH_2CH_2CH_2CH_2CH_2CH_2-C(=O)-H$	171	106	60	101
Benzaldehyde	$C_6H_5-C(=O)-H$	178	237	35	222
Ketones					
Acetone (2-propanone)	$CH_3-C(=O)-CH_3$	56	126	59	187
2-Pentanone	$CH_3-C(=O)-CH_2CH_2CH_3$	102	144	58 (b.p. 167°C)	112
3-Pentanone	$CH_3CH_2-C(=O)-CH_2CH_3$	102	156	69 (b.p. 165°C)	139
Cyclopentanone	cyclopentanone ring with C=O	131	146	56	210
Cyclohexanone	cyclohexanone ring with C=O	156	162	90	166
Acetophenone	$C_6H_5-C(=O)-CH_3$	202	238	60	198

Source: Compiled by Zvi Rappoport, *CRC Handbook of Tables for Organic Compound Identification*, 3rd ed., The Chemical Rubber Co.: Cleveland (1967).

CHEMICALS AND EQUIPMENT

1. Acetone (reagent grade)
2. 10% ammonia solution
3. Benzaldehyde
4. *Bis*(2-ethoxyethyl) ether
5. Chromic acid reagent
6. Cyclohexanone
7. 2,4-Dinitrophenylhydrazine reagent
8. 1,2-Dimethoxyethane
9. Ethanol
10. Hydroxylamine hydrochloride
11. I_2-KI test solution
12. Isovaleraldehyde
13. Methanol
14. 10% NaOH, sodium hydroxide
15. Pyridine
16. Semicarbazide reagent
17. Sodium acetate
18. Tollens' reagent (solution A and solution B)
19. Hirsch funnel
20. Hot plate
21. Neoprene adapter (no. 2)
22. Rubber stopper (no. 6, one-hole), with glass tubing
23. 50-mL side-arm filter flask
24. 250-mL side-arm filter flask
25. Vacuum tubing (heavy walled)
26. Unknown

name | section | date

partner | grade

26 EXPERIMENT 26

Pre-Lab Questions

1. Give an example of a compound with the following functional groups:
   a. an aldehyde

   b. an aromatic aldehyde

   c. a ketone

   d. a methyl ketone

2. Can 2,4-dinitrophenylhydrazine reagent distinguish between an aldehyde and a ketone?

3. What do you expect to see for a positive Tollens' test?

4. How do you use derivatives in this experiment?

name ______ section ______ date ______

partner ______ grade ______

26 EXPERIMENT 26

Report Sheet

Test	*Isovaler-aldehyde*	*Benzaldehyde*	*Cyclohexanone*	*Acetone*	*Unknown*
Chromic acid					
Tollens'					
Iodoform					
2,4-Dinitrophenyl-hydrazine					

Optional

Derivative	*Observed m.p.*	*Literature m.p.*
2,4-DNP		
Oxime		
Semicarbazone		

Unknown no. ______. The unknown compound is ______.

Post-Lab Questions

1. A compound of molecular formula C_3H_6O forms a yellow precipitate with 2,4-dinitrophenylhydrazine reagent but does not give a silver mirror with Tollens' reagent. Based on this information, draw a structural formula for the compound.

2. What should you see when the following compounds are mixed together with the given test solution?

   a. $CH_3CH_2CH_2-\underset{\underset{O}{\|}}{C}-CH_3$

   with 2,4-dinitrophenylhydrazine

   with KI-I_2 reagent

   with Tollens' reagent

   b. $C_6H_{11}-\overset{\overset{O}{\|}}{C}-H$

   with 2,4-dinitrophenylhydrazine

   with chromic acid

   with Tollens' reagent

3. Both 2-pentanone and 3-pentanone (see Table 26.4) have a boiling point of 102°C. Suggest a way that these two compounds could be distinguished and identified.

4. Use laboratory tests from this experiment and show how you could distinguish between the compounds listed below. The tests should give a clear, positive test result for the listed compound that is unique to it and thus, different from the others.

Test	*Cyclopentane*	*Cyclopentanone*	*Pentanal*

Properties of Carboxylic Acids and Esters

BACKGROUND

Carboxylic acids are structurally like aldehydes and ketones in that they contain the carbonyl group. However, an important difference is that carboxylic acids contain a hydroxyl group attached to the carbonyl carbon.

$$-\overset{\displaystyle O}{\overset{\|}{C}}-OH \quad \text{The carboxylic acid group}$$

This combination gives the group its most important characteristic; it behaves as an acid.

As a family, carboxylic acids are weak acids that ionize only slightly in water. As aqueous solutions, typical carboxylic acids ionize to the extent of only one percent or less.

$$R-\overset{\displaystyle O}{\overset{\|}{C}}-OH + H_2O \rightleftharpoons R-\overset{\displaystyle O}{\overset{\|}{C}}-O^- + H_3O^+$$

At equilibrium, most of the acid is present as un-ionized molecules. Dissociation constants, K_a, of carboxylic acids, where R is an alkyl group, are 10^{-5} or less. Water solubility depends to a large extent on the size of the R group. Only a few low-molecular-weight acids (up to four carbons) are very soluble in water.

Although carboxylic acids are weak, they are capable of reacting with bases stronger than water. Thus, while benzoic acid shows limited water solubility, it reacts with sodium hydroxide to form the soluble salt sodium benzoate. (Sodium benzoate is a preservative in some soft drinks.)

$$C_6H_5-COOH + NaOH \longrightarrow C_6H_5-COO^-Na^+ + H_2O$$

Benzoic acid
Insoluble

Sodium benzoate
Soluble

Sodium carbonate, Na_2CO_3, and sodium bicarbonate, $NaHCO_3$, solutions can neutralize carboxylic acids also.

The combination of a carboxylic acid and an alcohol gives an ester; water is eliminated. Ester formation is an equilibrium process, catalyzed by an acid catalyst.

$$CH_3CH_2CH_2\overset{\overset{\displaystyle O}{\|}}{C}{-}OH + CH_3CH_2OH \overset{H^+}{\rightleftharpoons} H_2O + CH_3CH_2CH_2\overset{\overset{\displaystyle O}{\|}}{C}{-}OCH_2CH_3$$

Butyric acid — Ethanol (Ethyl alcohol) — Ethyl butyrate (Ester)

Esterification ⟶

⟵ Hydrolysis

The reaction typically gives 60% to 70% of the maximum yield. The reaction is a reversible process. An ester reacting with water, giving the carboxylic acid and alcohol, is called *hydrolysis;* it is acid catalyzed. The base-promoted decomposition of esters yields an alcohol and a salt of the carboxylic acid; this process is called *saponification.* Saponification means "soap making," and the sodium salt of a fatty acid (e.g., sodium stearate) is a soap.

$$CH_3CH_2CH_2\overset{\overset{\displaystyle O}{\|}}{C}{-}OCH_2CH_3 + NaOH \longrightarrow CH_3CH_2CH_2\overset{\overset{\displaystyle O}{\|}}{C}{-}O^-Na^+ + CH_3CH_2OH$$

Saponification

A distinctive difference between carboxylic acids and esters is in their characteristic odors. Carboxylic acids are noted for their sour, disagreeable odors. On the other hand, esters have sweet and pleasant odors often associated with fruits, and fruits smell the way they do because they contain esters. These compounds are used in the food industry as fragrances and flavoring agents. For example, the putrid odor of rancid butter is due to the presence of butyric acid, while the odor of pineapple is due to the presence of the ester, ethyl butyrate. Only carboxylic acids of low molecular weight have odor at room temperature. Higher-molecular-weight carboxylic acids form strong hydrogen bonds, are solid, and have a low vapor pressure. Thus few molecules reach our noses. Esters, however, do not form hydrogen bonds among themselves; they are liquid at room temperature, even when the molecular weight is high. Thus they have high vapor pressure and many molecules can reach our noses, providing odor.

OBJECTIVES

1. To study the physical and chemical properties of carboxylic acids: solubility, acidity, aroma.
2. To prepare a variety of esters and note their odors.
3. To demonstrate saponification.

PROCEDURE

Carboxylic Acids and Their Salts

Characteristics of acetic acid

1. Place into a clean, dry test tube (100 × 13 mm) 2 mL of water and 10 drops of glacial acetic acid. Note its odor by wafting (moving your hand quickly over the open end of the test tube) the vapors toward your nose. Of what does it remind you?
2. Take a glass rod and dip it into the solution. Using wide-range indicator paper (pH 1–12), test the pH of the solution by touching the paper with the wet glass rod. Determine the value of the pH by comparing the color of the paper with the chart on the dispenser.
3. Now, add 2 mL of 2 M NaOH to the solution. Cork the test tube and sharply tap it with your finger. Remove the cork and determine the pH of the solution as before; if not basic, continue to add more base (dropwise) until the solution is basic. Note the odor and compare it with the odor of the solution before the addition of base.
4. By dropwise addition of 3 M HCl, carefully reacidify the solution from step 3 (above); test the solution as before with pH paper until the solution tests acid. Does the original odor return to its original strength?

Characteristics of benzoic acid

1. Your instructor will weigh out 0.1 g of benzoic acid for sample size comparison. With your microspatula, take some sample equivalent to the preweighed sample (an exact quantity is not important here). Add the solid to a test tube (100 × 13 mm) along with 2 mL of water. Is there any odor? Mix the solution by sharply tapping the test tube with your finger. How soluble is the benzoic acid?
2. Now add 1 mL of 2 M NaOH to the solution from step 1 (above), cork, and mix by sharply tapping the test tube with your finger. What happens to the solid benzoic acid? Is there any odor?
3. By dropwise addition of 3 M HCl, carefully reacidify the solution from step 2 (above); test as before with pH paper until acidic. As the solution becomes acidic, what do you observe?

Esterification

1. Into five clean, dry test tubes (100 × 13 mm), add 10 drops of liquid carboxylic acid or 0.1 g of solid carboxylic acid and 10 drops of alcohol according to the scheme in Table 27.1. Note the odor of each reactant.
2. Add 5 drops of concentrated sulfuric acid to each test tube and mix the contents thoroughly by sharply tapping the test tube with your finger.

CAUTION

Sulfuric acid causes severe burns. Flush any spill with lots of water. Use gloves with this reagent.

Table 27.1 *Acids and Alcohols*

Test Tube No.	*Carboxylic Acid*	*Alcohol*
1	Formic	Isobutyl
2	Acetic	Benzyl
3	Acetic	Isopentyl
4	Acetic	Ethyl
5	Salicylic	Methyl

3. Place the test tubes in a warm water bath at 60°C for 15 min. Remove the test tubes from the water bath, cool, and add 2 mL of water to each. Note that there is a layer on top of the water in each test tube. With a Pasteur pipet, take a few drops from this top layer and place on a watch glass. Note the odor. Match the ester from each test tube with one of the following odors: banana, peach, raspberry, nail polish remover, wintergreen.

Saponification

This part of the experiment can be done while the esterification reactions are being heated.

1. Place into a test tube (150 × 18 mm) 10 drops of methyl salicylate and 5 mL of 6 M NaOH. Heat the contents in a boiling water bath for 30 min. Record on the Report Sheet what has happened to the ester layer (1).
2. Cool the test tube to room temperature by placing it in an ice-water bath. Determine the odor of the solution and record your observation on the Report Sheet (2). Is the odor as strong as before?
3. Carefully add 6 M HCl to the solution, 1 mL at a time, until the solution is acidic. After each addition, mix the contents and test the solution with litmus. When the solution is acidic, what do you observe? What is the name of the compound formed? Answer these questions on the Report Sheet (3).

CHEMICALS AND EQUIPMENT

1. Glacial acetic acid
2. Benzoic acid
3. Formic acid
4. Salicylic acid
5. Benzyl alcohol
6. Ethanol (ethyl alcohol)
7. 2-Methyl-1-propanol (isobutyl alcohol)
8. 3-Methyl-1-butanol (isopentyl alcohol)
9. Methanol (methyl alcohol)
10. Methyl salicylate
11. 3 M HCl
12. 6 M HCl
13. 2 M NaOH
14. 6 M NaOH
15. Concentrated H_2SO_4
16. pH paper (broad range pH 1–12)
17. Litmus paper
18. Pasteur pipet
19. Hot plate

name section date

partner grade

27 EXPERIMENT 27

Pre-Lab Questions

1. Write structures for the following carboxylic acids and alcohols that will be used in this laboratory:

a. acetic acid

b. formic acid

c. salicylic acid

d. isobutyl alcohol (2-methylpropanol)

e. benzyl alcohol

f. isopentyl alcohol (3-methylbutanol)

g. ethyl alcohol (ethanol)

h. methyl alcohol (methanol)

2. What carboxylic acid was used to form sodium acetate?

3. Ethyl formate has the flavor of rum. Draw the structure and give the name of the alcohol and carboxylic acid needed to synthesize this ester.

4. What product(s) forms when an ester is treated with base? What is this reaction called? Give an example.

name | section | date

partner | grade

27

EXPERIMENT 27

Report Sheet

Carboxylic acids and their salts

Characteristics of Acetic Acid			
Property	*Water Solution*	*NaOH Solution*	*HCl Solution*
Odor			
Solubility			
pH			

Characteristics of Benzoic Acid			
Property	*Water Solution*	*NaOH Solution*	*HCl Solution*
Odor			
Solubility			
pH			

Esterification

Test Tube	*Acid*	*Odor*	*Alcohol*	*Odor*	*Name of Ester*	*Odor*
1	Formic		Isobutyl			
2	Acetic		Benzyl			
3	Acetic		Isopentyl			
4	Acetic		Ethyl			
5	Salicylic		Methyl			

Saponification

1. What has happened to the ester layer?

2. What has happened to the odor of the ester?

3. What forms on reacidification of the solution? Name the compound.

4. Write the chemical equation for the saponification of methyl salicylate.

Post-Lab Questions

1. a. Formic acid is completely soluble in water. Draw the structural formula of the molecule as it is in the water.

b. Sodium hydroxide, NaOH, is added and the pH raised to 12. What happens to the acid in this solution? Draw the structural formula now.

c. Hydrochloric acid, HCl, is now added and the pH lowered to 2. Draw the structural formula of the formic acid in this solution.

2. Write structures for the five esters formed in this experiment.

a.

b.

c.

d.

e.

3. Butyric acid has a putrid odor, like rancid butter. If you got some of this material on your hands, how could you remove the odor? (Butyric acid has limited solubility in water.)

Properties of Amines and Amides

BACKGROUND

Amines and amides are two classes of organic compounds that contain nitrogen. Amines behave as organic bases and may be considered derivatives of ammonia. Amides are compounds that have a carbonyl group connected to a nitrogen atom and are neutral. In this experiment, you will learn about the physical and chemical properties of some members of the amine and amide families.

If the hydrogens of ammonia are replaced by alkyl or aryl groups, amines result. Depending on the number of carbon atoms bonded directly to nitrogen, amines are classified as either primary (one carbon atom), secondary (two carbon atoms), or tertiary (three carbon atoms) (Table 28.1).

There are a number of similarities between ammonia and amines that carry beyond the structure. Consider odor. The smell of amines resembles that of ammonia but is not as sharp. However, amines can be quite pungent. Anyone handling or working with raw fish knows how strong the amine odor can be—raw fish contains low-molecular-weight amines such as dimethylamine and trimethylamine. Other amines associated with decaying flesh have names suggestive of their odors: putrescine and cadaverine.

$NH_2CH_2CH_2CH_2CH_2NH_2$
Putrescine
(1,4-Diaminobutane)

$NH_2CH_2CH_2CH_2CH_2CH_2NH_2$
Cadaverine
(1,5-Diaminopentane)

The solubility of low-molecular-weight amines in water is high. In general, if the total number of carbons attached to nitrogen is four or less, the amine is water soluble; amines with a carbon content greater than four are water insoluble. However, all amines are soluble in organic solvents such as diethyl ether or dichloromethane.

Because amines are organic bases, water solutions show weakly basic properties. If the basicity of aliphatic amines and aromatic amines is compared to that of ammonia, aliphatic amines are stronger than ammonia,

Table 28.1 *Types of Amines*

	Primary Amines	*Secondary Amines*	*Tertiary Amines*
NH_3 Ammonia	CH_3NH_2 Methylamine	$(CH_3)_2NH$ Dimethylamine	$(CH_3)_3N$ Trimethylamine
	C_6H_5–NH_2 Aniline	C_6H_5–NH–CH_3 N-Methylaniline	C_6H_5–N(CH_3)–CH_3 N,N-Dimethylaniline

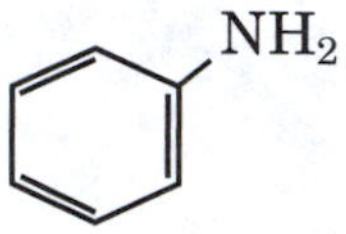

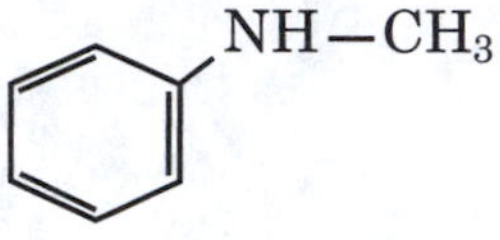

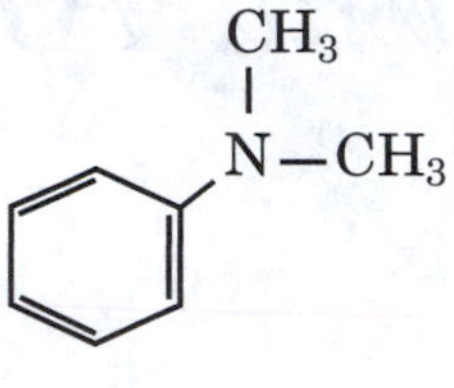

while aromatic amines are weaker. Amines characteristically react with acids to form ammonium salts; the nonbonded electron pair on nitrogen bonds the hydrogen ion.

$$\underset{\text{Amine}}{R\ddot{N}H_2} + HCl \longrightarrow \underset{\text{Ammonium Salt}}{RNH_3^+Cl^-}$$

If an amine is insoluble, reaction with an acid produces a water-soluble salt. Because ammonium salts are water soluble, many drugs containing amines are prepared as ammonium salts. After working with fish in the kitchen, a convenient way to rid one's hands of fish odor is to rub a freshly cut lemon over the hands. The citric acid found in the lemon reacts with the amines found on the fish; a salt forms that can be easily rinsed away with water.

Amides are carboxylic acid derivatives. The amide group is recognized by the nitrogen connected to the carbonyl group. Amides are neutral compounds; the electrons are delocalized into the carbonyl (resonance) and thus are not available to bond to a hydrogen ion.

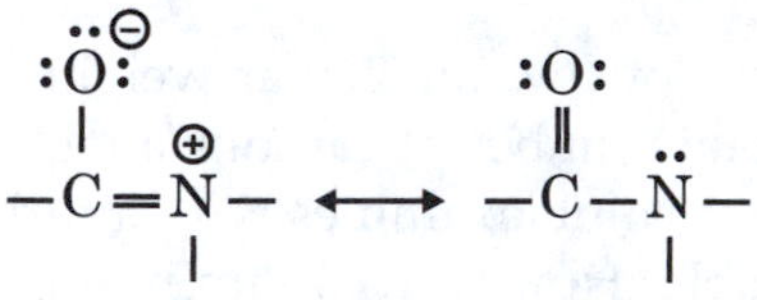

$$-\underset{}{C}(-\ddot{O}:^{\ominus})=\overset{\oplus}{N}- \longleftrightarrow -C(=\ddot{O})-\ddot{N}- \qquad CH_3-C(=O)-NH_2 \qquad C_6H_5-C(=O)-NH_2$$

Amide group (resonance) 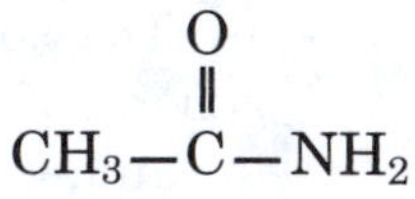 — Acetamide 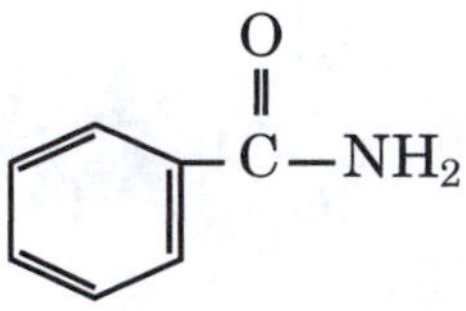 — Benzamide

Under suitable conditions, amide formation can take place between an amine and a carboxylic acid, an acyl halide, or an acid anhydride. Along with ammonia, primary and secondary amines yield amides with carboxylic acids or derivatives. Table 28.2 relates the nitrogen base with the amide class (based on the number of alkyl or aryl groups on the nitrogen of the amide).

Table 28.2 *Classes of Amides*

Nitrogen Base	*(reacts to form)*	*Amide* $(-\overset{\overset{\large O}{\|}}{C}-\overset{|}{N}-)$
Ammonia		Primary amide (no R groups)
Primary amine		Secondary amide (one R group)
Secondary amine		Tertiary amide (two R groups)

$$CH_3NH_2 + CH_3-\overset{\overset{O}{\|}}{C}-OH \longrightarrow CH_3-\overset{\overset{O}{\|}}{C}-O^-(CH_3NH_3^+) \xrightarrow{\Delta} CH_3-\overset{\overset{O}{\|}}{C}-NHCH_3 + H_2O$$

$$CH_3NH_2 + CH_3-\overset{\overset{O}{\|}}{C}-Cl \longrightarrow CH_3-\overset{\overset{O}{\|}}{C}-NHCH_3 + HCl$$

Hydrolysis of amides can take place in either acid or base. Primary amides hydrolyze in acid to ammonium salts and carboxylic acids. Neutralization of the acid and ammonium salts releases ammonia, which can be detected by odor or by litmus.

$$R-\overset{\overset{O}{\|}}{C}-NH_2 + HCl + H_2O \longrightarrow R-\overset{\overset{O}{\|}}{C}-OH + NH_4Cl$$

$$NH_4Cl + NaOH \longrightarrow NH_3 + NaCl + H_2O$$

Secondary and tertiary amides would release the corresponding alkyl ammonium salts which, when neutralized, would yield the amine.

In base, primary amides hydrolyze to carboxylic acid salts and ammonia. The presence of ammonia (or amine from corresponding amides) can be detected similarly by odor or litmus. The carboxylic acid would be generated by neutralization with acid.

$$R-\overset{\overset{O}{\|}}{C}-NH_2 + NaOH \longrightarrow R-\overset{\overset{O}{\|}}{C}-O^-Na^+ + NH_3$$

$$R-\overset{\overset{O}{\|}}{C}-O^-Na^+ + HCl \longrightarrow R-\overset{\overset{O}{\|}}{C}-OH + NaCl$$

OBJECTIVES

1. To show some physical and chemical properties of amines and amides.
2. To demonstrate the hydrolysis of amides.

PROCEDURE

CAUTION

Amines are toxic chemicals. Avoid excessive inhaling of the vapors and use gloves to avoid direct skin contact. Anilines are more toxic than aliphatic amines and are readily absorbed through the skin. Wash any amine or aniline spill with large quantities of water. Diethyl ether (ether) is extremely flammable. Be certain there are *NO* open flames in the immediate area. Discard all solutions in properly labeled organic waste containers.

Properties of Amines

1. Place 5 drops of liquid or 0.1 g of solid from the compounds listed in the following table into labeled clean, dry test tubes (100 × 13 mm).

Test Tube No.	*Nitrogen Compound*
1	6 M NH_3
2	Triethylamine
3	Aniline
4	N,N-Dimethylaniline
5	Acetamide

2. Carefully note the odors of each compound. **Do not inhale deeply. Merely wave your hand across the mouth of the test tube toward your nose in order to note the odor.** Record your observations on the Report Sheet.
3. Add 2 mL of distilled water to each of the labeled test tubes. Mix thoroughly by sharply tapping the test tube with your finger. Note on the Report Sheet whether the amines are soluble or insoluble.
4. Take a glass rod, and test each solution for its pH. Carefully dip one end of the glass rod into a solution and touch a piece of pH paper. Between each test, be sure to clean and dry the glass rod. Record the pH by comparing the color of the paper with the chart on the dispenser.
5. Carefully add 2 mL of 6 M HCl to each test tube. Mix thoroughly by sharply tapping the test tube with your finger. Compare the odor and solubility of this solution with previous observations.
6. Place 5 drops of liquid or 0.1 g of solid from the compounds listed in the table into labeled clean, dry test tubes (100 × 13 mm). Add 2 mL of diethyl ether (ether) to each test tube. Stopper with a cork and mix thoroughly by shaking. Record the observed solubilities.
7. Carefully place on a watch glass, side by side, without touching, a drop of triethylamine and a drop of concentrated HCl. Record your observations. (Do this in the hood.)

Hydrolysis of Acetamide

1. Dissolve 0.5 g of acetamide in 5 mL of 6 M H_2SO_4 in a large test tube (150 × 18 mm). Heat the solution in a boiling-water bath for 5 min.
2. Hold a small strip of moist pH paper just inside the mouth of the test tube without touching the sides; note any changes in color; record the pH reading. Remove the test tube from the water bath, holding it with a test tube holder. Carefully note any odor.
3. Place the test tube in an ice-water bath until cool to the touch. Now *carefully add, dropwise with shaking,* 6 M NaOH to the cool solution until basic. (You will need more than 7 mL of base.) Hold a piece of moist pH paper just inside the mouth of the test tube without touching the sides. Record the pH reading. Carefully note any odor.

CHEMICALS AND EQUIPMENT

1. Acetamide
2. 6 M NH_3, ammonia water
3. Aniline
4. N,N-Dimethylaniline
5. Triethylamine
6. Diethyl ether (ether)
7. 6 M NaOH
8. Concentrated HCl
9. 6 M HCl
10. 6 M H_2SO_4
11. pH papers
12. Hot plate

name | section | date

partner | grade

EXPERIMENT 28

Pre-Lab Questions

1. Draw and label an example of a primary amine, a secondary amine, and a tertiary amine.

2. Give an example of an amide and circle the functional group.

3. Write the chemical reaction between methyl amine, CH_3NH_2, and HCl.

4. You can remove the "fishy" odor from your hands after working with fish by rubbing the area with either lemon or lime juice. Why does this work? Why can you not do this just by rinsing with water?

name | section | date

partner | grade

28 EXPERIMENT 28

Report Sheet

Properties of amines

	Odor		Solubility			pH
	Original Sol.	*with HCl*	H_2O	*Ether*	*HCl*	H_2O
6 M NH_3						
Triethylamine						
Aniline						
N,N-Dimethylaniline						
Acetamide						

Triethylamine and concentrated hydrochloric acid observation:

Write the chemical equation for the reaction of triethylamine with concentrated hydrochloric acid.

Hydrolysis of acetamide

Solution	*pH Reading*	*Odor Noted*
1. Acid		
2. Base		

Post-Lab Questions

1. Effective mosquito repellents contain DEET (N,N-diethyl-3-methylbenzamide). If you were to synthesize this compound, with what carboxylic acid and amine would you start?

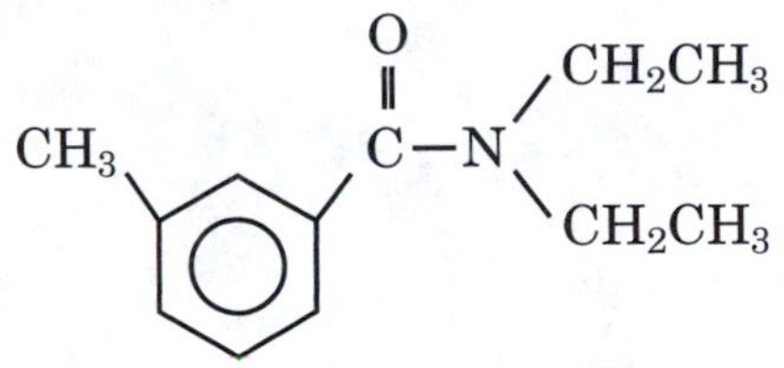

DEET

2. Metadone, a narcotic analgesic, is dispensed as its hydrochloride salt. Explain the use of the salt rather than the amine.

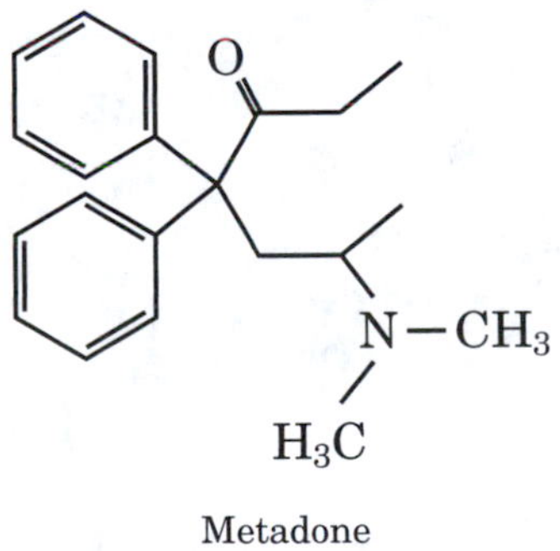

Metadone

3. Nicotine is an alkaloid, meaning alkali- or base-like. What is there in the molecule that would make it react as a base?

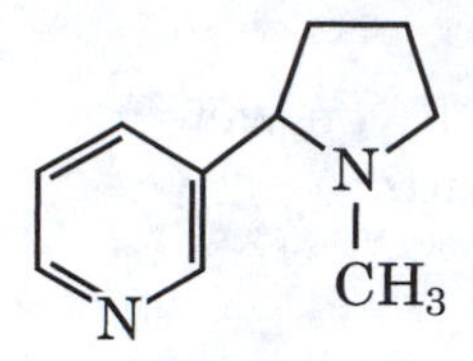

Nicotine

4. Write the equations that account for what happens in the hydrolysis of the acetamide solution in (a) acid and in (b) base.

a.

b.

Polymerization Reactions

BACKGROUND

Polymers are giant molecules made of many (poly-) small units. The starting material, which is a single unit, is called the monomer. Many of the most important biological compounds are polymers. Cellulose and starch are polymers of glucose units, proteins are made of amino acids, and nucleic acids are polymers of nucleotides. Since the 1930s, a large number of synthetic polymers have been manufactured. They contribute to our comfort and gave rise to the previous slogan of DuPont Co.: "Better living through chemistry." Synthetic fibers such as nylon and polyesters, plastics such as the packaging materials made of polyethylene and polypropylene films, polystyrene, and polyvinyl chloride, just to name a few, all became household words. Synthetic polymers are parts of buildings, automobiles, machinery, toys, appliances, etc.; we encounter them daily in our life.

In this experiment we will focus on man-made polymers and the basic mechanism by which some of them are formed. The two most important types of reactions that are employed in polymer manufacturing are the addition and condensation polymerization reactions. The first is represented by the polymerization of styrene and the second by the formation of nylon.

Styrene is a simple organic monomer that, by virtue of its containing a double bond, can undergo addition polymerization.

$$H_2C{=}CH(C_6H_5) + H_2C{=}CH(C_6H_5) \longrightarrow H_3C{-}CH(C_6H_5){-}CH{=}CH(C_6H_5)$$

The reaction is called an *addition reaction* because two monomers are added to each other with the elimination of a double bond. This is also called a chain growth polymerization reaction. However, the reaction as such does not go without the help of an unstable molecule, called an *initiator,* that starts the reaction. Benzoyl peroxide or *t*-butyl benzoyl peroxide are such initiators. Look at the structures of these two molecules. Notice that each structure contains an oxygen-to-oxygen bond, —O—O—.

This is the peroxide bond and is the weakest bond in the molecule. Benzoyl peroxide splits into two halves under the influence of heat or ultraviolet light and thus produces two free radicals. A *free radical* is a molecular fragment that has one unpaired electron. Thus, when the central bond was broken in the benzoyl peroxide, each of the shared pair of electrons went with one half of the molecule, each containing an unpaired electron.

$$C_6H_5-\overset{O}{\overset{\|}{C}}-O-O-\overset{O}{\overset{\|}{C}}-C_6H_5 \xrightarrow{\Delta} 2\, C_6H_5-\overset{O}{\overset{\|}{C}}-O^{\bullet}$$

benzoyl peroxide

Similarly, *t*-butyl benzoyl peroxide also gives two free radicals:

$$C_6H_5-\overset{O}{\overset{\|}{C}}-O-O-C(CH_3)_2-CH_3 \xrightarrow{\Delta} C_6H_5-\overset{O}{\overset{\|}{C}}-O^{\bullet} + {}^{\bullet}O-C(CH_3)_2-CH_3$$

t-butyl benzoyl peroxide

The reaction that generates these first free radicals is called **initiation**.

The dot indicates the unpaired electron. The free radical reacts with styrene and initiates the reaction:

$$C_6H_5-\overset{O}{\overset{\|}{C}}-O^{\bullet} + H_2C=CH-C_6H_5 \longrightarrow C_6H_5-\overset{O}{\overset{\|}{C}}-O-CH_2-\underset{C_6H_5}{CH^{\bullet}}$$

styrene

After this, the styrene monomers are added to the growing chain one by one until giant molecules containing hundreds and thousands of styrene-repeating units are formed. This part of the polymerization process is called **propagation**. Note the distinction between the monomer and the repeating unit. The monomer is the starting material, and the repeating unit is part of the polymer chain. Chemically they are not identical. In the case of styrene, the monomer contains a double bond, while the repeating unit (in the brackets in the following structure) does not.

$$CH_3-\underset{C_6H_5}{CH}-\left[CH_2-\underset{C_6H_5}{CH}-\right]_n CH_2-\underset{C_6H_5}{CH_2}$$

polystyrene

Because the initiators are unstable compounds, care should be taken not to keep them near flames or heat them directly. If a bottle containing a peroxide initiator is dropped, a minor explosion can even occur.

Eventually, as all the monomers get used up, the polymerization process slows. Any free radicals present start to pair up and the reaction ends. This part of the process is called **termination**.

The second type of reaction is called a *condensation reaction* because we condense two monomers into a longer unit, and at the same time we eliminate—expel—a small molecule. This is also called a step growth polymerization reaction. Nylon 6–6 is made of adipoyl chloride and hexamethylene diamine:

$$n\,Cl-\overset{\overset{\displaystyle O}{\|}}{C}-(CH_2)_4-\overset{\overset{\displaystyle O}{\|}}{C}-Cl + n\,H_2N-(CH_2)_6-NH_2 \longrightarrow$$

adipoyl chloride hexamethylene diamine

$$Cl-\overset{\overset{\displaystyle O}{\|}}{C}-(CH_2)_4-\overset{\overset{\displaystyle O}{\|}}{C}-\left[NH-(CH_2)_6-NH-\overset{\overset{\displaystyle O}{\|}}{C}-(CH_2)_4-\overset{\overset{\displaystyle O}{\|}}{C}-\right]_n NH-(CH_2)_6-NH_2 + n\,HCl$$

repeating unit

We form an amide linkage between the adipoyl chloride and the amine with the elimination of HCl. The polymer, a polyamide, is called nylon 6–6 because there are six carbons from the diamine and six carbons from the acyl chloride. Another nylon, nylon 6–10, is made from hexamethylene diamine (a 6-carbon diamine) and sebacoyl chloride (a 10-carbon atom acyl chloride). We use an acyl chloride rather than a carboxylic acid to form the amide bond because the former is more reactive. NaOH is added to the polymerization reaction in order to neutralize the HCl that is released every time an amide bond is formed.

The length of the polymer chain formed in both reactions depends on environmental conditions. Usually the chains formed can be made longer by heating the products longer. This process is called *curing*.

OBJECTIVES

1. To acquaint students with the conceptual and physical distinction between monomer and polymer.
2. To perform addition and condensation polymerization and solvent casting of films.

PROCEDURE

Preparation of Polystyrene

1. Set up your hot plate in the hood. Place 50 g of sea sand in a 150-mL beaker. Position a thermometer (0–200°C) in the sand bath so that it does not touch the bottom of the beaker. Heat the sand bath to 140°C.
2. Place approximately 2.5 mL styrene in a 150 × 18 mm Pyrex test tube. Add 3 drops of *t*-butyl benzoyl peroxide (*t*-butyl peroxide benzoate) initiator. Mix the solution.

3. Place the test tube in a test tube holder. Immerse the test tube in the sand bath. Heat the mixture in the test tube to about 140°C. The mixture will turn yellow.

CAUTION

Make sure the test tube points away from your face. Do not touch either the test tube or the beaker with your hand.

4. When bubbles appear, remove the test tube from the sand bath. The polymerization reaction is exothermic and thus it generates its own heat. Overheating would create sudden boiling. When the bubbles disappear, put the test tube back in the sand bath. But every time the mixture starts to boil you must remove the test tube.
5. Continue the heating until the mixture in the test tube has a syrupy consistency.
6. Immerse a glass rod in the hot mixture. Swirl it around a few times. Remove the glass rod immediately. A chunk of polystyrene will be attached to the glass rod that will solidify upon cooling. The remaining polystyrene will solidify on the walls of the test tube when you remove it from the sand bath. Turn off the hot plate and let the sand bath cool to room temperature. While it is cooling, add 10 drops of xylene to the test tube and dissolve some of the polystyrene by warming it in the sand bath.
7. Pour a few drops of the warm xylene solution on a microscope slide and let the solvent evaporate. A thin film of polystyrene will be obtained. This is one of the techniques—the so-called solvent-casting technique—used to make films from bulk polymers.
8. Discard the remaining xylene solution into a special jar labeled "Waste." Discard the test tube with the polystyrene in a special box labeled "Glass."
9. Investigate the consistency of the solidified polystyrene on your glass rod, removing the solid mass by prying it off with a spatula.

Preparation of Nylon

1. Set up a 50-mL reaction beaker and clamp above it a cylindrical paper roll (from toilet paper) or a stick.
2. Add 2.0 mL of 20% NaOH solution to the beaker. Then, with a volumetric pipet, add 10 mL of a 5% aqueous solution of hexamethylene diamine.
3. Take 10 mL of 5% adipoyl chloride solution in cyclohexane using a pipet or syringe. Layer the cyclohexane solution slowly on top of the aqueous solution in the beaker. Two layers will form and nylon will be produced at the interface (Figure 29.1).
4. With a copper hook, first scrape off the nylon formed on the walls of the beaker.
5. Slowly lift and pull the film from the center. If you pull it too fast, the nylon rope will break.
6. Wind it around the paper roll or stick two to three times. **Do not touch it with your hands.**

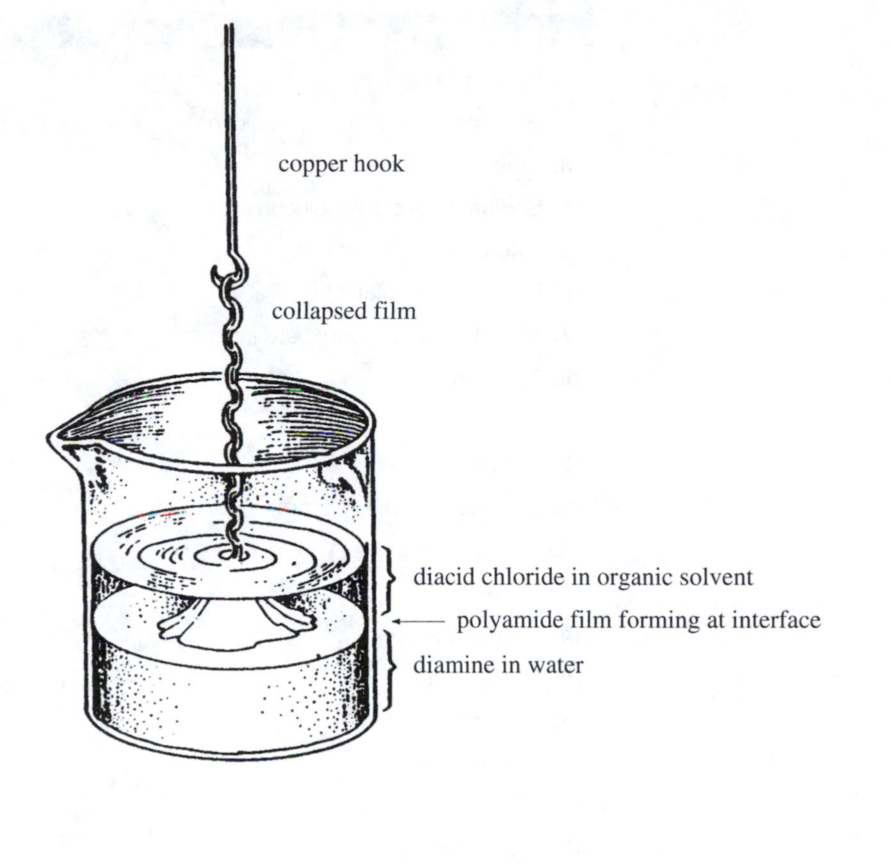

Figure 29.1
Preparation of nylon.

7. Slowly rotate the roll or the stick and wind at least a 1-m nylon rope.
8. Cut the rope and transfer the wound rope into a beaker filled with water (or 50% ethanol). Watch as the thickness of the rope collapses. Dry the rope between two filter papers.
9. There are still monomers left in the beaker. Mix the contents vigorously with a glass rod. See if beads of nylon form or only a clump of the polymer results.
10. Pour the mixture into a cold water bath and wash it. Dry the nylon between two filter papers. Note the consistency of your products.
11. Dissolve a small amount of nylon in 80% formic acid. Place a few drops of the solution onto a microscope slide and evaporate the solvent under the hood.
12. Compare the appearance of the solvent-cast nylon film with that of the polystyrene.

CHEMICALS AND EQUIPMENT

1. Styrene
2. Hexamethylene diamine solution
3. Adipoyl chloride solution
4. Sodium hydroxide solution
5. Xylene
6. Formic acid solution
7. *t*-Butyl peroxide benzoate initiator
8. 50% Ethanol
9. Sea sand
10. Hot plate
11. Test tube (150 × 18 mm) Pyrex
12. Test tube holder
13. Paper roll or stick
14. Copper hook
15. 10-mL pipets
16. Spectroline pipet filler
17. Beaker tongs
18. Filter paper
19. Microscope slides

name | section | date

partner | grade

29 EXPERIMENT 29

Pre-Lab Questions

1. What are the three steps in the polymerization process involving free radicals?

2. Why is the reaction that produces nylon called a *condensation* reaction?

3. Why don't you directly heat *t*-butyl benzoyl peroxide?

4. What is the weakest bond in any peroxide molecule?

5. What is formed when benzoyl peroxide is heated?

name section date

partner grade

Report Sheet

1. Compare the flexibility of the polystyrene and nylon.

2. Describe the difference in polarity between polystyrene and nylon. (*Hint:* Consider the different solvents that were used to dissolve each polymer.)

3. Is there any difference in the appearance of the solvent-cast films of nylon and polystyrene?

4. Describe the differences in texture between the polystyrene and the nylon.

Post-Lab Questions

1. The polymerization of styrene is an exothermic reaction. However, you still needed to heat to 140°C. Why was this done?

2. Is there any advantage to adding NaOH to the aqueous phase in the experiment? (*Hint:* Think of the Le Chatelier Principle.)

3. Nylon 6–6: to what do these numbers refer?

4. A polymer can be made from pentamethylene diamine and sebacoyl chloride:

$$H_2N{-}(CH_2)_5{-}NH_2 \text{ and } Cl{-}\overset{\displaystyle O}{\overset{\|}{C}}{-}(CH_2)_8{-}\overset{\displaystyle O}{\overset{\|}{C}}{-}Cl$$

a. Draw the structure of the polymer formed. Show the repeating unit.

b. Is this a free radical or a condensation reaction?

c. What molecule has been eliminated in this reaction?

d. What name would you apply to this polymer (see question 3, on previous page)?

Preparation of Acetylsalicylic Acid (Aspirin)

BACKGROUND

One of the most widely used nonprescription drugs is aspirin. Throughout the world, more than 50 million pounds are consumed annually. It is no wonder there is such wide use when one considers the medicinal applications for aspirin. It is an effective analgesic (pain killer) that can reduce the mild pain of headaches, toothache, neuralgia (nerve pain), muscle pain, and joint pain (from arthritis and rheumatism). Aspirin behaves as an antipyretic drug (it reduces fever) and an antiinflammatory agent capable of reducing the swelling and redness associated with inflammation. It is an effective agent in preventing strokes and heart attacks due to its ability to act as an anticoagulant by preventing platelet aggregation.

Early studies showed the active agent that gave these properties to be salicylic acid. It is an organic compound found in nature and one that has had a long history of use. Hippocrates (460–377 B.C.) wrote that by chewing leaves of willow (*Salix*) or forming a bitter powder extracted from willow bark, aches and pains were eased and fever was reduced. Native North Americans and the Hottentots of southern Africa also used willow for these beneficial effects. Salicylic acid also was discovered in a European species of dropwort, *Spiraea*. However, salicylic acid contains the phenolic and the carboxylic acid groups. As a result, the compound was too harsh for the linings of the mouth, esophagus, and stomach. Prolonged contact with the stomach lining, in particular, caused some hemorrhaging. The Bayer Company in Germany patented the ester, acetylsalicylic acid, and marketed the product as "aspirin" in 1899. (Aspirin is derived from the words acetyl and *Spiraea*.) Bayer's studies showed that this ester was less of an irritant. The acetylsalicylic acid was hydrolyzed in the small intestine to salicylic acid, which then was absorbed into the bloodstream. It became so widely advertised and common that the U.S. Supreme Court ruled against the Bayer Company in 1917 that the name "aspirin" was no longer owned by the company. Aspirin is now part of the common language.

The relationship between salicylic acid and aspirin is shown in the following formulas:

Salicylic acid (ring with COOH and OH substituents) — Acetylsalicylic acid (Aspirin) (ring with COOH and O–C(=O)–CH_3 substituents)

Aspirin still has side effects. Hemorrhaging of the stomach walls can occur even with normal dosages. These side effects can be reduced through the addition of coatings or through the use of buffering agents. Magnesium hydroxide, magnesium carbonate, and aluminum glycinate, when mixed into the formulation of the aspirin (e.g., Bufferin), reduce the irritation.

This experiment will acquaint you with a simple synthetic problem in the preparation of aspirin. The preparative method uses acetic anhydride and an acid catalyst, like sulfuric or phosphoric acid, to speed up the reaction with salicylic acid.

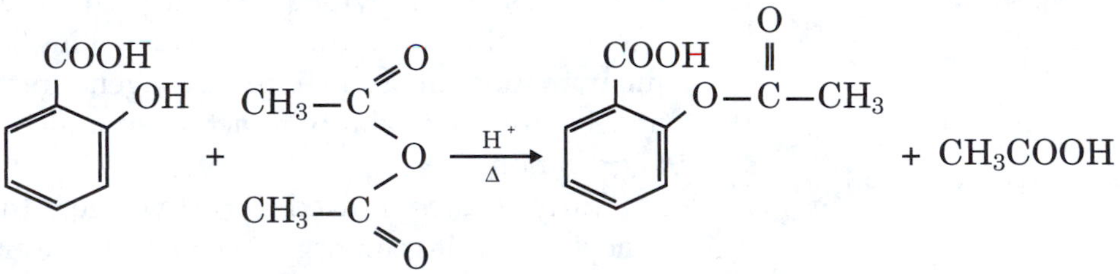

$$\text{Salicylic acid} + (CH_3CO)_2O \text{ (Acetic anhydride)} \xrightarrow[\Delta]{H^+} \text{Aspirin} + CH_3COOH \text{ (Acetic acid)}$$

If any salicylic acid remains unreacted, its presence can be detected with a 1% iron(III) chloride solution. Salicylic acid has a phenol group in the molecule. The iron(III) chloride gives a violet color with any molecule possessing a phenol group (see Experiment 25). Notice the aspirin no longer has the phenol group. Thus a pure sample of aspirin will not give a purple color with 1% iron(III) chloride solution.

OBJECTIVES

1. To illustrate the synthesis of the drug aspirin.
2. To use a chemical test to determine the purity of the preparation.

PROCEDURE

Preparation of Aspirin

1. Prepare a bath using a 400-mL beaker filled about half way with water. Heat to boiling.

2. Take 2.0 g of salicylic acid and place it in a 125-mL Erlenmeyer flask. Use this quantity of salicylic acid to calculate the theoretical or expected yield of aspirin (1). Carefully add 3 mL of acetic anhydride to the flask and, while swirling, add 3 drops of concentrated sulfuric acid.

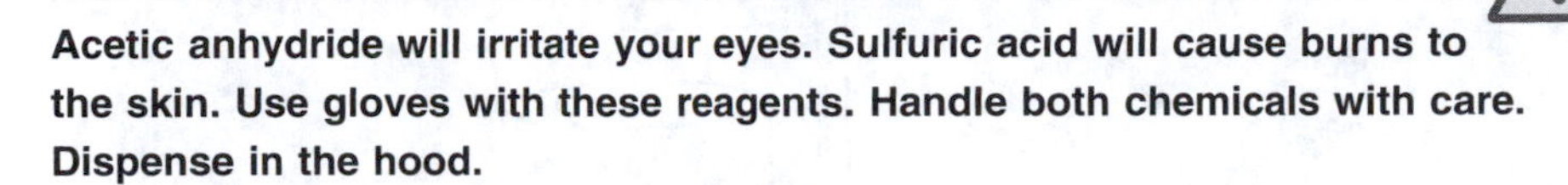

Acetic anhydride will irritate your eyes. Sulfuric acid will cause burns to the skin. Use gloves with these reagents. Handle both chemicals with care. Dispense in the hood.

3. Mix the reagents and then place the flask in the boiling-water bath; heat for 30 min. (Figure 30.1). The solid will completely dissolve. Swirl the solution occasionally.
4. Remove the Erlenmeyer flask from the bath and let it cool to approximately room temperature. Then, slowly pour the solution into a 150-mL beaker containing 20 mL of ice water, mix thoroughly, and place the beaker in an ice bath. Use a glass rod to mix the solution while in the ice bath, vigorously rubbing the glass rod (called scratching) along the bottom of the beaker (be careful not to poke a hole through the beaker). The water destroys any unreacted acetic anhydride and will cause the insoluble aspirin to precipitate from the solution. Scratching the beaker helps to precipitate the aspirin, also.
5. Collect the crystals by filtering under suction with a Büchner funnel. The assembly is shown in Figure 30.2. (Also see Figure 26.1, p. 319.)
6. Obtain a 250-mL filter flask and connect the side arm of the filter flask to a water aspirator with heavy-walled vacuum rubber tubing. (The thick walls of the tubing will not collapse when the water is turned on and the pressure is reduced.)
7. The Büchner funnel is inserted into the filter flask through either a filtervac, a neoprene adapter, or a one-hole rubber stopper, whichever is available. Filter paper is then placed into the Büchner funnel. Be sure that the paper lies flat and covers all the holes. Wet the filter paper with water.

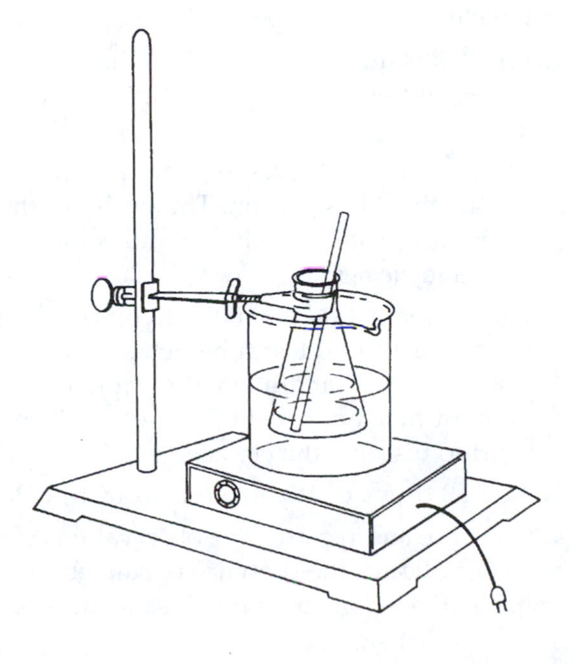

Figure 30.1
Assembly for the synthesis of aspirin.

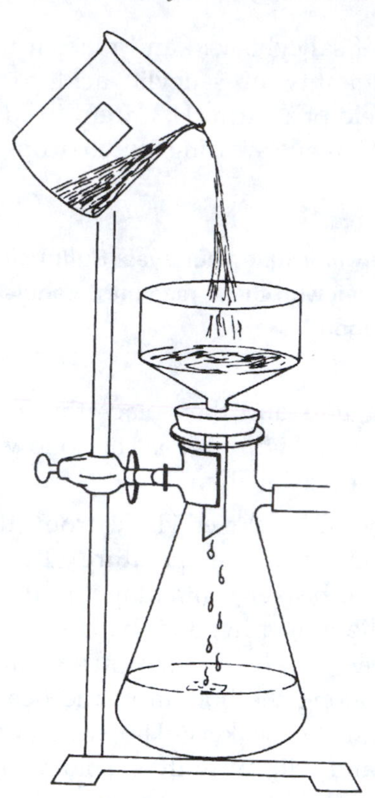

Figure 30.2
Filtering using the Büchner funnel.

8. Turn on the water aspirator to maximum water flow. Pour the solution into the Büchner funnel.
9. Wash the crystals with two 5-mL portions of cold water, followed by one 10-mL portion of cold ethyl acetate.
10. Continue suction through the crystals for at least 5 min. to help dry them. Disconnect the rubber tubing from the filter flask before turning off the water aspirator.
11. Using a spatula, place the crystals between several sheets of paper toweling or filter paper and press dry the solid.
12. Weigh a 50-mL beaker (2). Add the crystals and reweigh (3). Calculate the weight of crude aspirin (4). Determine the percent yield (5).

Determine the Purity of the Aspirin

1. The aspirin you prepared is not pure enough for use as a drug and is *not* suitable for ingestion. The purity of the sample will be tested with 1% iron(III) chloride solution and compared with a commercial aspirin and salicylic acid.
2. Label three test tubes (100 × 13 mm) 1, 2, and 3; place a few crystals of salicylic acid into test tube no. 1, a small sample of your aspirin into test tube no. 2, and a small sample of a crushed commercial aspirin into test tube no. 3. Add 5 mL of distilled water to each test tube and swirl to dissolve the crystals.
3. Add 10 drops of 1% aqueous iron(III) chloride to each test tube.
4. Compare and record your observations (6). The formation of a purple color indicates the presence of salicylic acid. The intensity of the color qualitatively tells how much salicylic acid is present.

CHEMICALS AND EQUIPMENT

1. Acetic anhydride
2. Concentrated sulfuric acid, H_2SO_4
3. Commercial aspirin tablets
4. Ethyl acetate
5. 1% Iron(III) chloride
6. Salicylic acid
7. Boiling stones
8. Büchner funnel, small (65 mm OD)
9. 250-mL filter flask
10. Filter paper
11. Filtervac or neoprene adapter
12. Hot plate

name

section

date

partner

grade

30

EXPERIMENT 30

Pre-Lab Questions

1. What is the active ingredient derived from aspirin that gives the observed therapeutic properties? Where is this compound released?

2. Should aspirin test positive with a 1% iron(III) chloride solution? Explain your answer.

3. What functional groups in salicylic acid are responsible for the side reaction that involves irritation of the stomach lining?

4. How do buffering agents reduce the side effects of aspirin? Name some of the buffering agents used in formulation.

name _______________ section _______________ date _______________

partner _______________ grade _______________

30 EXPERIMENT 30

Report Sheet

1. Theoretical yield:

$$\_\_\_\_\_\_\_\_\ \text{g salicylic acid} \times \frac{180\ \text{g aspirin}}{1\ \text{mole}} \times \frac{1\ \text{mole}}{138\ \text{g salicylic acid}}$$

= _______________ g aspirin

2. Weight of 50-mL beaker _______________ g

3. Weight of your aspirin and beaker _______________ g

4. Weight of your aspirin: (3) − (2) _______________ g

5. Percent yield: [(4)/(1)] × 100 = % _______________ %

6. Iron(III) chloride test

No.	*Sample*	*Color*	*Intensity*
1	Salicylic acid		
2	Your aspirin		
3	Commercial aspirin		

Post-Lab Questions

1. The experiment required 3 drops of concentrated sulfuric acid. Why was this acid added? What would happen if this reagent were omitted?

2. A student thought he carefully followed the directions in the experimental procedure, and he expected 2.63 g of product. However, his isolated product weighed 3.05 g. What would account for the excess weight?

3. The above student also found that his product tested positive with 1% iron(III) chloride solution. What does this test indicate?

4. Another student also expected 2.63 g of product, but isolated only 2.01 g. What is the percentage yield? Show your work.

5. Below are structures of common aspirin substitutes. Why do ibuprofen and naproxen not give a positive test with 1% iron(III) chloride? Why does acetaminophen give a positive test?

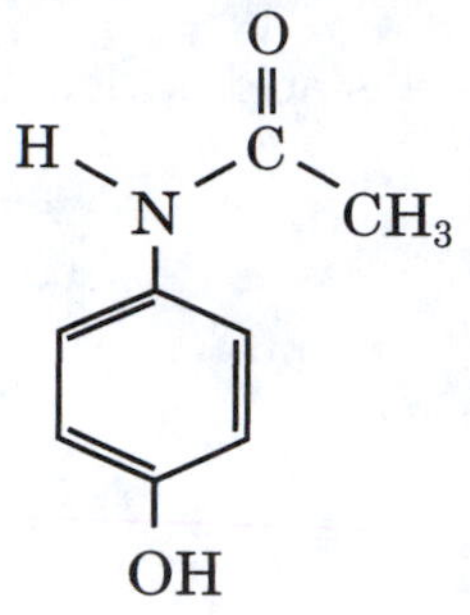

Acetaminophen 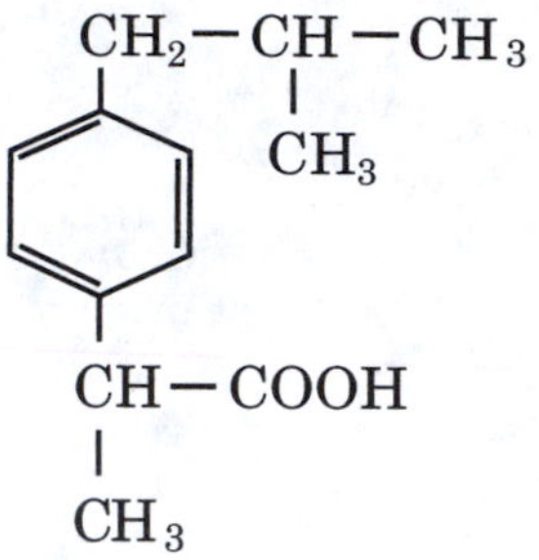 Ibuprofen 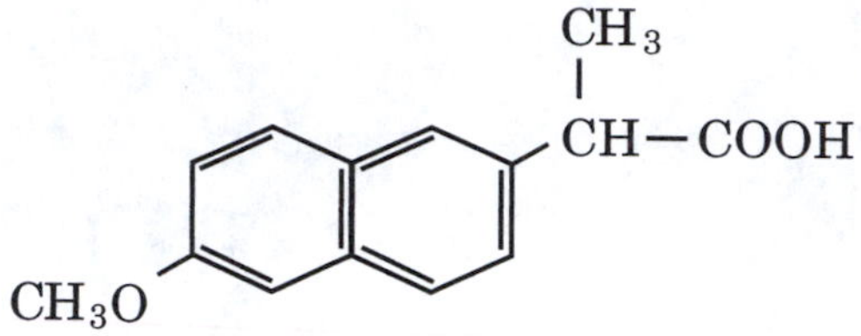 Naproxen

6. Bottles containing old aspirin tablets often smell of vinegar. The presence of what chemical causes this smell? How does this chemical form?

Isolation of Caffeine from Tea Leaves

BACKGROUND

Many organic compounds are obtained from natural sources through extraction. This method takes advantage of the solubility characteristics of a particular organic substance with a given solvent. In the experiment here, caffeine is readily soluble in hot water and is thus separated from the tea leaves. Caffeine is one of the main substances that make up the water solution called tea. Besides being found in tea leaves, caffeine is present in coffee, kola nuts, and cocoa beans. As much as 5% by weight of the leaf material in tea plants consists of caffeine.

The caffeine structure is shown below. It is classed as an alkaloid, meaning that with the nitrogen present, the molecule has base characteristics (alkali-like). In addition, the molecule has the purine ring system, a framework that plays an important role in living systems.

Purine

Caffeine

Caffeine is the most widely used of all the stimulants. Small doses of this chemical (50 to 200 mg) can increase alertness and reduce drowsiness and fatigue. The "No-Doz" tablet lists caffeine as the main ingredient. In addition, it affects blood circulation because the heart is stimulated and blood vessels are relaxed (vasodilation). It also acts as a diuretic. There are side effects. Large doses of over 200 mg can result in insomnia, restlessness, headaches, and muscle tremors ("coffee nerves"). Continued, heavy use may bring on physical dependence. (Do you know somebody who cannot function in the morning until they have that first cup of coffee?)

Tea leaves consist primarily of cellulose; this is the principal structural material of all plant cells. Fortunately, the cellulose is insoluble in water, so that by using a hot water extraction, more soluble caffeine can be separated. Also dissolved in water are complex substances called tannins. These are colored phenolic compounds of high molecular weight (500 to 3000) that have acidic behavior. If a basic salt such as Na_2CO_3 is added to the water solution, the tannins can react to form a salt. These salts are insoluble in organic solvents, such as chloroform or dichloromethane, but are soluble in water.

Although caffeine is soluble in water (2 g/100 g of cold water), it is more soluble in the organic solvent dichloromethane (14 g/100 g). Thus caffeine can be extracted from the basic tea solution with dichloromethane, but the sodium salts of the tannins remain behind in the aqueous solution. Evaporation of the dichloromethane yields crude caffeine; the crude material can be purified by sublimation (see Experiment 13).

OBJECTIVES

1. To demonstrate the isolation of a natural product.
2. To learn the techniques of extraction.
3. To use sublimation as a purification technique.

PROCEDURE

The isolation of caffeine from tea leaves follows the scheme below:

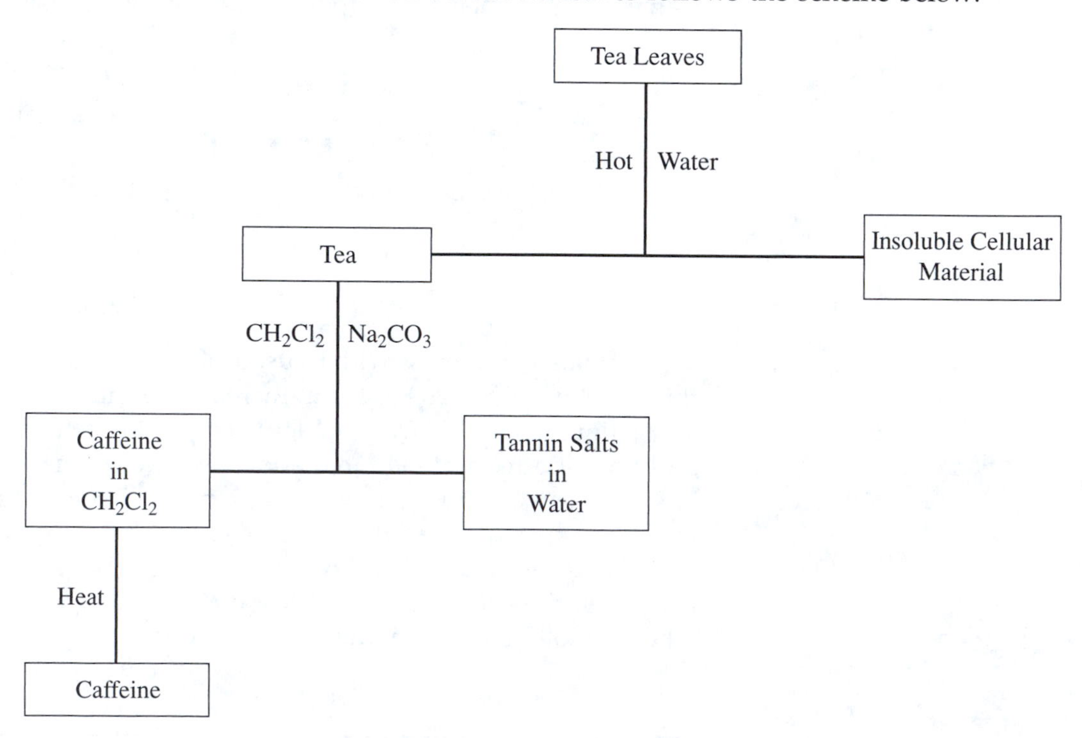

1. Carefully open two commercial tea bags (try not to tear the paper) and weigh the contents to the nearest 0.001 g. Record this weight (1). Place the tea leaves back into the bags, close, and secure the bags with staples.

2. Into a 150-mL beaker, place the tea bags so that they lie flat on the bottom. Add 30 mL of distilled water and 2.0 g of anhydrous Na_2CO_3; heat the contents with a hot plate, keeping a *gentle* boil, for 20 min. While the mixture is boiling, keep a watch glass on the beaker. Hold the tea bags under water by occasionally pushing them down with a glass rod.
3. Decant the hot liquid into a 50-mL Erlenmeyer flask. Wash the tea bags with 10 mL of hot water, carefully pressing them with a glass rod; add this wash water to the tea extract. (If any solids are present in the tea extract, filter them by gravity to remove.) Cool the combined tea extract to room temperature. The tea bags may be discarded.
4. Transfer the cool tea extract to a 125-mL separatory funnel that is supported on a ring stand with a ring clamp.

CAUTION

Dichloromethane is toxic and must be carefully handled. Avoid inhaling vapors. Avoid spilling any on your skin. Spills should be washed with soap and flushed with water. Do all work in the hood.

5. Carefully add 5.0 mL of dichloromethane to the separatory funnel. Stopper the funnel and lift it from the ring clamp; hold the funnel with two hands as shown in Figure 31.1. By holding the stopper in place with one hand, invert the funnel. *Make certain the stopper is held tightly and no liquid is spilled; make sure the liquid is not in contact with the stopcock; open the stopcock, being sure to point the opening away from you and your neighbors.* Built-up pressure caused by gases accumulating inside will be released. Now, close the stopcock and gently mix the contents by inverting the funnel two or three times. Again, release any pressure by opening the stopcock as before.
6. Return the separatory funnel to the ring clamp, remove the stopper, and allow the aqueous layer to separate from the dichloromethane layer (Figure 31.2). You should see two distinct layers form after a few minutes, with the dichloromethane layer at the bottom. Sometimes an emulsion may form at the juncture of the two layers. The emulsion often can be broken by gently swirling the contents or by gently stirring the emulsion with a glass rod.

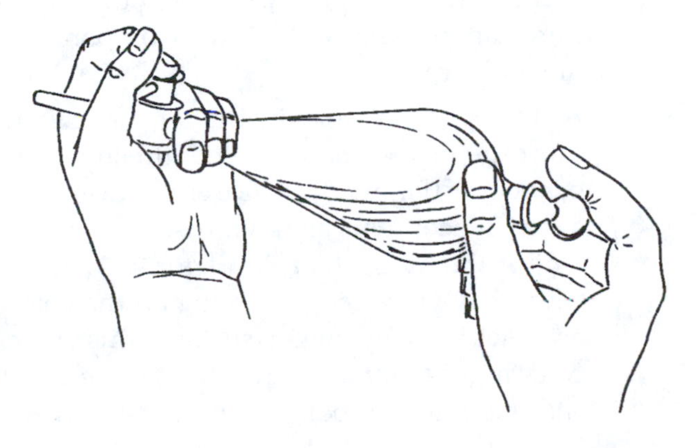

Figure 31.1
Using the separatory funnel.

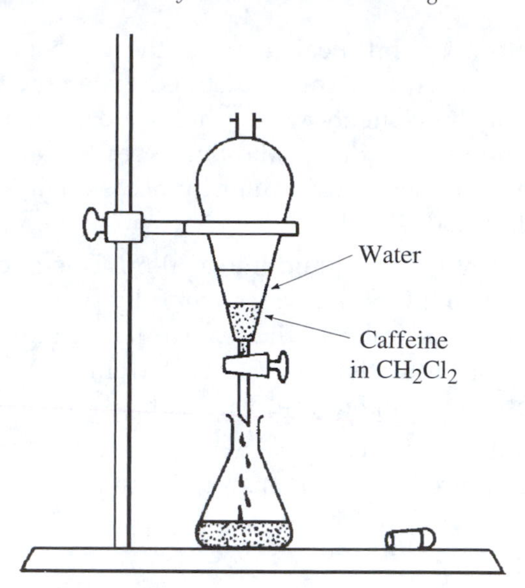

Figure 31.2
Separation of the aqueous layer and the dichloromethane layer in the separatory funnel.

7. Carefully drain the lower layer into a 25-mL Erlenmeyer flask. Try not to include any water with the dichloromethane layer; careful manipulation of the stopcock will prevent this.
8. Repeat the extraction with an additional 5.0 mL of dichloromethane. Combine the separated bottom layer with the dichloromethane layer obtained from step 7.
9. Add 0.5 g of anhydrous Na_2SO_4 to the combined dichloromethane extracts. Swirl the flask. The anhydrous salt is a drying agent and will remove any water that may still be present.
10. Weigh a 25-mL side-arm filter flask containing one or two boiling stones. Record this weight (2). By means of gravity filtration, filter the dichloromethane–salt mixture into the preweighed flask. Rinse the salt on the filter paper with an additional 2.0 mL of dichloromethane.
11. Remove the dichloromethane by evaporation *in the hood.* Be careful not to overheat the solvent; it may foam over. The solid residue that remains after the solvent is gone is the crude caffeine. Reweigh the cooled flask (3). Calculate the weight of the crude caffeine by subtraction (4) and determine the percent yield (5).
12. Take a melting point of your solid. First, scrape the caffeine from the bottom and sides of the flask with a microspatula and collect a sample of the solid in a capillary tube (review Experiment 13 for the technique). Pure caffeine melts at 238°C. Compare your melting point (6) to the literature value.

Optional

13. At the option of your instructor, the caffeine may be purified further. The caffeine may be sublimed directly from the flask with a cold finger condenser (Figure 31.3). Carefully insert the cold finger condenser into a no. 2 neoprene adapter (use a drop of glycerine as a lubricant). Adjust the tip of the cold finger to 1 cm from the bottom of the flask. Clean any glycerine remaining on the cold finger with a Kimwipe and acetone; the cold finger surface must be clean and dry. Connect the cold finger to a faucet by latex tubing (water *in* the upper tube; water *out* the lower tube). Connect the side-arm filter flask to a water

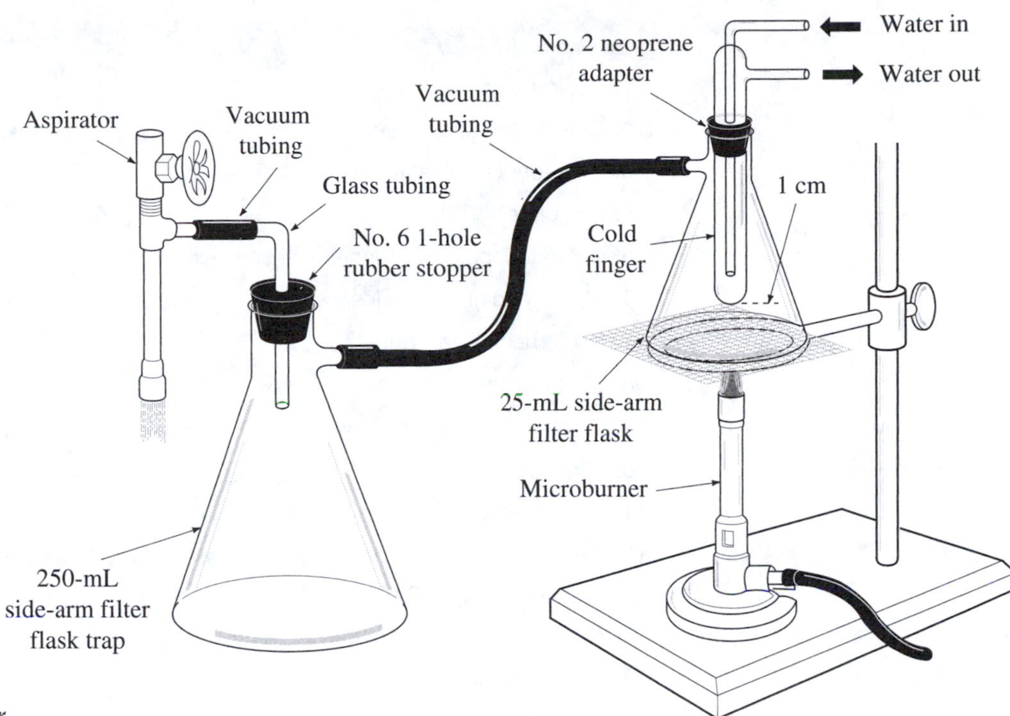

Figure 31.3
Sublimation apparatus connected to an aspirator.

aspirator with vacuum tubing, installing a trap between the aspirator and the sublimation setup (Figure 31.3). When you turn the water on, press the cold finger into the filter flask until a good seal is made. Gently heat the bottom of the filter flask that holds the caffeine with a microburner (support the *base* of the microburner); move the flame back and forth and along the sides of the flask. **Do not allow the sample to melt.** If the sample melts, *stop* heating and allow to cool before continuing. When the sublimation is complete, disconnect the heat and allow the system to cool; leave the aspirator connected and the water running.

14. When the system has reached room temperature, carefully disconnect the aspirator from the side-arm filter flask by removing the vacuum tubing from the side-arm. Turn off the water to the cold finger. *Carefully* remove the cold finger from the flask along with the neoprene adapter without dislodging any crystals. Scrape the sublimed caffeine onto a preweighed piece of weighing paper (7). Reweigh (8); determine the weight of caffeine (9). Calculate the percent recovery (10). Determine the melting point (11).

15. Collect the caffeine in a sample vial, and submit it to your instructor.

CHEMICALS AND EQUIPMENT

1. Boiling stones
2. Cold finger condenser
3. Filter paper (Whatman no. 7.0), fast flow
4. Hot plate
5. 125-mL separatory funnel with stopper
6. Melting-point capillaries
7. No. 2 neoprene adapter
8. 25-mL side-arm filter flask
9. Small sample vials
10. Tea bags
11. Tubing: latex, 2 ft.; vacuum, 2 ft.
12. 250-mL trap: 250-mL side-arm filter flask fitted with a no. 6 one-hole rubber stopper containing a piece of glass tubing (10 cm long × 37 mm OD)
13. Anhydrous sodium sulfate, Na_2SO_4
14. Anhydrous sodium carbonate, Na_2CO_3
15. Dichloromethane, CH_2Cl_2
16. Stapler
17. Weighing paper

name | section | date

partner | grade

31 EXPERIMENT 31

Pre-Lab Questions

1. Name some sources of caffeine.

2. Below is the structure of caffeine. Circle and label (a) an amide functional group; (b) a tertiary amine; and (c) the purine ring system.

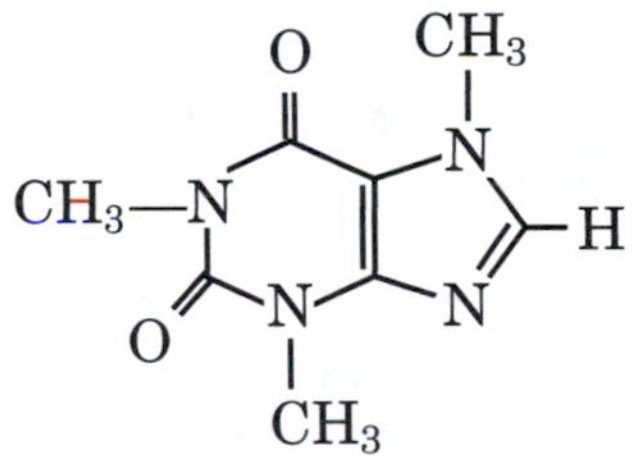

CH_3—N ... O, CH_3, N, H, N, N, O, CH_3

3. Caffeine is an alkaloid. What does this mean? What is there in the structure of the molecule that would give it this property?

4. Besides caffeine, what else is extracted into the tea? How are they removed?

5. How do you test for the purity of a sample of caffeine?

name   section   date

partner   grade

31 EXPERIMENT 31

Report Sheet

1. Weight of tea in 2 tea bags ________ g

2. Weight of 25-mL side-arm filter flask and boiling stones ________ g

3. Weight of flask, boiling stones, and crude caffeine ________ g

4. Weight of caffeine: (3) − (2) ________ g

5. Percent yield: [(4)/(1)] × 100 = % ________ %

6. Melting point of your crude caffeine ________ °C

7. Weight of weighing paper ________ g

8. Weight of sublimed caffeine and paper ________ g

9. Weight of caffeine: (8) − (7) ________ g

10. Percent recovery: [(9)/(4)] × 100 = % ________ %

11. Melting point of sublimed caffeine ________ °C

Post-Lab Questions

1. Look at Figure 31.2 in the **Procedure** section. Note the position of the layers in the separatory funnel. What does this say about the relative densities of the water and the dichloromethane?

2. Why was sodium carbonate, Na_2CO_3, used to remove the tannins?

3. What was the purpose of adding the anhydrous sodium sulfate to the dichloromethane solution?

4. If the above step was omitted in the isolation procedure, how would this affect the yield?

5. A student used 7.32 g of tea leaves in the experiment. How much caffeine is expected, assuming all the caffeine is extracted? (*Hint:* See the **Background** section.)

Carbohydrates

BACKGROUND

Carbohydrates are a major food source. Rice, potatoes, bread, corn, candy, and fruits are rich in carbohydrates. A carbohydrate can be classified as a monosaccharide (for example, glucose or fructose), a disaccharide (sucrose or lactose), which consists of two joined monosaccharides, or a polysaccharide (starch or cellulose), which consists of thousands of monosaccharide units linked together. If you look at the functional groups present, carbohydrates are polyhydroxy aldehydes or ketones or compounds that yield polyhydroxy aldehydes or ketones upon hydrolysis. Monosaccharides exist mostly as cyclic structures containing hemiacetal (or hemiketal) groups. These structures in solutions are in equilibrium with the corresponding open-chain structures bearing aldehyde or ketone groups. Glucose, blood sugar, is an example of a polyhydroxy aldehyde (Figure 32.1).

Disaccharides and polysaccharides exist as cyclic structures containing functional groups such as hydroxyl groups, acetal (or ketal) groups, and hemiacetal (or hemiketal) groups. Most of the di-, oligo-, or polysaccharides have two distinct ends. The end that has a hemiacetal (or hemiketal) on its terminal is called the reducing end, and the one that does not contain a hemiacetal (or hemiketal) terminal is the nonreducing end. The name "reducing" is given because hemiacetals (and to a lesser extent hemiketals) can reduce an oxidizing agent such as Fehling's reagent (and has an "unlocked" ring).

Figure 32.2 is an example of a disaccharide with a hemiacetal and an acetal in it.

Not all disaccharides or polysaccharides contain a reducing end. An example is sucrose, which does not have a hemiacetal (or hemiketal) group on either of its ends (Figure 32.3).

Polysaccharides, such as amylose or amylopectin, do have a hemiacetal group on one of their terminal ends, but practically, they are nonreducing substances because there is only one reducing group for every 2,000–10,000 monosaccharidic units. In such a low concentration, the reducing group does not give a positive test with Benedict's or Fehling's reagent.

On the other hand, when a nonreducing disaccharide (sucrose) or a polysaccharide such as amylose is hydrolyzed, the glycosidic linkages

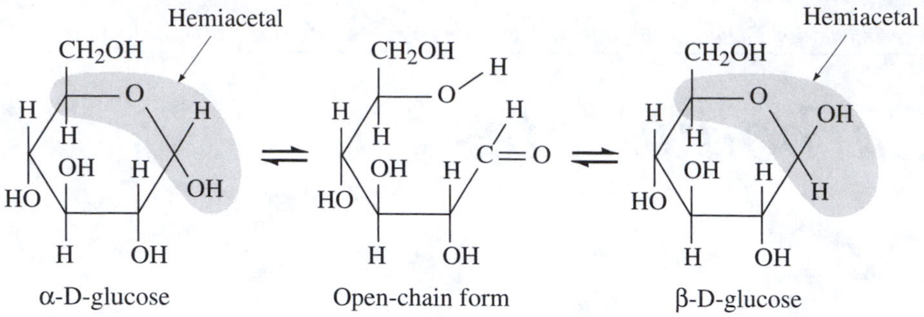

Figure 32.1
The structures of D-glucose.

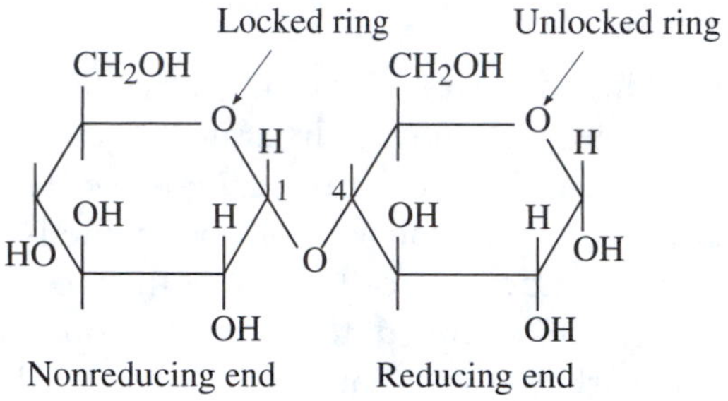

Figure 32.2
The structure of maltose, a disaccharide.

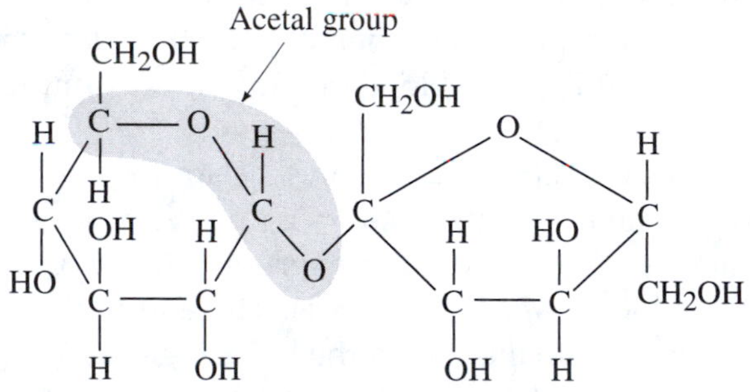

Figure 32.3
The structure of sucrose.

(acetal) are broken and reducing ends are created. Hydrolyzed sucrose (a mixture of D-glucose and D-fructose) will give a positive test with Benedict's or Fehling's reagent as well as hydrolyzed amylose (a mixture of glucose and glucose-containing oligosaccharides). The hydrolysis of sucrose or amylose can be achieved by using a strong acid such as HCl or with the aid of biological catalysts (enzymes; see Experiment 41).

Starch can form an intense, brilliant, dark blue or violet colored complex with iodine. The straight chain component of starch, the amylose, gives a blue color, while the branched component, the amylopectin, yields a purple color. In the presence of iodine the amylose forms helixes inside of which the iodine molecules assemble as long polyiodide chains. The helix-forming branches of amylopectin are much shorter than those of amylose. Therefore, the polyiodide chains are also much shorter in the amylopectin–iodine complex than in the amylose–iodine complex. The result is a different color (purple). When starch is hydrolyzed and broken down to small carbohydrate units, the iodine will not give a dark blue (or purple) color. The iodine test is used in this experiment to indicate the completion of the hydrolysis.

In this experiment you will investigate some chemical properties of carbohydrates in terms of their functional groups.

Reducing and Nonreducing Properties of Carbohydrates

1. **Aldoses (polyhydroxy aldehydes).** All aldoses are reducing sugars because they contain free aldehyde functional groups. The aldehydes are oxidized by mild oxidizing agents (e.g., Benedict's or Fehling's reagent) to the corresponding carboxylates. For example,

$$\underset{\text{(from Fehling's reagent)}}{R{-}CHO + 2Cu^{2+}} \xrightarrow{NaOH} R{-}COO^{-}Na^{+} + \underset{\text{Red precipitate}}{Cu_2O\downarrow}$$

2. **Ketoses (polyhydroxy ketones).** All ketoses are reducing sugars because they have a ketone functional group next to an alcohol functional group. The reactivity of this specific ketone (also called an α-hydroxyketone) is attributed to its ability to form an α-hydroxyaldehyde in basic media according to the following equilibrium equations:

$$\underset{\text{Ketose}}{\begin{array}{c} CH_2OH \\ | \\ C{=}O \\ | \\ H{-}C{-}OH \\ | \end{array}} \underset{}{\overset{\text{Base}}{\rightleftharpoons}} \underset{\text{Enediol}}{\begin{array}{c} CHOH \\ \| \\ C{-}OH \\ | \\ H{-}C{-}OH \\ | \end{array}} \overset{\text{Base}}{\rightleftharpoons} \underset{\text{Aldose}}{\begin{array}{c} CHO \\ | \\ H{-}C{-}OH \\ | \\ H{-}C{-}OH \\ | \end{array}}$$

3. **Hemiacetal functional group (potential aldehydes).** Carbohydrates with hemiacetal functional groups can reduce mild oxidizing agents such as Fehling's reagent because hemiacetals can easily form aldehydes through the following equilibrium equation:

$$\begin{array}{ccc} H & & OR' \\ & C & \\ R & & OH \end{array} \rightleftharpoons R{-}\overset{\overset{\displaystyle O}{\|}}{C}{-}H + R'OH$$

Sucrose, on the other hand, is a nonreducing sugar because it does not contain a hemiacetal functional group. Although starch has a hemiacetal functional group at one end of its molecule, it is considered a nonreducing sugar because the effect of the hemiacetal group in a very large starch molecule becomes insignificant to give a positive Benedict's test.

Hydrolysis of Acetal Groups

Disaccharides and polysaccharides can be converted into monosaccharides by hydrolysis. The following is an example:

$$\underset{\substack{\text{Lactose} \\ \text{(milk sugar)}}}{C_{12}H_{22}O_{11}} + H_2O \xrightarrow{\text{catalyst}} \underset{\text{Glucose}}{C_6H_{12}O_6} + \underset{\text{Galactose}}{C_6H_{12}O_6}$$

OBJECTIVES

1. To become familiar with the reducing or nonreducing nature of carbohydrates.
2. To experience the acid-catalyzed hydrolysis of acetal groups and the relation to digestion.
3. To compare monosaccharides, disaccharides, and polysaccharides.

PROCEDURE

Reducing or Nonreducing Carbohydrates

1. Place approximately 2 mL (approximately 40 drops) of Fehling's solution (20 drops each of solution part A and solution part B) into each of five labeled test tubes.
2. Add 10 drops of each of the following carbohydrates to the corresponding test tubes as shown in the following table.

Test Tube No.	*Name of Carbohydrate*
1	Glucose
2	Fructose
3	Sucrose
4	Lactose
5	Starch

3. Place the test tubes in a boiling-water bath for 5 min. A 600-mL beaker containing about 200 mL of tap water and a few boiling stones is used as the bath. Record your results on your Report Sheet. Which of those carbohydrates are reducing carbohydrates?

CAUTION

Remember to use boiling stones; they prevent bumping. Handle the hot test tubes with a test tube holder and the hot beaker with beaker tongs.

Hydrolysis of Carbohydrates

Hydrolysis of sucrose (acid versus base)

1. Place 3 mL of 2% sucrose solution in each of two labeled test tubes. To the first test tube (no. 1), add 3 mL of water and 3 drops of dilute sulfuric acid solution (3 M H_2SO_4). To the second test tube (no. 2), add 3 mL of water and 3 drops of dilute sodium hydroxide solution (3 M NaOH).

CAUTION

To avoid burns from the acid or the base, use gloves when dispensing these reagents.

2. Heat the test tubes in a boiling-water bath for about 5 min. Then cool both solutions to room temperature.
3. To the contents of test tube no. 1, add dilute sodium hydroxide solution (3 M NaOH) (about ten drops) until red litmus paper turns blue.
4. Test a few drops of each of the two solutions (test tubes no. 1 and no. 2) with Fehling's reagent, following the procedure that is described for carbohydrates above. Record your results on your Report Sheet.

Acid-catalyzed hydrolysis of starch

1. Place 5.0 mL of starch solution in a 150 × 15-mm test tube and add 1.0 mL of dilute sulfuric acid (3 M H_2SO_4). Mix it by gently shaking the test tube. Heat the solution in a boiling-water bath for about 5 min.
2. Using a clean medicine dropper, transfer about 3 drops of the starch solution into a white spot plate and then add 2 drops of iodine solution. Observe the color of the solution. If the solution gives a positive test with iodine solution (the solution should turn blue), the hydrolysis is not complete and you should continue heating.
3. Transfer about 3 drops of the boiling solution at 5-min. intervals for an iodine test. (*Note: Rinse the medicine dropper very thoroughly before each test.*) When the solution no longer gives the characteristic blue color with iodine solution, stop heating and record the time needed for the completion of hydrolysis on the Report Sheet.

CHEMICALS AND EQUIPMENT

1. Bunsen burner
2. Medicine droppers
3. White spot plate
4. Boiling stones
5. Fehling's reagent
6. 3 M NaOH
7. 2% starch solution
8. 2% sucrose
9. 2% fructose
10. 2% glucose
11. 2% lactose
12. 3 M H_2SO_4
13. 0.01 M iodine in KI

name | section | date

partner | grade

32

EXPERIMENT 32

Pre-Lab Questions

1. Which functional group(s) are present in reducing carbohydrates?

2. Circle and label the hemiacetal functional group and the acetal functional group in each of the following carbohydrates:

a. sucrose

b. lactose

3. Which carbohydrate in question 2 is a reducing sugar?

4. Name the functional group that links two monosaccharides in a disaccharide.

name | section | date

partner | grade

32 EXPERIMENT 32

Report Sheet

Reducing or nonreducing carbohydrates

Test Tube No.	*Substance*	*Color Observation*	*Reducing or Nonreducing Carbohydrates*
1	Glucose		
2	Fructose		
3	Sucrose		
4	Lactose		
5	Starch		

Hydrolysis of carbohydrates

	Hydrolysis of sucrose (acid versus base catalysis)		
Sample	*Condition of Hydrolysis*	*Color Observation*	*Fehling's Test (positive or negative)*
1	Acidic (H_2SO_4)		
2	Basic (NaOH)		

Sample No.	Acid-catalyzed hydrolysis of starch		
	Heating Time (min.)	*Color Observation*	*Iodine Test (positive or negative)*
1	5		
2	10		
3	15		
4	20		
5	25		
6	30		

Post-Lab Questions

1. How does the iodine test distinguish between amylose and amylopectin?

2. Why is sucrose a nonreducing sugar?

3. How can you tell when the hydrolysis of starch is complete? Why does the test work this way? What is the monosaccharide that results at the end?

4. Why does amylose give a negative test with Fehling's solution?

Fermentation of a Carbohydrate: Ethanol from Sucrose

BACKGROUND

Ethanol is the least toxic of the alcohols, and as such, has been used as a beverage. It is an alcohol that has been known since ancient times. How ancient people first discovered the process of making ethanol (ethyl alcohol or grain alcohol) by fermentation is lost in time. However, the process has been known for more than 5000 years; evidence exists from Egyptian tombs that beer preparation can be dated to the third millennium B.C. No doubt the basic chemistry of alcohol fermentation came about completely by accident.

An impure form of ethanol results from the fermented juice of fruits, such as grapes and apples, and grains, such as hops and barley. These beverages are known as wine, cider, beer, and ale. Alchemists further learned that from these juices relatively concentrated ethanol solutions could be produced using the technique of distillation.

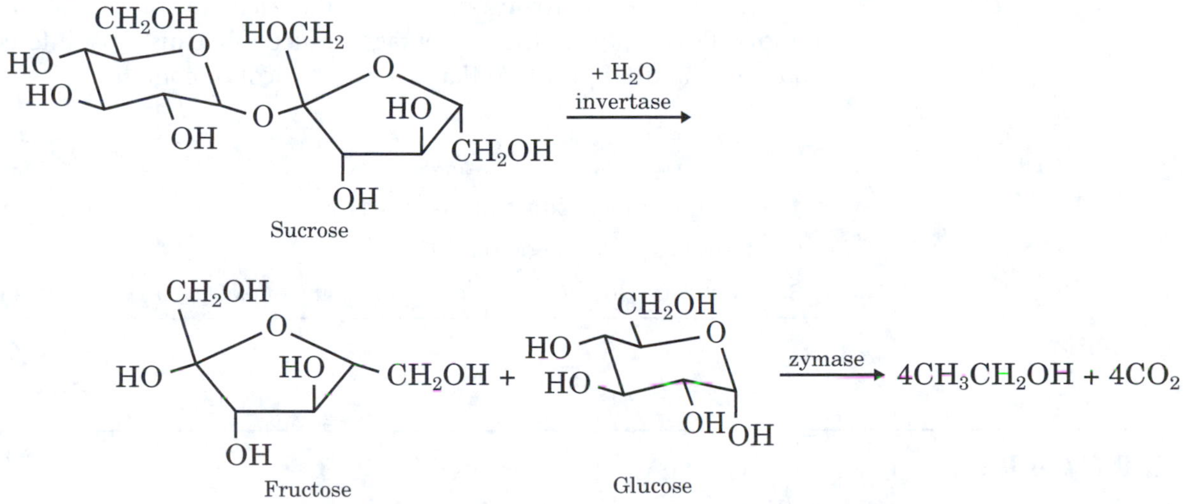

Sucrose $\xrightarrow[\text{invertase}]{+ H_2O}$ Fructose + Glucose $\xrightarrow{\text{zymase}}$ $4CH_3CH_2OH + 4CO_2$

In fermentation, any carbohydrate, either a complex polysaccharide or simple sugar, can be used as a starting material. In the present experiment, sucrose, a disaccharide, is converted to ethanol by the action of yeast. Sucrose has the formula $C_{12}H_{22}O_{11}$ and is composed of a glucose molecule and a fructose molecule. Enzymes in yeast convert these sugars to ethanol and carbon dioxide. First, *invertase* catalyzes the hydrolysis of sucrose to glucose and fructose; second, *zymase* converts the hydrolyzed sugars to ethanol and CO_2. Addition of phosphate salts to the nutrient medium increases the rate of fermentation.

The fermentation process produces solutions of only 10–15% ethanol. This is because the end-product ethanol inhibits the action of the enzymes in the yeast. In order to get more concentrated solutions, distillation is necessary. It is a method that is based on differences in boiling points of the components of a mixture. If one component is nonvolatile, or is a liquid that boils much higher or lower than the component desired, a simple distillation is used. (See Experiment 4 for an application of this technique.) Further purification can be done by fractional distillation. Here the liquids to be distilled are heated in a boiling flask and the vapors are passed through a fractioning column. This column allows for many condensations and vaporizations to take place before pure vapors finally enter a cooling condenser. The purified material is collected in a clean container.

The ethanol prepared in this way is a constant boiling mixture of 95% ethanol and 5% water. This constant boiling mixture is called an *azeotrope*. Distillation is unable to concentrate and purify further. Pure, or 100%, ethanol is called *absolute* ethanol and is obtained by adding benzene to the water-ethanol mixture. Benzene-water-ethanol forms a low-boiling azeotrope that distills before pure ethanol. Thus, after this three-component azeotrope distills, a 100% solution of ethanol is left. Treatment with calcium oxide removes the last traces of water, and a final distillation gives pure ethanol.

Proof is the usual designation of the alcohol content of wines and liquors. Proof is twice the percentage of alcohol. Thus, absolute ethanol is 200 proof, and 80 proof vodka has a 40% alcohol content.

OBJECTIVES

1. To demonstrate a fermentation process.
2. To isolate the ethanol produced.
3. To determine the percent composition of the ethanol solution recovered.

PROCEDURE

This is a two-part procedure that is to be carried out in two successive weeks: **Fermentation** the first week and **Distillation** the second week.

Fermentation

1. Weigh out 20.0 g of sucrose and place it into a 250-mL Erlenmeyer flask.
2. Add 100 mL of water and gently shake until all the sucrose has dissolved.

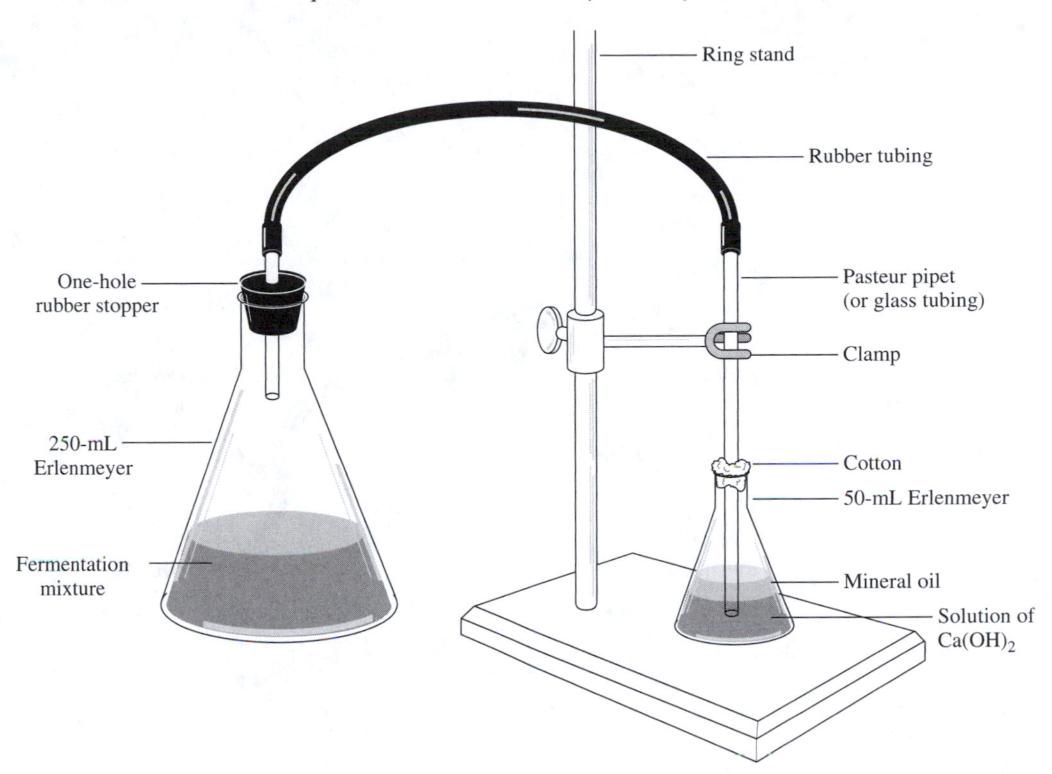

Figure 33.1
Fermentation setup.

3. To this solution add 0.50 g disodium hydrogen phosphate (Na_2HPO_4), 1.0 g potassium phosphate (K_3PO_4), and 2.0 g *dried* baker's yeast. Vigorously shake the contents to mix them thoroughly.
4. Figure 33.1 shows the setup for the fermentation experiment. The Erlenmeyer flask is fitted with a one-hole rubber stopper containing a short piece of glass tubing. Latex tubing (approx. 12 in.) is attached to the glass tubing; at the other end is a short length of glass tubing or a Pasteur pipet. The open end is placed into a 50-mL Erlenmeyer flask (or a test tube [150 × 15 mm]) containing 30 mL of calcium hydroxide ($Ca(OH)_2$) solution; be sure the glass opening is at least 1 cm below the surface of the liquid. Carefully cover the top of the calcium hydroxide solution with 1 cm of mineral (paraffin) oil. Hold the glass tubing vertically in place with a clamp attached to a ring stand; plug the opening of the Erlenmeyer loosely with a small piece of glass wool or cotton. This setup excludes air (and oxygen) from the system (which allows anaerobic oxidation) and prevents further oxidation (by aerobic oxidation) of the ethanol (to acetic acid).
5. Label the fermentation setup with your name. Your laboratory instructor will instruct you where to place your experiment for the week. The fermentation process requires from 5–7 days at 30–35°C to be complete.

Isolation by Fractional Distillation

1. Carefully remove the rubber stopper from the 250-mL Erlenmeyer flask. Do not shake the flask; avoid disturbing the sediment on the bottom.

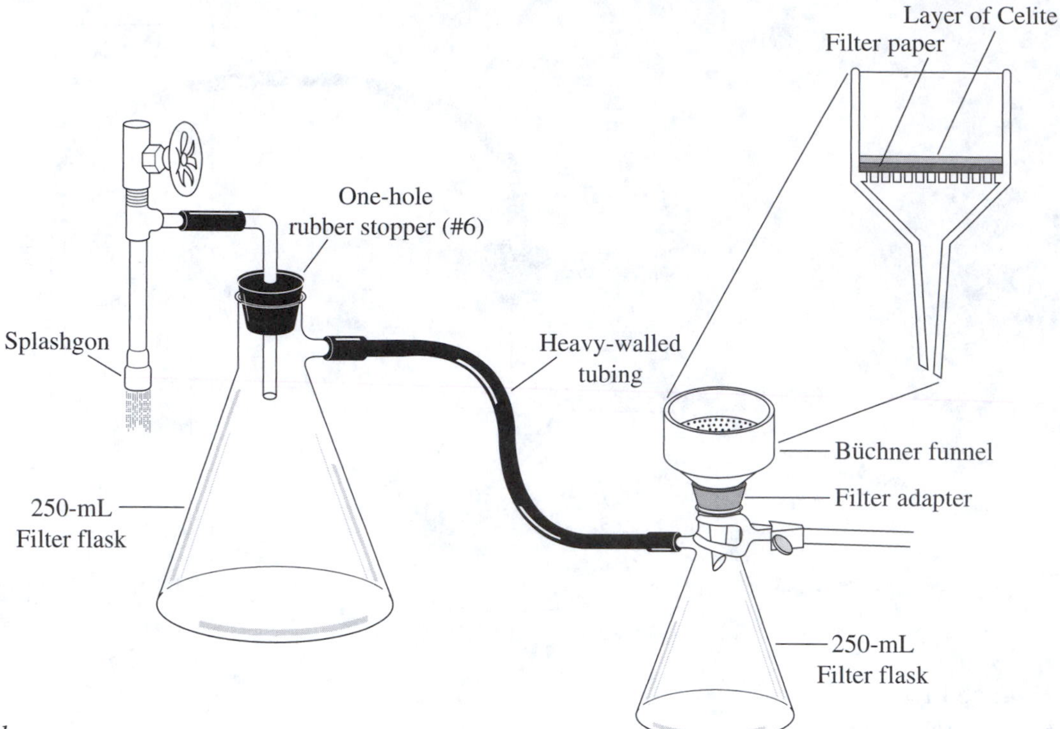

Figure 33.2
Vacuum filtration with a Büchner funnel.

2. Prepare a vacuum filtration assembly using two 250-mL side-arm filter flasks, a 5.5-cm Büchner funnel (with a Filtervac or neoprene adapter), and two lengths (each 12 in.) of vacuum tubing, as assembled in Figure 33.2. Place a piece of filter paper into the Büchner funnel so that it *covers all the holes and lies flat*. Into a 250-mL beaker, place 100 mL of water and one tablespoon of Celite (Filter Aid, which is diatomaceous earth). Stir vigorously and pour the mixture into the Büchner funnel while the water is running and a vacuum is applied. A thin layer of the Celite Filter Aid will form on the filter paper. Discard the water collected in the filter flask.
3. Carefully decant the liquid in the fermentation flask above the sediment through the Celite Filter Aid, using suction. This technique traps the small yeast particles in the Celite Filter Aid but lets through water, ethanol, and any other liquid impurities. This liquid filtrate will be distilled.
4. Obtain a distillation setup and assemble the glassware as shown in Figure 33.3. Use a small dab of silicone grease on all the standard-taper joints as you connect them.
   a. Use a round-bottom distilling flask that will be filled approximately one-half to two-thirds full; a 250-mL round-bottom flask should do. Add 2–3 boiling stones to the flask.
   b. Use a heating mantle for the heat source and a Variac to control the heat.
   c. Pack the fractionating column *loosely*, but uniformly, with soap-free stainless steel cleaning pad material.

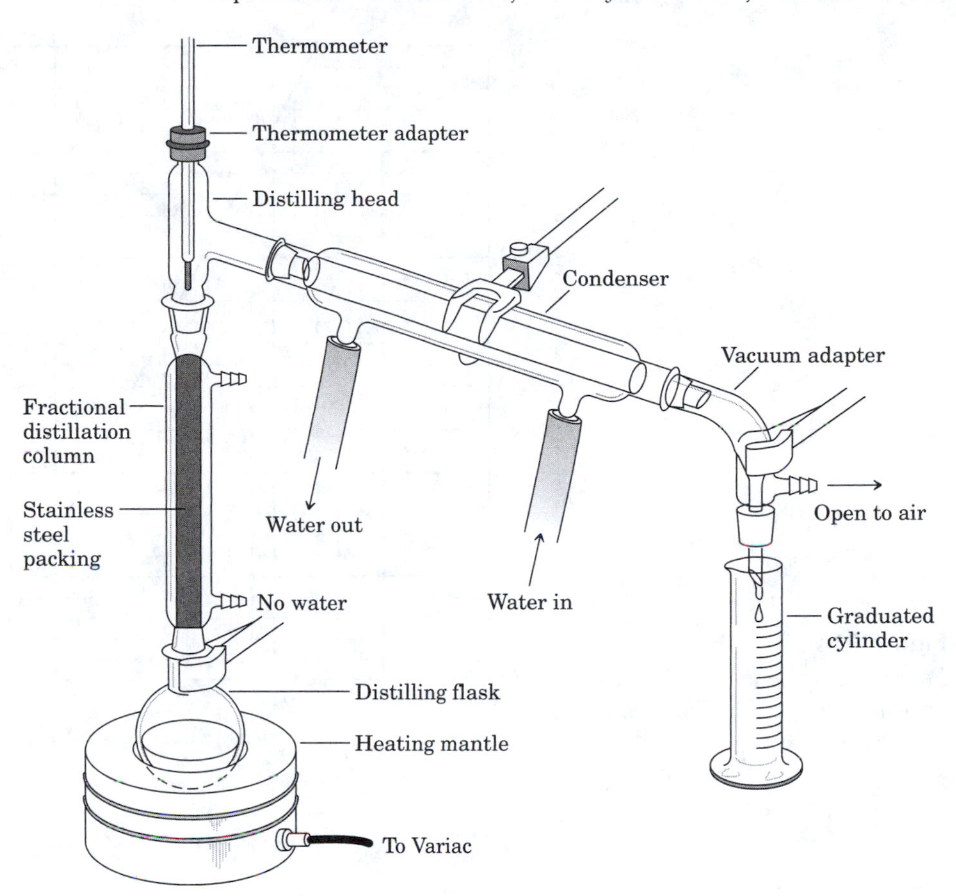

Figure 33.3
Fractional distillation apparatus.

5. Gradually turn up the heat until the liquid in the distillation flask begins to boil. As the vapors rise in the column, you will see liquid condensing; this ring of condensate will rise in the column. Control the setting on the Variac so that the condensate rises slowly through the column and at an even rate. (If the rate is too fast, the column will flood.) The temperature readings at the distillation head will rise; when the temperature reaches 78°C, begin to collect the liquid that distills (1). Discard any liquid distilling before this temperature is reached.
6. Collect liquid distilling between 78 and 90°C. Collect 10 to 15 mL of distillate (2).
7. Turn off the heat source and remove the heating mantle from the distillation flask.
8. Weigh a 50-mL beaker to the nearest 0.001 g (3). With a 10-mL volumetric pipet, transfer 10 mL of distillate to the beaker ($V_{distillate}$). Reweigh the beaker and liquid (5), and by difference, determine the weight of the distillate (6). Determine the density (7), and by referring to Figure 33.4, determine the percent composition of the ethanol (8).
9. Clean-up: Solid waste can be disposed of in solid waste containers; liquid waste can be flushed down the sink with plenty of water.

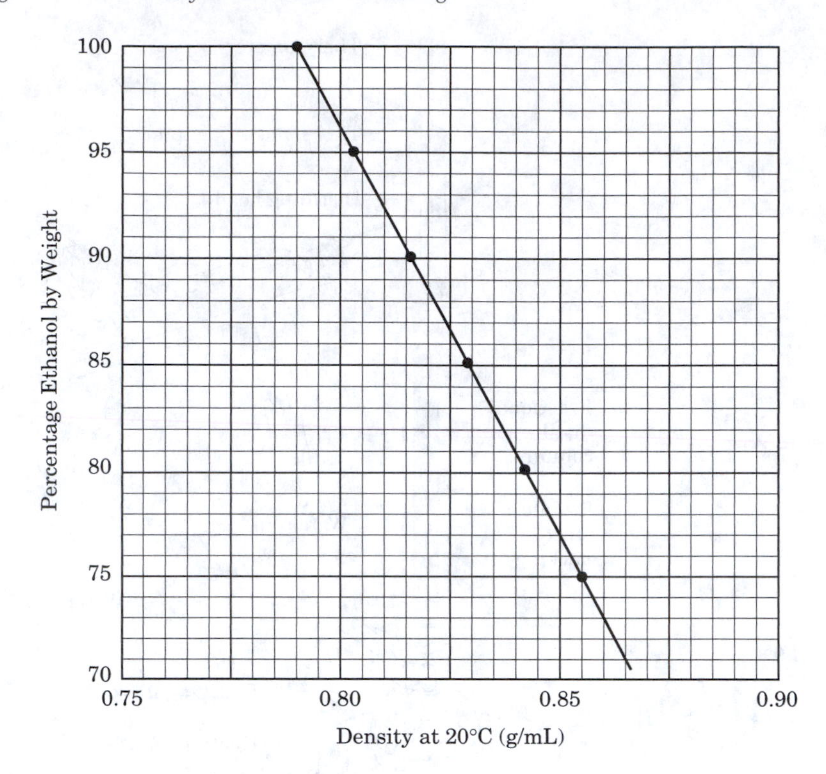

Figure 33.4
Percent composition of aqueous ethanol as a function of density.

CHEMICALS AND EQUIPMENT

1. Calcium hydroxide solution, $Ca(OH)_2$
2. Disodium hydrogen phosphate, Na_2HPO_4
3. Dried baker's yeast
4. Mineral oil
5. Potassium phosphate, K_3PO_4
6. Sucrose
7. Boiling stones
8. Celite (Filter Aid)
9. Clamps
10. Glass wool (or cotton)
11. Heating mantle (or Thermowell)
12. Kit for fractional distillation
13. Ring stands
14. Silicone grease
15. Steel wool, coarse
16. Variac

name

section

date

partner

grade

33 EXPERIMENT 33

Pre-Lab Questions

1. What is the product obtained through *anaerobic* fermentation of a carbohydrate?

2. What must be excluded in order to prevent *aerobic* oxidation?

3. Wine is about 24 proof. What is the percent alcohol in this wine?

4. How can you get a more concentrated solution of alcohol from the fermentation process?

5. Why doesn't the fermentation process produce a concentration of ethanol greater than 10–15%?

6. Give a complete, balanced equation for the conversion of glucose ($C_6H_{12}O_6$) into ethanol.

name

section

date

partner

grade

33 EXPERIMENT 33

Report Sheet

1. Distilling temperature range, °C ____________
2. Volume of distillate collected ____________
3. Weight of 50-mL beaker: W_1 ____________
4. Volume of distillate: $V_{distillate}$ 10.00 mL
5. Weight of beaker and distillate: W_2 ____________
6. Weight of distillate: (5) − (3) ____________
7. Density of distillate
   $d = (W_2 - W_1)/V_{distillate} = [(5) - (3)]/(4)$ ____________
8. Percentage ethanol ____________

Post-Lab Questions

1. What can be used as a starting material for a successful fermentation?

2. What are the catalysts necessary for fermentation to take place? Where are they found?

3. Calcium hydroxide traps a gas formed in the fermentation. What is the gas? What solid forms as a result?

4. In the setup you used (refer to Figure 33.1), the system is essentially closed in order to prevent air from entering. If you forgot to use the trap, would you still get ethanol as the end product? Explain your answer.

5. In a final distillation in the purification of ethanol, only a product of 95% ethanol can be obtained. Explain why this occurs.

6. From the percentage ethanol in your experiment, determine the *proof*.

Preparation and Properties of a Soap

BACKGROUND

A soap is the sodium or potassium salt of a long-chain fatty acid. The fatty acid usually contains 12 to 18 carbon atoms. Solid soaps usually consist of sodium salts of fatty acids, whereas liquid soaps consist of the potassium salts of fatty acids.

A soap such as sodium stearate consists of a nonpolar, hydrophobic end (the hydrocarbon chain of the fatty acid) and a polar, hydrophilic end (the ionic carboxylate).

$$CH_3CH_2CH_2CH_2CH_2CH_2CH_2CH_2CH_2CH_2CH_2CH_2CH_2CH_2CH_2CH_2CH_2{-}\overset{\overset{\displaystyle O}{\|}}{C}{-}O^-Na^+$$

Nonpolar; hydrophobic (Dissolves in oils) — Polar; hydrophilic (Dissolves in water)

Because "like dissolves like," the nonpolar end (hydrophobic, or "water-hating," part) of the soap molecule can dissolve the greasy dirt, and the polar or ionic end (hydrophilic, or "water-loving," part) of the molecule is attracted to water molecules. Therefore, the dirt from the surface being cleaned will be pulled away and suspended in water. Thus soap acts as an *emulsifying agent*, a substance used to disperse one liquid (oil molecules) in the form of finely suspended particles or droplets in another liquid (water molecules).

Treatment of fats or oils with strong bases, such as lye (NaOH) or potash (KOH), causes them to undergo saponification to form glycerol and the salt of a long-chain fatty acid (soap).

$$\begin{array}{l} CH_2-O-\overset{O}{\overset{\|}{C}}-C_{17}H_{35} \\ | \\ CH-O-\overset{O}{\overset{\|}{C}}-C_{17}H_{35} \\ | \\ CH_2-O-\overset{O}{\overset{\|}{C}}-C_{17}H_{35} \end{array} + 3NaOH \xrightarrow{\Delta} \begin{array}{c} CH_2OH \\ | \\ CHOH \\ | \\ CH_2OH \end{array} + 3C_{17}H_{35}\overset{O}{\overset{\|}{C}}-O^-Na^+$$

Tristearin — Glycerol — Sodium stearate (a soap)

Because soaps are salts of strong bases and weak acids, they should be weakly alkaline in aqueous solution. However, a soap with free alkali can cause damage to skin, silk, or wool. Therefore, a test for basicity of the soap is quite important.

Soap has been largely replaced by synthetic detergents during the last two decades, because soap has two serious drawbacks. One is that soap becomes ineffective in hard water. Hard water contains appreciable amounts of Ca^{2+} or Mg^{2+} salts.

$$\underset{\text{Soap}}{2C_{17}H_{35}COO^-Na^+} + M^{2+} \longrightarrow \underset{\text{Scum}}{[C_{17}H_{35}COO^-]_2\,M^{2+}\downarrow} + 2Na^+$$

$$M = (Ca^{2+} \text{ or } Mg^{2+})$$

The other is that, in an acidic solution, soap is converted to free fatty acid and therefore loses its cleansing action.

$$\underset{\text{Soap}}{C_{17}H_{35}COO^-Na^+} + H^+ \longrightarrow \underset{\text{Fatty acid}}{C_{17}H_{35}COOH\downarrow} + Na^+$$

OBJECTIVES

1. To prepare a simple soap.
2. To investigate some properties of a soap.

PROCEDURE

Preparation of a Soap

Measure 23 mL of a vegetable oil into a 250-mL Erlenmeyer flask. Add 20 mL of ethanol (ethyl alcohol) (to act as a solvent) and 20 mL of 25% sodium hydroxide solution (25% NaOH). While stirring the mixture constantly with a glass rod, the flask with its contents is heated gently in a boiling-water bath. A 600-mL beaker containing about 200 mL of tap water and a few boiling stones can serve as a water bath (Figure 34.1).

CAUTION

Alcohol is flammable! No open flames should be in the laboratory.

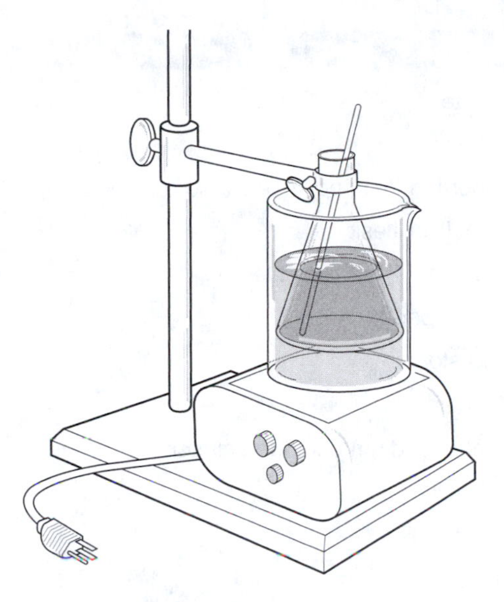

Figure 34.1
Experimental setup for soap preparation.

After heating for about 20 min., the odor of alcohol will disappear, indicating the completion of the reaction. A pasty mass containing a mixture of the soap, glycerol, and excess sodium hydroxide is obtained. Use an ice-water bath to cool the flask with its contents. To precipitate or "salt out" the soap, add 150 mL of a saturated sodium chloride solution to the soap mixture while stirring vigorously. This process increases the density of the aqueous solution; therefore, soap will float out from the aqueous solution. Filter the precipitated soap with the aid of suction (see Figure 33.2) and wash it with 10 mL of ice-cold water. Observe the appearance of your soap and record your observation on the Report Sheet.

Properties of a Soap

1. *Emulsifying properties.* Shake 5 drops of mineral oil in a test tube containing 5 mL of water. A temporary emulsion of tiny oil droplets in water will be formed. Repeat the same test, but this time add a small piece of the soap you have prepared before shaking. Allow both solutions to stand for a short time. Compare the appearance and the relative stabilities of the two emulsions. Record your observations on the Report Sheet.
2. *Hard water reactions.* Place about one-third spatula full of the soap you have prepared in a 50-mL beaker containing 25 mL of water. Warm the beaker with its contents to dissolve the soap. Pour 5 mL of the soap solution into each of five test tubes (nos. 1, 2, 3, 4, and 5). Test no. 1 with 2 drops of a 5% solution of calcium chloride (5% $CaCl_2$), no. 2 with 2 drops of a 5% solution of magnesium chloride (5% $MgCl_2$), no. 3 with 2 drops of a 5% solution of iron(III) chloride (5% $FeCl_3$), and no. 4 with tap water. Tube no. 5 will be used for a basicity test, which will be performed later. Record your observations on the Report Sheet.
3. *Alkalinity (basicity).* Test soap solution no. 5 with a wide-range pH paper. What is the approximate pH of your soap solution? Record your answer on the Report Sheet.

CHEMICALS AND EQUIPMENT

1. Hot plate
2. Ice cubes
3. Büchner funnel in no. 7 one-hole rubber stopper
4. 500-mL filter flask
5. Filter paper, 7 cm diameter
6. pHydrion paper
7. Boiling stones
8. 95% ethanol
9. Saturated sodium chloride solution
10. 25% NaOH
11. Vegetable oil
12. 5% $FeCl_3$
13. 5% $CaCl_2$
14. Mineral oil
15. 5% $MgCl_2$

name | section | date

partner | grade

34 EXPERIMENT 34

Pre-Lab Questions

1. Define the following terms:
   a. Hydrophobic

   b. Hydrophilic

   c. Emulsifying agent

2. The use of soap in water containing Ca^{2+} and Mg^{2+} salts is not very efficient; a "scum" forms. Explain why this occurs.

3. Why is it difficult to use soap in an acid solution?

4. If you were to design a "new" soap, a synthetic detergent, what should it look like?

name | section | date

partner | grade

34 EXPERIMENT 34

Report Sheet

Preparation

Appearance of your soap ____________________

Properties

Emulsifying Properties

Which mixture, oil–water or oil–water–soap, forms a more stable emulsion?

Hard Water Reaction

No. 1 + $CaCl_2$ ____________________

No. 2 + $MgCl_2$ ____________________

No. 3 + $FeCl_3$ ____________________

No. 4 + tap water ____________________

Alkalinity

pH of your soap solution (no. 5) ____________________

Post-Lab Questions

1. Write the chemical equation for the reaction that occurs when $CaCl_2$ is added to your soap solution.

2. What is the advantage of a detergent over a soap?

3. Explain what happens when you "salt out" a solution.

4. Commercial soap is not as basic as your preparation. How could you reduce the pH of your preparation?

5. Stearic acid is insoluble in water, but sodium stearate (a soap) is soluble. What causes the difference in solubility?

Preparation of a Hand Cream

BACKGROUND

Hand creams are formulated to carry out a variety of cosmetic functions. Among these are softening the skin and preventing dryness; elimination of natural waste products (oils) by emulsification; and cooling the skin by radiation, thus helping to maintain body temperature. In addition, hand creams must have certain ingredients that aid spreadability and provide body. In many cases added fragrance improves the odor, and in some special cases medications combat assorted ills.

The basic hand cream formulations all contain water to provide moisture and lanolin, which helps its absorption by the skin. The latter is a yellowish wax. Chemically, wax is made of esters of long-chain fatty acids and long-chain alcohols. Lanolin is usually obtained from sheep wool; it has the ability to absorb 25–30% of its own weight of water and to form a fine emulsion. Mineral oil, which consists of high-molecular-weight hydrocarbons, provides spreadability. In order to allow nonpolar substances, such as lanolin and mineral oil, to be uniformly dispersed in a polar medium, water, one needs strong emulsifying agents. An emulsifying agent must have nonpolar, hydrophobic portions to interact with the oil and also polar, hydrophilic portions to interact with water. A mixture of stearic acid and triethanolamine, through acid–base reaction, yields the salt that has the requirements to act as an emulsifying agent.

Besides the above five basic ingredients, some hand creams also contain alcohols, such as propylene glycol (1,2-propanediol), and esters, such as methyl stearate, to provide the desired texture of the hand cream.

In this experiment you will prepare four hand creams using the combination of ingredients as shown in Table 35.1.

OBJECTIVES

1. To learn the method of preparing a hand cream.
2. To appraise the function of the ingredients in the hand cream.

Table 35.1 *Recipes to Prepare Hand Creams*

Ingredients	*Sample 1*	*Sample 2*	*Sample 3*	*Sample 4*	
Water	25 mL	25 mL	25 mL	25 mL	
Triethanolamine	1 mL	1 mL	1 mL	—	Beaker 1
Propylene glycol	0.5 mL	0.5 mL	—	0.5 mL	
Stearic acid	5 g	5 g	5 g	5 g	
Methyl stearate	0.5 g	0.5 g	—	0.5 g	
Lanolin	4 g	4 g	4 g	4 g	Beaker 2
Mineral oil	5 mL	—	5 mL	5 mL	

PROCEDURE

Preparation of the Hand Creams

For each sample in Table 35.1, assemble the ingredients in two beakers. Beaker 1 contains the polar ingredients, and beaker 2 contains the nonpolar contents.

1. To prepare sample 1, put the nonpolar ingredients in a 50-mL beaker (beaker 2) and heat it in a water bath. The water bath can be a 400-mL beaker half-filled with tap water and heated with a Bunsen burner (Figure 35.1). Carefully hold the beaker with crucible tongs in the boiling water until all ingredients melt. As an alternative, use hot plates instead of a Bunsen burner.
2. In the same water bath, heat the 100-mL beaker (beaker 1) containing the polar ingredients for about 5 min. Remove the beaker and set it on the bench top.
3. Into the 100-mL beaker containing polar ingredients, slowly pour the contents of the 50-mL beaker that holds the molten nonpolar ingredients (Figure 35.2). Stir the mixture for 5 min. until you have a smooth uniform paste.
4. Repeat the same procedure in preparing the other three samples.

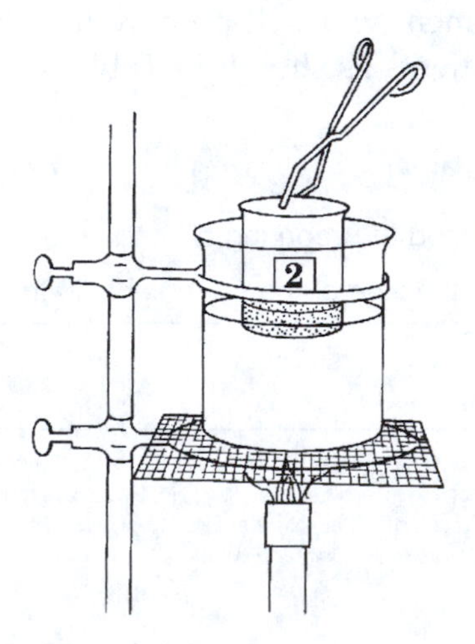

Figure 35.1
Heating ingredients.

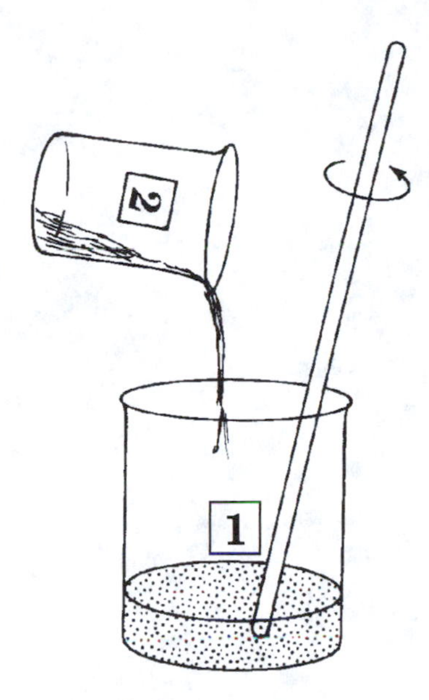

Figure 35.2
Mixing hand cream ingredients.

Characterization of the Hand Cream Preparations

1. Test the pH of the hand creams prepared using a wide-range pH paper.
2. Rubbing a small amount of the hand cream between your fingers, test for smoothness and homogeneity. Also note the appearance. Record your observations on the Report Sheet.
3. Dispose of your hand cream preparations in the waste containers provided. **DO NOT** place them in the sink.

CHEMICALS AND EQUIPMENT

1. Bunsen burner
2. Lanolin
3. Stearic acid
4. Methyl stearate
5. Mineral oil
6. Triethanolamine
7. Propylene glycol
8. pHydrion paper

name | section | date

partner | grade

35

EXPERIMENT 35

Pre-Lab Questions

1. Write the structural formula of propylene glycol, 1,2-propanediol. Why is it classed as an alcohol? Is it a polar molecule?

2. An emulsion agent was used in the hand cream. Explain its purpose.

3. How is the structure of the emulsion agent similar to that of a soap?

4. Some hand creams use methyl stearate in their formulation. Why is it there? What functional group is present in the molecule? Write the structure of methyl stearate. Can it be effective as an emulsifying agent?

name | section | date

partner | grade

35 EXPERIMENT 35

Report Sheet

Characterization of the hand cream samples

Properties	*Sample 1*	*Sample 2*	*Sample 3*	*Sample 4*
pH				
Smoothness				
Homogeneity				
Appearance				

Post-Lab Questions

1. In comparing the properties of the hand creams you produced, ascertain the function of each of the missing ingredients in the hand cream:

   a. Mineral oil

   b. Triethanolamine

   c. Methyl stearate and propylene glycol

2. The emulsion agent, a salt, was prepared from two chemicals.

   a. Write the structure of stearic acid.

   b. Write the structure of triethanolamine.

   c. Write the structure of the salt.

   d. In the above salt, label the hydrophobic part of the salt and the hydrophilic part.

3. What evidence do you have from your preparations that suggests that the emulsifying agent is a necessary component of hand creams?

4. Was the pH of all your samples of hand cream the same? Which one differed? Explain why there was a difference.

5. Two ingredients are necessary in order for a substance to be called a hand cream. What are they and why are they included?

Extraction and Identification of Fatty Acids from Corn Oil

BACKGROUND

Fats are esters of glycerol and fatty acids. Liquid fats are often called oils. Whether a fat is solid or liquid depends on the nature of the fatty acids. Solid animal fats contain mostly saturated fatty acids, while vegetable oils contain high amounts of unsaturated fatty acids. To avoid arteriosclerosis, hardening of the arteries, diets low in saturated fatty acids as well as in cholesterol are recommended.

Note that even solid fats contain some unsaturated fatty acids, and oils contain saturated fatty acids as well. Besides the degree of unsaturation, the length of the fatty acid chain also influences whether a fat is solid or liquid. Short-chain fatty acids, such as are found in coconut oil, convey liquid consistency in spite of the low unsaturated fatty acid content. Two of the unsaturated fatty acids, linoleic and linolenic acids, are essential fatty acids because the body cannot synthesize them from precursors; they must be included in the diet.

The four unsaturated fatty acids most frequently found in vegetable oils are:

Oleic acid: $CH_3(CH_2)_7CH{=}CH(CH_2)_7COOH$

Linoleic acid: $CH_3(CH_2)_4CH{=}CHCH_2CH{=}CH(CH_2)_7COOH$

Linolenic acid: $CH_3CH_2CH{=}CHCH_2CH{=}CHCH_2CH{=}CH(CH_2)_7COOH$

Arachidonic acid:
$CH_3(CH_2)_4CH{=}CHCH_2CH{=}CHCH_2CH{=}CHCH_2CH{=}CH(CH_2)_3COOH$

All the C=C double bonds in the unsaturated fatty acids are *cis*- double bonds, which interrupt the regular packing of the aliphatic chains, and thereby convey a liquid consistency at room temperature. This physical property of the unsaturated fatty acid is carried over to the physical properties of triglycerides (oils).

To extract and isolate fatty acids from corn oil, first the ester linkages must be broken. This is achieved in the saponification reaction, in which a

triglyceride is converted to glycerol and the potassium salt of its fatty acids:

$$\begin{array}{l} CH_2-O-\overset{O}{\overset{\|}{C}}-C_{17}H_{35} \\ | \\ CH-O-\overset{O}{\overset{\|}{C}}-C_{17}H_{35} \\ | \\ CH_2-O-\overset{O}{\overset{\|}{C}}-C_{17}H_{35} \end{array} + 3KOH \longrightarrow \begin{array}{c} CH_2OH \\ | \\ CHOH \\ | \\ CH_2OH \end{array} + 3C_{17}H_{35}\overset{O}{\overset{\|}{C}}-O^-K^+$$

Triglyceride — Glycerol — Potassium stearate

In order to separate the potassium salts of fatty acids from glycerol, the products of the saponification mixture must be acidified. Subsequently, the fatty acids can be extracted by petroleum ether. To identify the fatty acids that were isolated, they must be converted to their respective methyl ester by a perchloric acid–catalyzed reaction:

$$C_{17}H_{35}COOH + CH_3OH \xrightarrow{HClO_4} C_{17}H_{35}\overset{O}{\overset{\|}{C}}-O-CH_3 + H_2O$$

The methyl esters of fatty acids can be separated by thin-layer chromatography (TLC). They can be identified by comparison of their rate of migration (R_f values) to the R_f values of authentic samples of methyl esters of different fatty acids (Figure 36.1).

R_f = distance traveled by fatty acid/distance traveled by the solvent front.

OBJECTIVES

1. To extract fatty acids from neutral fats.
2. To convert them to their methyl esters.
3. To identify them by thin-layer chromatography.

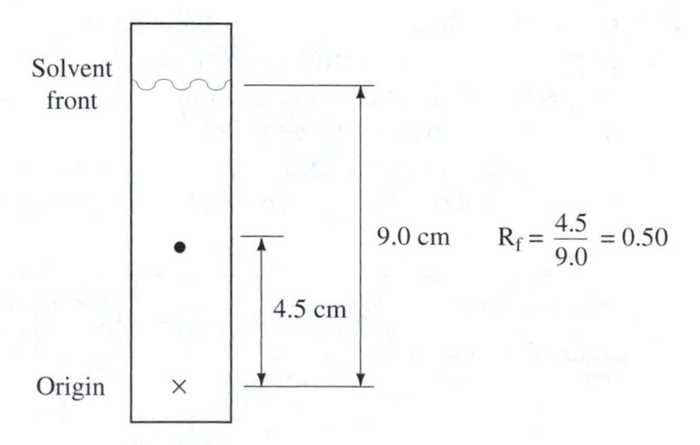

Figure 36.1
TLC chromatogram.

PROCEDURE

Extraction of Fatty Acids

1. Weigh a 50-mL Erlenmeyer flask and record the weight on your Report Sheet (1).
2. Add 2 mL of corn oil and weigh it again. Record the weight on your Report Sheet (2).
3. Add 5 mL of 0.5 M KOH in ethanol to the Erlenmeyer flask. Stopper it. Place the flask in a water bath at 55°C for 20 min.

CAUTION

Strong acid; use gloves when using the concentrated HCl.

4. When the saponification is completed, add 2.5 mL of concentrated HCl. Mix it by swirling the Erlenmeyer flask. Transfer the contents into a 50-mL separatory funnel. Add 5 mL of petroleum ether. Mix it thoroughly (see Figure 31.1). Drain the lower aqueous layer into a flask and the upper petroleum ether layer into a glass-stoppered test tube. Repeat the process by adding back the aqueous layer into the separatory funnel and extracting it with another portion of 5 mL of petroleum ether. Combine the extracts together in the glass-stoppered test tube.

Preparation of Methyl Esters

1. Place a plug of glass wool (the size of a pea) into the upper stem of a funnel, fitting it loosely. Add 10 g of anhydrous Na_2SO_4. Rinse the salt on to the glass wool with 5 mL of petroleum ether; discard the wash. Pour the combined petroleum ether extracts into the funnel and collect the filtrate in an evaporating dish. Add another portion (2 mL) of petroleum ether to the funnel and collect this wash, also in the evaporating dish.
2. Evaporate the petroleum ether under the hood by placing the evaporating dish in a water bath at 60°C. (Alternatively, if dry N_2 gas is available, the evaporation could be achieved by bubbling nitrogen through the extract. This also must be done under the hood.)
3. When dry, add 10 mL of CH_3OH:$HClO_4$ mixture (95:5). Place the evaporating dish in the water bath at 55°C for 10 min.

Identification of Fatty Acids

1. Transfer the methyl esters prepared above into a separatory funnel. Extract twice with 5 mL of petroleum ether. Combine the extracts.
2. Prepare another funnel with anhydrous Na_2SO_4 on top of the glass wool. Filter the combined petroleum ether extracts through the salt into a dry, clean evaporating dish. Evaporate the petroleum ether in the water bath at 60°C, as before. When dry, add 0.2 mL of petroleum ether and transfer the solution to a clean and dry test tube.
3. Take a 15 × 6.5-cm thin-layer chromatography (TLC) plate. Make sure you do not touch the face of the TLC plate with your fingers. Preferably use plastic gloves, or handle the plate by holding it only at the edges. This precaution must be observed throughout the operation because your fingers may contaminate the sample. With a pencil,

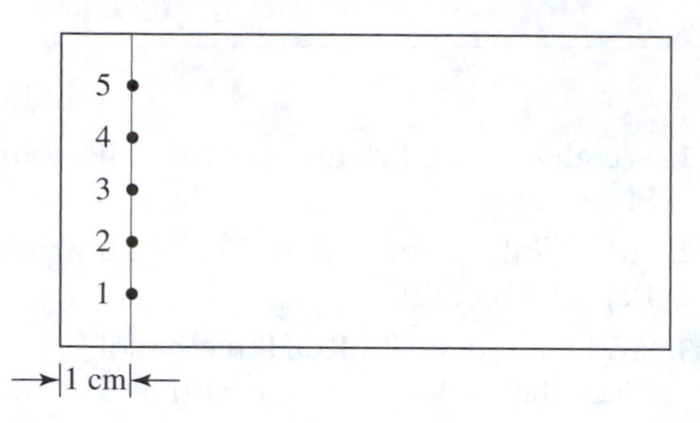

Figure 36.2
Spotting.

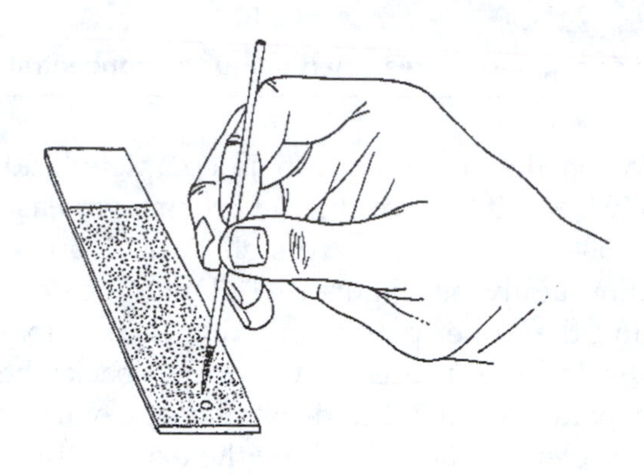

Figure 36.3
Spotting the thin-layer chromatography plate.

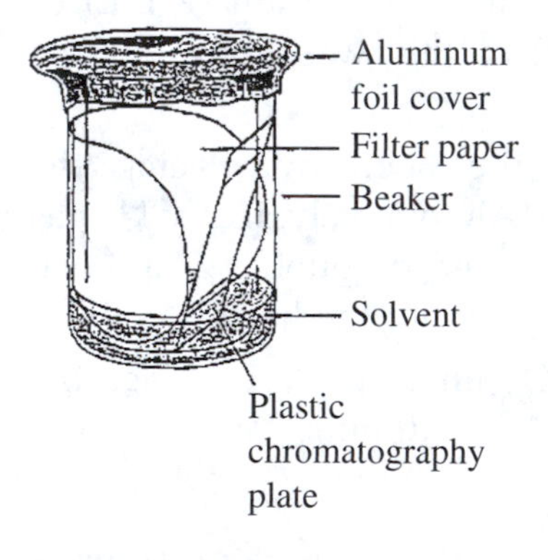

Figure 36.4
Chamber for developing the thin-layer chromatography plate.

lightly draw a line parallel to the 6.5-cm edge about 1 cm from the edge. Mark the positions of the five spots, equally spaced, where you will spot your samples (Figure 36.2).

4. For spots no. 1 and no. 5, use your isolated methyl esters obtained from corn oil. For spot no. 2, use methyl oleate; for spot no. 3, methyl linoleate; and for spot no. 4, methyl palmitate. Use a separate capillary tube for each sample. In spotting, apply each sample in the capillary by gently touching the plate until it spreads to a spot of 1 mm diameter (Figure 36.3). Dry the spots with a heat lamp. Pour about 15 mL of solvent (hexane:diethyl ether; 4:1) into a 500-mL beaker. Place the spotted TLC plate diagonally for ascending chromatography. Make certain that the spots applied are **above** the surface of the eluting solvent. Cover the beaker lightly with aluminum foil to avoid excessive solvent evaporation (Figure 36.4).
5. When the solvent front has risen to about 1–2 cm from the top edge, remove the plate from the beaker. Mark the advance of the solvent front with a pencil. Dry the plate with a heat lamp under the hood. Place the dried plate in a beaker containing a few iodine crystals. Cover the beaker tightly with aluminum foil. Warm the beaker slightly on a hot plate to enhance the sublimation of the iodine. The iodine vapor will interact with the esters and make the spots visible. After a few minutes' exposure to the iodine vapor, remove the chromatogram from the beaker and mark the spots **immediately** with a pencil. The spots may fade with exposure to air.

CAUTION

Do this in the hood. Iodine vapor is toxic. Do not remove the aluminum foil cover until after you have removed the beaker from the hot plate and the beaker has cooled to room temperature.

6. Record the distance the solvent front advanced on your Report Sheet to the nearest 0.5 mm (4). Record on your Report Sheet (5, 6, 7, 8, 9) the distance to the center of each iodine-stained spot from its origin to the nearest 0.5 mm. Calculate the R_f values of your samples (10, 11, 12, 13, 14).

CHEMICALS AND EQUIPMENT

1. Corn oil
2. Methyl palmitate
3. Methyl oleate
4. Methyl linoleate
5. Petroleum ether (b.p. 30–60°C)
6. 0.5 M KOH in ethanol
7. Concentrated HCl
8. Anhydrous Na_2SO_4
9. Methanol:perchloric acid mixture (95:5)
10. Hexane:diethyl ether mixture (4:1)
11. Iodine crystals, I_2
12. Aluminum foil
13. Polyethylene gloves
14. 15 × 6.5-cm silica gel TLC plate
15. Capillary tubes open on both ends
16. Heat lamp
17. Water bath
18. Ruler
19. Glass wool
20. Glass-stoppered test tube

name | section | date

partner | grade

Pre-Lab Questions

1. What is the stereochemistry of the double bonds in unsaturated fatty acids?

2. Does the stereochemistry of the double bonds have anything to do with the liquid consistency of fatty acids? Explain.

3. Fatty acids are soluble in solvents such as petroleum ether, but the salts of fatty acids are not. Explain this observation.

4. What is taking place in the saponification of the fats in corn oil?

5. Write the equation for the conversion of the fatty acid, oleic acid, into its methyl ester.

name ______ section ______ date ______

partner ______ grade ______

36 EXPERIMENT 36

Report Sheet

1. Weight of beaker ______ g
2. Weight of beaker and oil ______ g
3. Weight of oil: (2) − (1) ______ g

Distances on the chromatogram in mm

4. The solvent front ______ mm
5. Spot no. 1 a, b, c, d, e a______ b______ c______ d______ e______ mm
6. Spot no. 2 ______ mm
7. Spot no. 3 ______ mm
8. Spot no. 4 ______ mm
9. Spot no. 5 a, b, c, d, e a______ b______ c______ d______ e______ mm

Calculated R_f values

10. For spot no. 1 [(5)/(4)] a, b, c, d, e a______ b______ c______ d______ e______
11. For spot no. 2 [(6)/(4)] ______
12. For spot no. 3 [(7)/(4)] ______
13. For spot no. 4 [(8)/(4)] ______
14. For spot no. 5 [(9)/(4)] a, b, c, d, e a______ b______ c______ d______ e______
15. How many fatty acids were present in your corn oil?

16. How many fatty acids could you identify? Name the identifiable fatty acids in the corn oil.

Post-Lab Questions

1. The extraction of the fatty acids initially involves a *saponification*. What is happening here?

2. The preparation of the methyl esters is an esterification reaction. Because it is an equilibrium reaction, consider Le Chatelier's Principle and propose two ways that you could improve the yield of the methyl esters of the fatty acids.

3. Anhydrous sodium sulfate, Na_2SO_4, was used in the preparation procedure. Why was this reagent used?

4. The unsaturated fatty acid esters spotted most strongly with the iodine vapor. Explain this observation.

5. Which of your methyl ester samples moved fastest on the TLC plate? What factor accounts for this observation? (*Hint:* Look at chain length.)

Analysis of Lipids

BACKGROUND

Lipids are chemically heterogeneous mixtures. The only common property they have is their insolubility in water. We can test for the presence of various lipids by analyzing their chemical constituents. Foods contain a variety of lipids; most important among them are fats, complex lipids, and steroids. Fats are triglycerides, esters of fatty acids and glycerol. Complex lipids also contain fatty acids, but their alcohol may be either glycerol or sphingosine. They also contain other constituents such as phosphate, choline, or ethanolamine or mono- to oligosaccharides. An important representative of this group is lecithin, a glycerophospholipid, containing fatty acids, glycerol, phosphate, and choline. The most important steroid in foods is cholesterol. Different foods contain different proportions of these three groups of lipids.

Structurally, cholesterol contains the steroid nucleus that is the common core of all steroids.

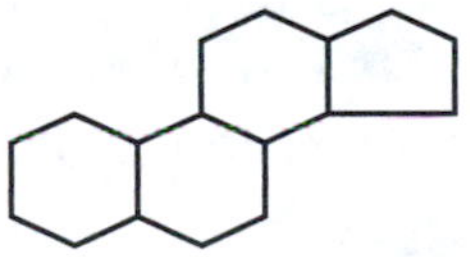

Steroid nucleus

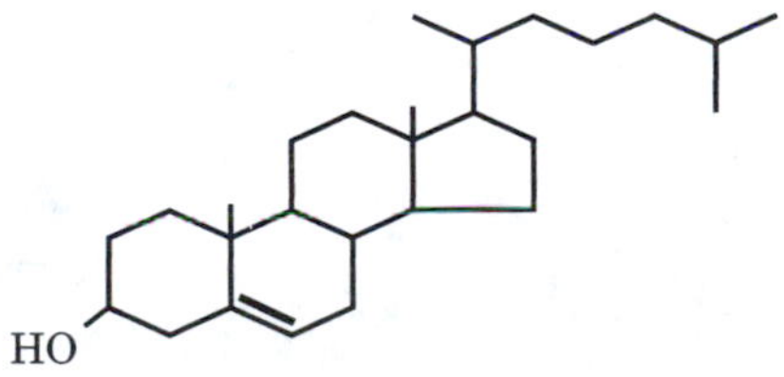

HO

Cholesterol

There is a special colorimetric test, the Lieberman-Burchard reaction, which uses acetic anhydride and sulfuric acid as reagents, that gives a characteristic green color in the presence of cholesterol. This color is due to the —OH group of cholesterol and the unsaturation found in the adjacent fused ring. The color change is gradual: first it appears as a pink coloration, changing later to lilac, and finally to deep green.

When lecithin is hydrolyzed in acidic medium, both the fatty acid ester bonds and the phosphate ester bonds are broken and free fatty acids and inorganic phosphate are released. Using a molybdate test, we can detect the presence of phosphate in the hydrolysate by the appearance of a purple color. Although this test is not specific for lecithin (other phosphate-containing lipids will give a positive molybdate test), it differentiates clearly between fat and cholesterol on the one hand (negative test), and phospholipid on the other (positive test).

$$\begin{array}{l} H_2C-O-C(=O)-(CH_2)_nCH_3 \\ HC-O-C(=O)-(CH_2)_nCH_3 \\ H_2C-O-P(=O)(O^-)-O-CH_2CH_2N^+(CH_3)_3 \end{array} + 4H_2O \overset{H^+}{\rightleftharpoons} \begin{array}{c} CH_2OH \\ CHOH \\ CH_2OH \end{array} + 2CH_3(CH_2)_nC(=O)-OH + HO-P(=O)(O^-)-OH + HOCH_2CH_2N^+(CH_3)_3$$

A second test that differentiates between cholesterol and lecithin is the acrolein reaction. When lipids containing glycerol are heated in the presence of potassium hydrogen sulfate, the glycerol is dehydrated, forming acrolein, which has an unpleasant odor. Further heating results in polymerization of acrolein, which is indicated by the slight blackening of the reaction mixture. Both the pungent smell and the black color indicate the presence of glycerol, and thereby fat and/or lecithin. Cholesterol gives a negative acrolein test.

$$\begin{array}{c} CH_2OH \\ | \\ CHOH \\ | \\ CH_2OH \end{array} \xrightarrow{\Delta} \begin{array}{c} O \\ \| \\ C-H \\ | \\ CH \\ \| \\ CH_2 \end{array} + 2H_2O$$

OBJECTIVE

To investigate the lipid composition of common foods such as corn oil, butter, and egg yolk.

PROCEDURE

Each test will require the following samples: (1) pure cholesterol, (2) pure glycerol, (3) lecithin preparation, (4) corn oil, (5) butter, (6) egg yolk.

Phosphate Test

CAUTION

6 M nitric acid is a strong acid. Handle it with care. Use gloves.

1. Take six clean and dry test tubes (150 × 18 mm). Label them 1 through 6. Add about 0.2 g of sample to each test tube. Hydrolyze the compounds by adding 3 mL of 6 M nitric acid to each test tube.
2. Prepare a water bath by boiling about 100 mL of tap water in a 250-mL beaker on a hot plate. Place the test tubes in the boiling-water bath for 5 min. Do not inhale the vapors. Cool the test tubes. Neutralize the acid by adding 3 mL of 6 M NaOH. Mix. During the hydrolysis, a

precipitate may form, especially in the egg yolk sample. The samples in which a precipitate appeared must be filtered. Place a piece of cheese cloth on top of a 25-mL Erlenmeyer flask. Pour the turbid hydrolysate in the test tube on the cheese cloth and collect the liquid filtering through it.

3. Transfer 2 mL of each neutralized (and filtered) sample into clean and labeled test tubes (150 × 18 mm). Add 3 mL of a molybdate solution to each test tube and mix the contents. Heat the test tubes in a boiling-water bath for 5 min. Cool them to room temperature.

CAUTION

The molybdate reagent contains sulfuric acid. Wear gloves and avoid spills when dispensing this reagent.

4. Add 0.5 mL of an ascorbic acid solution to each test tube and mix the contents thoroughly. Wait 20 min. for the development of the purple color. Record your observations on the Report Sheet. While you are waiting, you can perform the rest of the colorimetric tests.

The Acrolein Test for Glycerol

1. Place 1 g of potassium hydrogen sulfate, $KHSO_4$, in each of seven clean and dry test tubes (100 × 13 mm). Label them 1 through 7. Add a few grains of pure lecithin and cholesterol to two of the test tubes. Add a drop, about 0.1 g, of glycerol, corn oil, butter, and egg yolk to the other four test tubes. To the seventh test tube add a few crystals of sucrose.
2. Set up your Bunsen burner in the hood.

CAUTION

Do this test in the hood. It is important for this test to be done under the hood because of the pungent odor of the acrolein. When asked to smell the test tubes, do not inhale the fumes directly. Smell the test tubes by moving them sideways under your nose or by wafting, or waving, the vapors toward your nose with a cupped hand.

3. Gently heat each test tube, one at a time, over a Bunsen burner flame, shaking it continuously from side to side. When the mixture melts it slightly blackens, and you will notice the evolution of fumes. Stop the heating. Carefully smell the test tubes; pay attention to the **CAUTION**. A pungent odor, resembling burnt hamburgers, is a positive test for glycerol. Sucrose in the seventh test tube also will be dehydrated and will give a black color. However, its smell is different, and thus is not a positive test for acrolein. Do not overheat the test tubes, for the residue will become hard, making it difficult to clean the test tubes. Record your observations on the Report Sheet.

Lieberman-Burchard Test for Cholesterol

1. Take six clean and dry test tubes (150 × 18 mm). Label them 1 through 6. Place a few grains of your cholesterol and lecithin in two of the test tubes. Similarly, add about 0.1-g samples of glycerol, corn oil, butter, and egg yolk to the other four test tubes.

CAUTION

Do this test in the hood. Wear gloves and avoid spills. Acetic anhydride has a very strong, pungent odor. Carefully handle the concentrated sulfuric acid; any spill on your person must be flushed with water for at least 10 min.

2. Transfer 3 mL of chloroform and 1 mL of acetic anhydride to each test tube. Finally, add 1 drop of concentrated sulfuric acid to each mixture. Mix the contents and record the color changes, if any. Wait 5 min. Record again the color of your solutions. Record your observations on the Report Sheet.

CHEMICALS AND EQUIPMENT

1. 6 M NaOH
2. 6 M HNO_3
3. Molybdate reagent
4. Ascorbic acid solution
5. $KHSO_4$
6. Chloroform
7. Acetic anhydride
8. Sulfuric acid, H_2SO_4
9. Cholesterol
10. Lecithin
11. Glycerol
12. Corn oil
13. Butter
14. Egg yolk
15. Hot plate
16. Cheese cloth
17. Sucrose

name section date

partner grade

37 EXPERIMENT 37

Pre-Lab Questions

1. Why would lipids be insoluble in water?

2. How do fats and steroids differ?

3. Below are structures of several compounds derived from cholesterol. Circle the steroid nucleus in each.

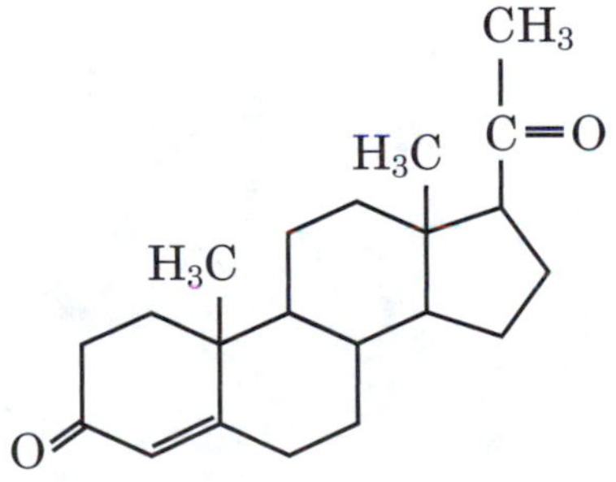

CH_3, C=O, H_3C, H_3C, O — Progesterone

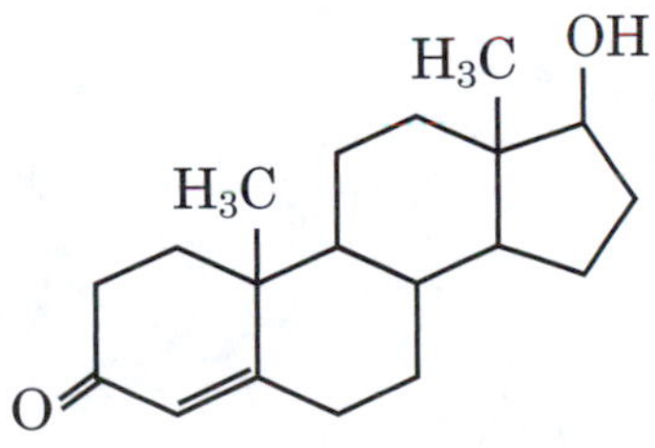

H_3C, OH, H_3C, O — Testosterone

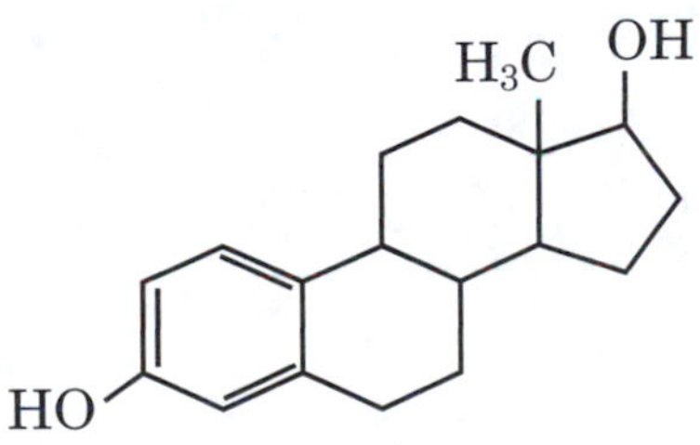

H_3C, OH, HO — Estradiol

4. Would any of the above give a positive Lieberman-Burchard test? Explain your answer.

5. Cholesterol is an alcohol. Alcohols dehydrate by losing the elements of HOH and forming a carbon-to-carbon double bond. Draw the structure cholesterol forms upon dehydration. Would this dehydration compound give a positive Lieberman-Burchard test?

name

section

date

partner

grade

37 EXPERIMENT 37

Report Sheet

Tests	*Cholesterol*	*Lecithin*	*Glycerol*	*Corn Oil*	*Butter*	*Egg Yolk*	*Sucrose*
1. Phosphate **a.** Color							
b. Conclusions							
2. Acrolein **a.** Odor							
b. Color							
c. Conclusions							
3. Lieberman-Burchard **a.** Initial color							
b. Color after 5 min.							
c. Conclusions							

Post-Lab Questions

1. Below are the structures of two complex lipids: (a) a glycolipid; and (b) a sphingolipid. Would either of these compounds give you a positive test with molybdate solution? Explain your answer.

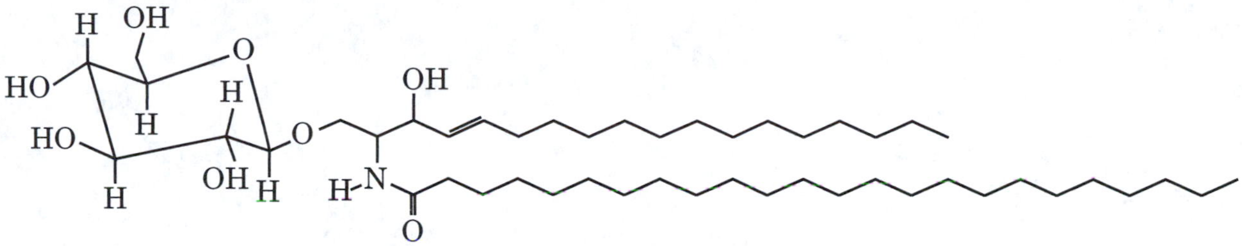

a. Glucocerebroside

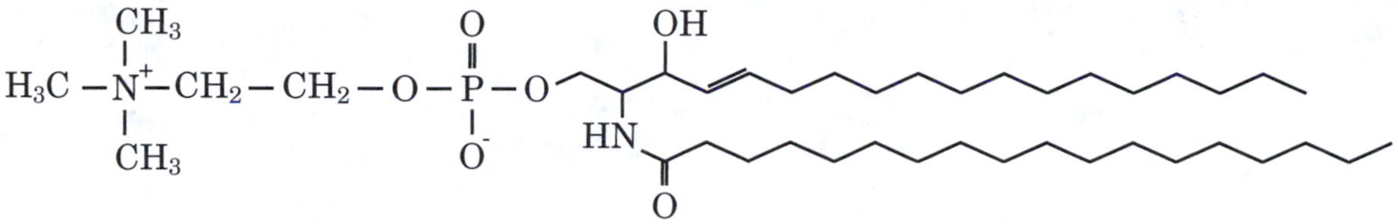

b. Sphingomyelin

2. Samples were initially hydrolyzed with nitric acid before the molybdate test was carried out. Explain why this hydrolysis was done.

3. From your results, what is present in corn oil? Is it a pure triglyceride?

4. Based on the intensity of the color in your tests for cholesterol, which food showed the most cholesterol present? Which food showed the least cholesterol present?

Separation of Amino Acids by Paper Chromatography

BACKGROUND

Amino acids are the building blocks of peptides and proteins. They possess two functional groups—the carboxylic acid group gives the acidic character, and the amino group provides the basic character. The common structure of all amino acids is

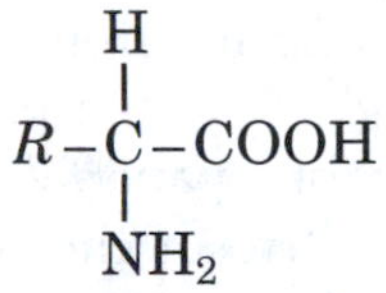

$$R-\underset{\displaystyle NH_2}{\overset{\displaystyle H}{\underset{|}{\overset{|}{C}}}}-COOH$$

The *R* represents the side chain that is different for each of the amino acids that are commonly found in proteins. However, all 20 amino acids have a free carboxylic acid group and a free amino (primary amine) group except proline, which has a cyclic side chain and a secondary amino group.

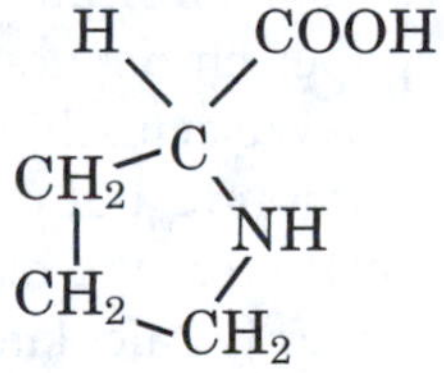

H COOH
C
CH_2 NH
CH_2 CH_2

Proline

We use the properties provided by these groups to characterize the amino acids. The common carboxylic acid and amino groups provide the acid–base nature of the amino acids. The different side chains, and the solubilities provided by these side chains, can be utilized to identify the different amino acids by their rate of migration in paper chromatography.

In this experiment, we use paper chromatography to identify aspartame, an artificial sweetener, and its hydrolysis products from certain foods.

$$HOOC-CH_2-\underset{NH_2}{CH}-\overset{O}{\overset{\|}{C}}-NH-\underset{CH_2-C_6H_5}{CH}-\overset{O}{\overset{\|}{C}}-OCH_3$$

Aspartame

Aspartame is the methyl ester of the dipeptide aspartylphenylalanine. Upon hydrolysis with HCl it yields aspartic acid, phenylalanine, and methanol. When this artificial sweetener was approved by the Food and Drug Administration, opponents of aspartame claimed that it is a health hazard, because aspartame would be hydrolyzed and would yield poisonous methanol in soft drinks that are stored over long periods of time. The Food and Drug Administration ruled, however, that aspartame is sufficiently stable and fit for human consumption. Only a warning must be put on the labels of foods containing aspartame. This warning is for patients suffering from phenylketonurea who cannot tolerate phenylalanine.

To run a paper chromatography, we use Whatman no. 1 chromatography paper. We apply the sample (aspartame or amino acids) as a spot to a strip of chromatography paper. The paper is dipped into a mixture of solvents. The solvent moves up the paper by capillary action and carries the sample with it. Each amino acid may have a different migration rate depending on the solubility of the side chain in the solvent. Amino acids with similar side chains are expected to move with similar, though not identical, rates; those that have quite different side chains are expected to migrate with different velocities. Depending on the solvent system used, almost all amino acids and dipeptides can be separated from each other by paper chromatography.

We actually do not measure the rate of migration of an amino acid or a dipeptide, but rather, how far a particular amino acid travels on the paper relative to the migration of the solvent. This ratio is called the R_f value. In order to calculate the R_f values, one must be able to visualize the position of the amino acid or dipeptide. This is done by spraying the chromatography paper with a ninhydrin solution that reacts with the amino group of the amino acid. A purple color is produced when the paper is heated. (The proline not having a primary amine gives a yellow color with ninhydrin.) For example, if the purple spot of an amino acid appears on the paper 4.5 cm away from the origin and the solvent front migrates 9.0 cm (Figure 38.1), the R_f value for the amino acid is calculated

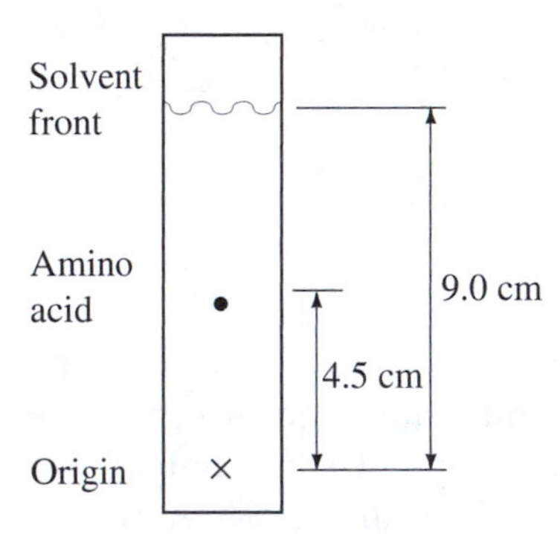

Figure 38.1
Paper chromatogram.

$$R_f = \frac{\textbf{distance traveled by the amino acid}}{\textbf{distance traveled by the solvent front}} = \frac{\textbf{4.5 cm}}{\textbf{9.0 cm}} = \textbf{0.50}$$

In the present experiment you will determine the R_f values of three amino acids: phenylalanine, aspartic acid, and leucine. You will also measure the R_f value of aspartame.

The aspartame you will analyze is actually a commercial sweetener, Equal, made by the NutraSweet Co., that contains silicon dioxide, glucose, cellulose, and calcium phosphate in addition to the aspartame. None of these other ingredients of Equal will give a purple or any other colored spot with ninhydrin. Other generic aspartame sweeteners may contain other nonsweetening ingredients. Occasionally, some sweeteners may contain a small amount of leucine, which can be detected by the ninhydrin test. You will also hydrolyze aspartame using HCl as a catalyst to see if the hydrolysis products will prove that the sweetener is truly aspartame. Finally, you will analyze some commercial soft drinks supplied by your instructor. The analysis of the soft drink can tell you if the aspartame was hydrolyzed at all during the processing and storing of the soft drink.

OBJECTIVES

1. To separate amino acids and a dipeptide by paper chromatography.
2. To identify hydrolysis products of aspartame.
3. To analyze the state of aspartame in soft drinks.

PROCEDURE

Part A

CAUTION

When heating a solution in a test tube, do not point the open end toward anyone. The liquid may boil suddenly and spatter the contents.

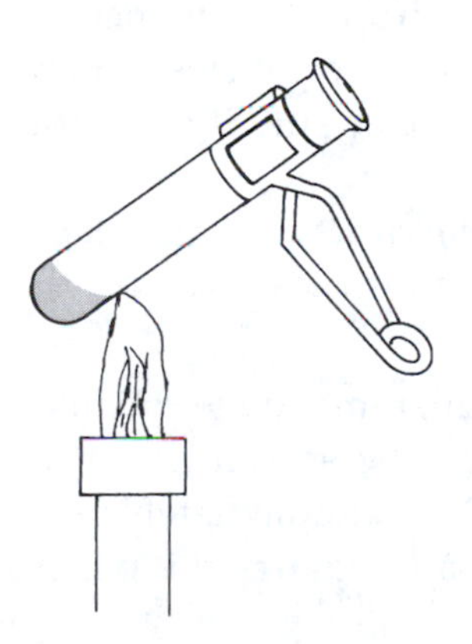

Figure 38.2
Position for holding a test tube in a Bunsen flame.

1. Dissolve 10 mg of the sweetener Equal in 1 mL of 3 M HCl in a test tube (100 × 13 mm). Heat the solution with a Bunsen burner, using a small flame (or with a microburner) to a boil for 30 sec. Do not heat the bottom of the test tube, but heat slightly above the surface level of the solution (Figure 38.2). As the solution boils, do not let the liquid evaporate completely. Set the solution aside to cool; this is the hydrolyzed aspartame.
2. Label six small test tubes (75 × 10 mm) as follows: (1) phenylalanine; (2) aspartic acid; (3) leucine; (4) aspartame; (5) hydrolyzed aspartame; (6) Diet Coca-Cola. Place about 0.5-mL samples in each test tube.
3. Use plastic gloves throughout in order not to contaminate the paper chromatogram. Take a strip of Whatman no. 1 chromatographic paper, 8 × 15 cm and 0.016 cm thick. With a pencil, lightly draw a line parallel to the 8-cm edge 1 cm from the edge. Mark the positions of 6 spots, placed equally, where you will spot your samples (Figure 38.3).

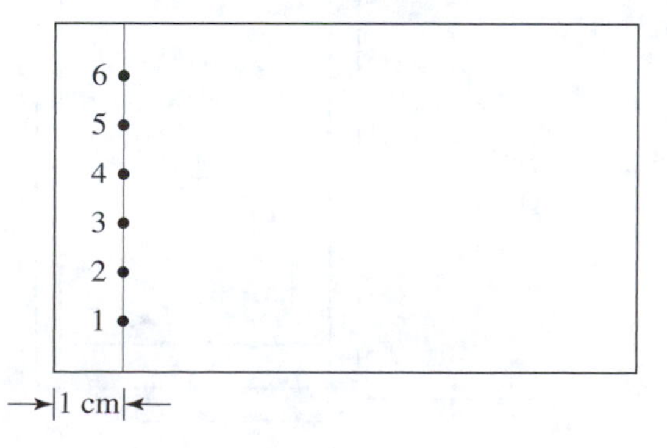

Figure 38.3
Spotting.

4. *Spotting*. For each sample, use a separate capillary tube. Apply a drop of sample to the paper until it spreads to a spot of 1-mm diameter. Dry the spot. (If a heat lamp is available, use it for drying.) Do the spotting in the following order:

1. Phenylalanine	one drop
2. Aspartic acid	one drop
3. Leucine	one drop
4. Aspartame (in Equal)	one drop
5. Hydrolyzed aspartame	six drops
6. Diet Coca-Cola	ten drops

When you do multiple drops of nos. 5 and 6, allow the paper to dry after each drop before applying the next drop. (Avoid putting a hole in the paper!) Do not allow the spots to spread to larger than 1 mm in diameter. (See Figure 36.3, p. 434, for the spotting technique.)

5. Pour about 15 mL of solvent mixture (butanol:acetic acid:water) into a large (1-L) beaker. Place a glass rod over the beaker. Using clear tape, affix the chromatographic paper to the rod so that when you lower the rod over the beaker, the paper will dip into the solvent **but the pencil marks and the spots will be above the solvent surface** (Figure 38.4). Cover the rod and beaker with aluminum foil. Place the beaker on a hot plate. Turn to a low setting (e.g., no. 2 out of 10 on the dial) and heat the beaker to about 35°C. Allow the solvent front to advance at least 6 cm (it will take 40–50 min.), but do not allow it to get closer than 1 cm from the edge. Watch the temperature and make sure it does not rise above 35°C.

   While you are waiting for the solvent front to complete its rise, you can do the short chromatography experiment in **Part B**. *Your instructor may choose to make this optional.*

6. When the solvent front has advanced at least 6 cm, remove the rod and the paper from the beaker. You must not allow the solvent front to advance up to or beyond the edge of the paper. Mark immediately *with a pencil* the position of the solvent front. Under a hood, dry the paper

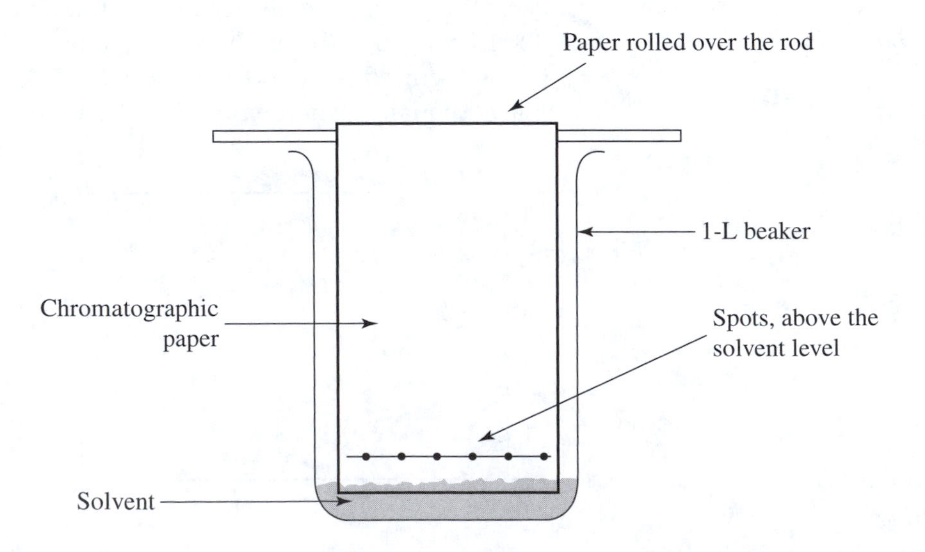

Figure 38.4
Chamber for developing the chromatogram.

with the aid of a heat lamp or hair dryer. With polyethylene gloves on your hands, spray the dry paper with ninhydrin solution. *Be careful not to spray ninhydrin on your hand and do not touch the sprayed areas with bare hands. If the ninhydrin spray touches your skin (which contains amino acids), your fingers will be discolored for a few days.* Place the sprayed paper into a drying oven set at 105–110°C for 2–3 min.

7. Remove the paper from the oven. Mark the center of the spots and calculate the R_f values of each spot. Record your observations on the Report Sheet.
8. If the spots on the chromatogram are faded, you can visualize them by exposing the chromatogram to iodine vapor. Place your chromatogram into a wide-mouth jar (or a large beaker, 1-L size) containing a few iodine crystals. Cap the jar (or cover the beaker with aluminum foil) and *warm it slightly* on a hot plate to enhance the sublimation of iodine. The iodine vapor will interact with the faded pigment spots and make them visible. After a few minutes' exposure to the iodine vapor, remove the chromatogram and mark the spots **immediately** with a pencil. The spots will fade again with exposure to air. Measure the distance of the center of the spots from the origin and calculate the R_f values.

CAUTION

Do this procedure in the hood. Iodine vapor is toxic. Remove the jar (or beaker) from the hot plate and allow the contents to cool to room temperature before you remove the paper.

Part B

1. Take a strip of Whatman no. 1 chromatographic paper, 4 × 12 cm and 0.016 cm thick. With a pencil, lightly draw a line parallel to the 4-cm edge 1 cm from the edge. Mark the positions of 3 spots, placed equally, where you will spot your samples.
2. Obtain three differently colored felt-tip pens: red, blue, and black. On the spots you have marked, touch the paper, leaving a different-colored dot at each place. The spot should be no more than 1 cm in diameter and not too heavily marked.
3. Pour distilled water into a 250-mL beaker to a depth of 1 cm. Place a glass rod over the beaker. Using clear tape, affix the chromatographic paper to the rod so that when you lower the rod over the beaker, the paper will dip into the water **but the pencil marks and the spots will be above the water surface**. Cover the rod and the beaker with aluminum foil.
4. When the water front has advanced to about 2 cm from the top edge, remove the rod and paper from the beaker. Set aside and allow the paper to air dry. Record how many colors you see for each felt-tip pen you used.

CHEMICALS AND EQUIPMENT

1. 0.1% solutions of aspartic acid, phenylalanine, and leucine
2. 0.5% solution of aspartame (Equal)
3. Diet Coca-Cola
4. 3 M HCl
5. 0.2% ninhydrin spray
6. Butanol:acetic acid:water–solvent mixture
7. Equal sweetener
8. Aluminum foil
9. Whatman no. 1 chromatographic paper: 15×8 cm; 12×4 cm
10. Ruler
11. Polyethylene gloves
12. Capillary tubes open on both ends
13. Heat lamp or hair dryer
14. Drying oven, 110°C
15. Wide-mouth jar
16. Iodine crystals
17. Felt-tip pens: red, blue, black

name | section | date

partner | grade

38 EXPERIMENT 38

Pre-Lab Questions

1. Draw the generalized structure of an amino acid. Circle the two functional groups and describe their chemical properties.

2. Aspartame is a dipeptide. Draw and label the two amino acids that make this compound.

3. Write the structure of the mono- and dimethyl ester of aspartic acid.

4. If an amino acid moved 2.0 cm on the chromatogram and the solvent front reached 8.0 cm, what is the R_f value of the amino acid?

5. Why is it necessary to wear gloves when working with ninhydrin spray?

name

section

date

partner

grade

38 EXPERIMENT 38

Report Sheet

1.

Sample	*Distance Traveled (mm)*	*Solvent Front (mm)*	R_f
Phenylalanine			
Aspartic acid			
Leucine			
Aspartame			
Hydrolyzed aspartame			
Diet Coca-Cola			

2. Identification

a. Name the amino acids you found in the hydrolysate of the sweetener Equal.

b. How many spots were stained with ninhydrin (1) in Equal and (2) in Diet Coca-Cola samples?

3. Felt-tip pen colors

Red:

Blue:

Black:

Post-Lab Questions

1. Why must you use a pencil, and not ink, to mark on the paper the origin of the spots of the amino acid preparations?

2. In this experiment, suppose you forgot to mark the position of the solvent front when you tested the hydrolysate of aspartame against the other amino acids.

   a. Is it still possible to determine how many amino acids were present in the hydrolysate? Explain your answer.

   b. Could you still identify what those amino acids were? Explain your answer.

   c. What can you not do?

3. There are polarity and molecular weight differences between aspartic acid and phenylalanine: Aspartic acid has a polar, acidic side chain and a smaller molecular weight; phenylalanine has a nonpolar side chain and a larger molecular weight. Based on the R_f values you obtained for these two amino acids in the solvent employed, which of these two properties influenced the rate of migration?

4. From your experiment, is there any evidence that the aspartame in the Diet Coca-Cola sample showed any hydrolysis into the two amino acids?

Acid–Base Properties of Amino Acids

BACKGROUND

In the body, amino acids exist as zwitterions.

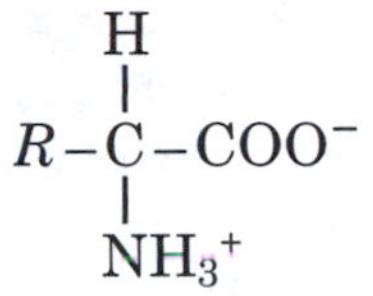

$$R-\underset{\underset{NH_3^+}{|}}{\overset{\overset{H}{|}}{C}}-COO^-$$

This is an amphoteric compound because it behaves as both an acid and a base in the Brønsted definition. As an acid, it can donate an H^+ and becomes the conjugate base:

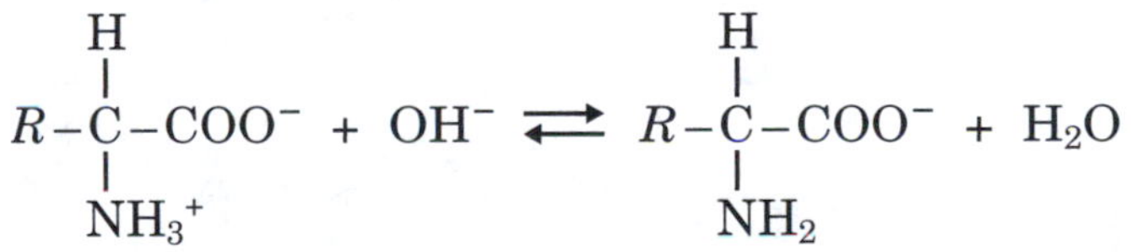

$$\underset{\text{Acid}}{R-\underset{\underset{NH_3^+}{|}}{\overset{\overset{H}{|}}{C}}-COO^-} + \underset{\text{Base}}{OH^-} \rightleftharpoons \underset{\text{Conj. base}}{R-\underset{\underset{NH_2}{|}}{\overset{\overset{H}{|}}{C}}-COO^-} + \underset{\text{Conj. acid}}{H_2O}$$

As a base, it can accept an H^+ ion and becomes the conjugate acid:

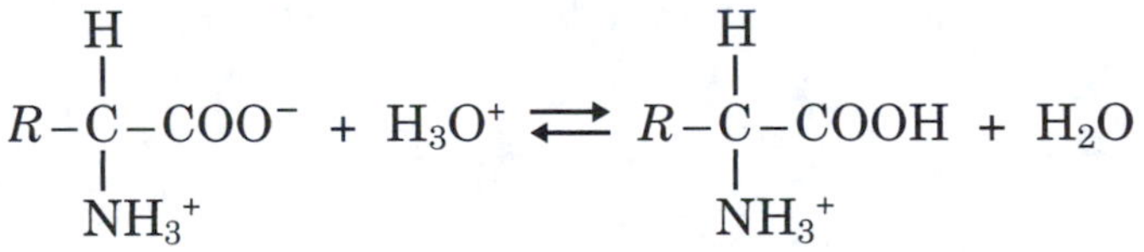

$$\underset{\text{Base}}{R-\underset{\underset{NH_3^+}{|}}{\overset{\overset{H}{|}}{C}}-COO^-} + \underset{\text{Acid}}{H_3O^+} \rightleftharpoons \underset{\text{Conj. acid}}{R-\underset{\underset{NH_3^+}{|}}{\overset{\overset{H}{|}}{C}}-COOH} + \underset{\text{Conj. base}}{H_2O}$$

To study the acid–base properties, one can perform a simple titration. We start our titration with the amino acid being in its acidic form at a low pH:

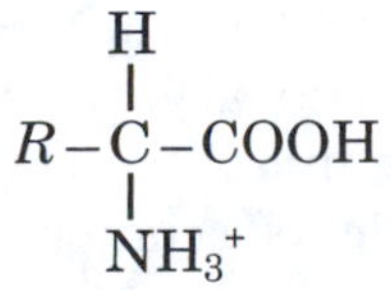

$$R-\underset{\underset{NH_3^+}{|}}{\overset{\overset{H}{|}}{C}}-COOH \quad (I)$$

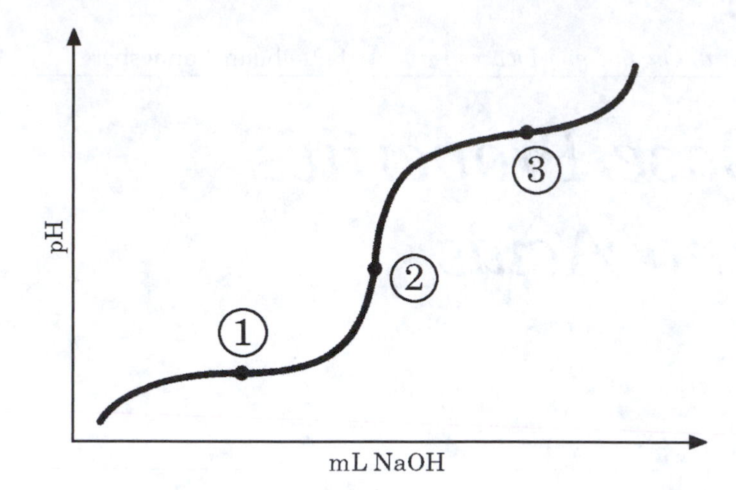

Figure 39.1
The titration curve of an amino acid.

As we add a base, OH^-, to the solution, the pH will rise. We record the pH of the solution by using a pH meter after each addition of the base. To obtain the titration curve, we plot the milliliters of NaOH added against the pH of the solution (Figure 39.1).

Note that there are two flat portions (called legs) on the titration curve where the pH does not increase appreciably with the addition of NaOH. At the midpoint of the first leg, ①, half of the original amino acid is in the acid form (I) and half is in the zwitterion form (II).

$$R-\underset{NH_3^+}{\overset{H}{\underset{|}{\overset{|}{C}}}}-COO^- \quad \text{(II)}$$

The point of inflection, ②, occurs when the amino acid is entirely in the zwitterion form (II). At the midpoint of the second leg, ③, half of the amino acid is in the zwitterion form and half is in the basic form (III).

$$R-\underset{NH_2}{\overset{H}{\underset{|}{\overset{|}{C}}}}-COO^- \quad \text{(III)}$$

From the pH at the midpoint of the first leg we obtain the pK value of the carboxylic acid group, as this is the group that is titrated with NaOH at this stage (the structure going from I to II). The pH of the midpoint of the second leg, ③, is equal to the pK of the $-NH_3^+$, because this is the functional group that donates its H^+ at this stage of the titration. The pH at the inflection point, ②, is equal to the isoelectric point. At the isoelectric point of a compound, the positive and negative charges balance each other. This occurs at the inflection point when all the amino acids are in the zwitterion form.

You will obtain a titration curve of an amino acid with a neutral side chain such as glycine, alanine, phenylalanine, leucine, or valine. If pH meters are available, you can read the pH directly from the instrument after each addition of the base. If a pH meter is not available, you can obtain the pH with the aid of indicator papers. From the titration curve obtained, you can determine the pK values and the isoelectric point.

OBJECTIVES

1. To study acid–base properties by titration.
2. To calculate pK values for the titratable groups.

PROCEDURE

1. With a 20-mL volumetric pipet, pipet 20 mL of 0.1 M amino acid solution (glycine, alanine, phenylalanine, leucine, or valine) that has been acidified with HCl to a pH of 1.5 into a 100-mL beaker. (Be sure to use a Spectroline pipet filler; see Experiment 2 for review of technique.)
2. If a pH meter is available, insert the clean and dry electrode of the pH meter into a standard buffer solution with known pH. Turn the knob of the meter to the pH mark and adjust it to read the pH of the buffer. Turn the knob of the pH meter to "Standby" position. Remove the electrode from the buffer, wash it with distilled water, and dry it. Insert the dry electrode into the amino acid solution. Turn the knob of the meter to "pH" position and record the pH of the solution. Fill a 50-mL buret with 0.25 M NaOH solution. (Refer to Experiment 19 for a review of titration and the use of the buret.) Add the NaOH solution from the buret in 1.0-mL increments to the beaker. After each increment, stir the contents with a glass rod and then read the pH of the solution. Record these on your Report Sheet. Continue the titration as described until you reach pH 12. Turn off your pH meter, wash the electrode with distilled water, wipe it dry, and store it in its original buffer.
3. If a pH meter is not available, perform the titration as above, but use pH indicator papers. After the addition of each increment and stirring, withdraw a drop of the solution with a Pasteur pipet. Touch the end of the pipet to a dry piece of the pH indicator paper. Compare the color of the indicator paper with the color on the charts supplied. Read the corresponding pH from the chart and record it on your Report Sheet.
4. Draw your titration curve. From the graph, determine your pK values and the isoelectric point of your amino acid. Record these on your Report Sheet.

CHEMICALS AND EQUIPMENT

1. 0.1 M amino acid solution (glycine, alanine, leucine, phenylalanine, or valine)
2. 0.25 M NaOH solution
3. pH meter and standard buffer (or pH indicator paper and Pasteur pipet)
4. 50-mL buret
5. 20-mL pipet
6. Spectroline pipet filler

name

section

date

partner

grade

39

EXPERIMENT 39

Pre-Lab Questions

1. Define *zwitterion*.

2. When is an amino acid at its *isoelectric point*?

3. The following is the structure of glycine: $HOOC—CH_2—NH_2$. Write the structure of this amino acid in its (a) basic form; (b) acid form; and (c) zwitterion form.

 a.

 b.

 c.

4. Show how glycine, as its zwitterion, can act as a buffer.

name _______________ section _______ date _______

partner _______________ grade _______

39 EXPERIMENT 39

Report Sheet

1. Amino acid used for titration ____________________

mL of 0.25 M NaOH added	*pH*	*mL of 0.25 M NaOH added*	*pH*
0		13.0	
1.0		14.0	
2.0		15.0	
3.0		16.0	
4.0		17.0	
5.0		18.0	
6.0		19.0	
7.0		20.0	
8.0		21.0	
9.0		22.0	
10.0		23.0	
11.0		24.0	
12.0		25.0	

2. Plot your data below to get the titration curve.

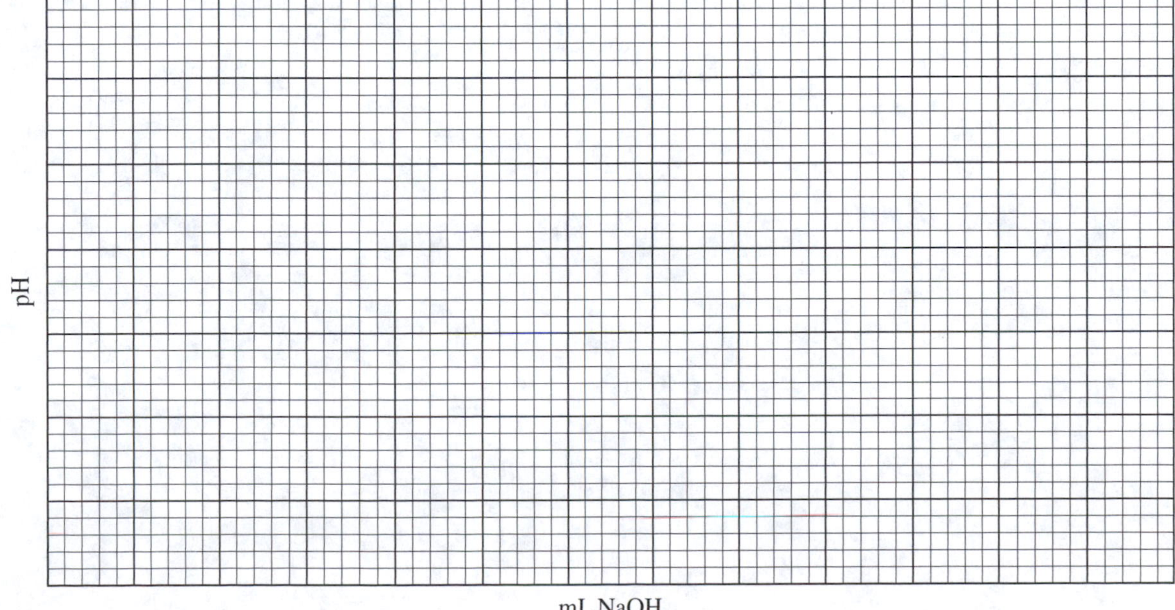

3. a. Indicate the positions of the midpoints of each leg and the position of the inflection point on your graph.

b. Record the pK values for the carboxylic acid group ______________________, and for the amino group ______________________.

c. Record the pH of the isoelectric point ______________________.

Post-Lab Questions

1. Look up the structure in your textbook of the amino acid you titrated. Write the structure you expect at pH (a) 2.5 and (b) 12.

2. Compare your isoelectric point value (pI value) with the value of the amino acid listed in your textbook. How does it compare? Into which class does your amino acid fall?

3. Which data point can you obtain with greater accuracy from your graph—the pK values from the ''legs'' or the isoelectric point from the point of inflection? Explain your answer.

4. From the pK_a value you obtained in your experiment, calculate the equilibrium constant, K_a.

5. You are at the isoelectric point for a solution of leucine. If the number of negatively charged carboxylate groups, $-COO^-$, is 1×10^{19}, how many positively charged amino groups, $-NH_3^+$, would there be?

Isolation and Identification of Casein

BACKGROUND

Casein is the most important protein in milk. It functions as a storage protein, fulfilling nutritional requirements. Casein can be isolated from milk by acidification to bring it to its isoelectric point. At the isoelectric point, the number of positive charges on a protein equals the number of negative charges. Proteins are least soluble in water at their isoelectric points because they tend to aggregate by electrostatic interaction. The positive end of one protein molecule attracts the negative end of another protein molecule, and the aggregates precipitate out of solution.

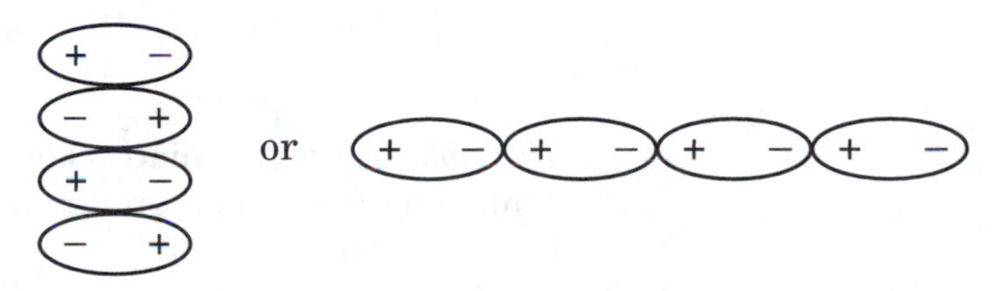

On the other hand, if a protein molecule has a net positive charge (at low pH or acidic condition) or a net negative charge (at high pH or basic condition), its solubility in water is increased.

$$\overset{+}{N}H_3 \sim COOH \xleftarrow[\text{low pH}]{H^+} \overset{+}{N}H_3 \sim COO^- \xrightarrow[\text{high pH}]{OH^-} NH_2 \sim COO^- + H_2O$$

More soluble | Least soluble (at isoelectric pH) | More soluble

In the first part of this experiment, you are going to isolate casein from milk, which has a pH of about 7. Casein will be separated as an insoluble precipitate by acidification of the milk to its isoelectric point (pH = 4.6). The fat that precipitates along with casein can be removed by dissolving it in alcohol.

In the second part of this experiment, you are going to prove that the precipitated milk product is a protein. The identification will be achieved by performing a few important chemical tests.

1. *The biuret test.* This is one of the most general tests for proteins. When a protein reacts with copper(II) sulfate, a positive test is the formation of a copper complex that has a violet color.

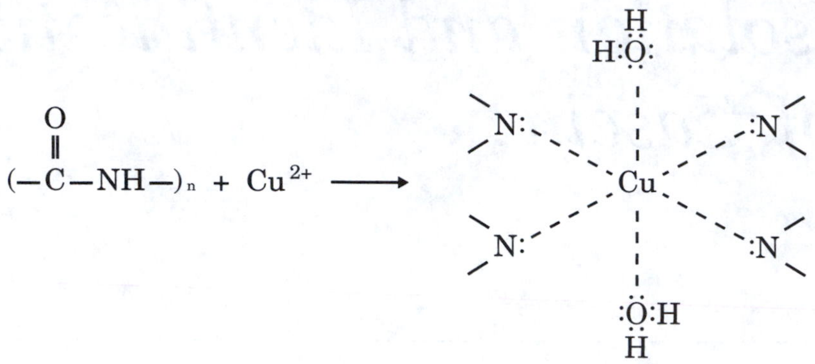

$$(-\overset{O}{\overset{\|}{C}}-NH-)_n + Cu^{2+} \longrightarrow \text{Protein–copper complex}$$

Protein   Blue color   Protein–copper complex (violet color)

This test works for any protein or compound that contains two or more of the following groups:

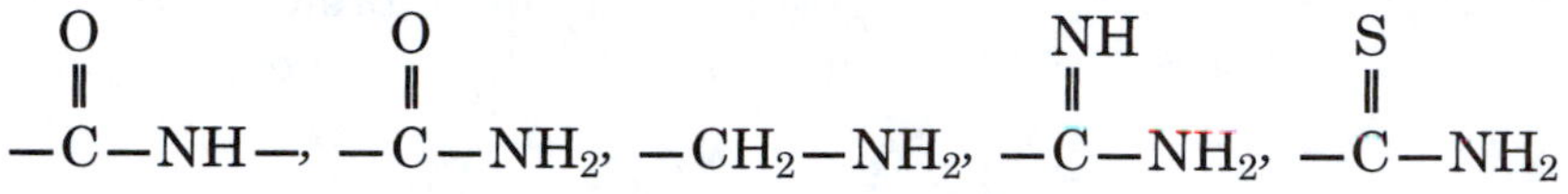

$$-\overset{O}{\overset{\|}{C}}-NH-,\ -\overset{O}{\overset{\|}{C}}-NH_2,\ -CH_2-NH_2,\ -\overset{NH}{\overset{\|}{C}}-NH_2,\ -\overset{S}{\overset{\|}{C}}-NH_2$$

2. *The ninhydrin test.* Amino acids with a free $-NH_2$ group and proteins containing free amino groups react with ninhydrin to give a purple–blue complex.

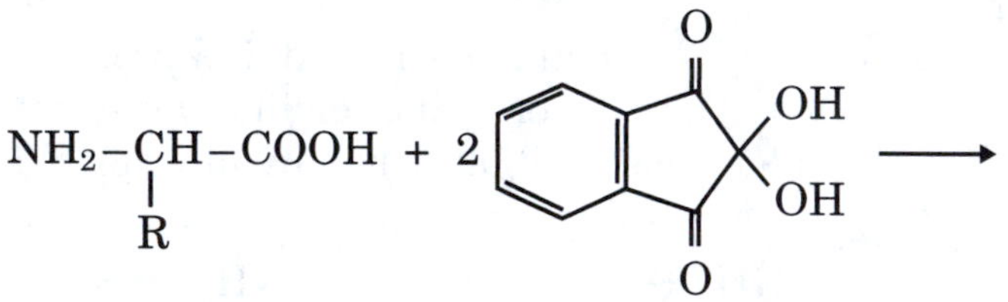

$$NH_2-\underset{R}{\underset{|}{CH}}-COOH + 2\ \text{Ninhydrin} \longrightarrow \text{Purple–blue complex} + RCHO + CO_2 + 3H_2O$$

Amino acid   Ninhydrin

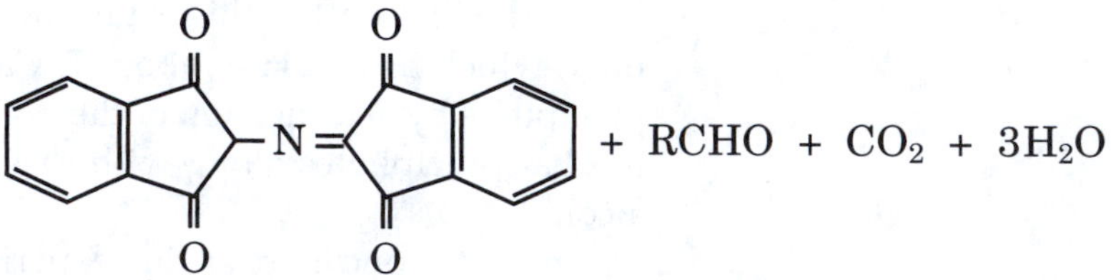

Purple–blue complex

3. *Heavy metal ions test.* Heavy metal ions precipitate proteins from solution. The ions that are most commonly used for protein precipitation are Zn^{2+}, Fe^{3+}, Cu^{2+}, Sb^{3+}, Ag^{+}, Hg^{2+}, Cd^{2+}, and Pb^{2+}. Among these metal ions, Hg^{2+}, Cd^{2+}, and Pb^{2+} are known for their notorious toxicity to humans. They can cause serious damage to proteins

(especially enzymes) by denaturing them. This can result in death. The precipitation occurs because proteins become cross-linked by heavy metals.

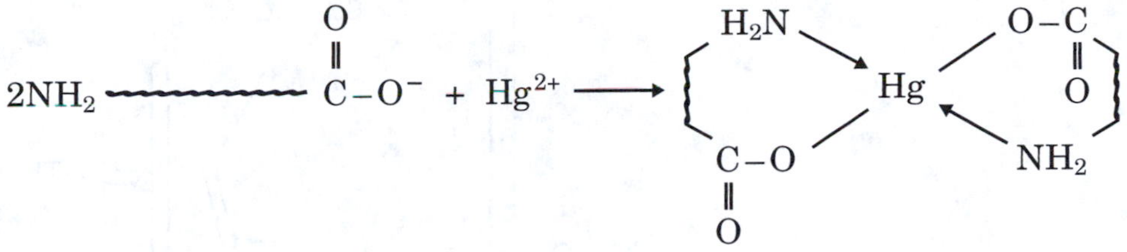

$$2\,NH_2 \sim\sim\sim C(=O)-O^- + Hg^{2+} \longrightarrow \text{Hg complex}$$

Insoluble precipitate

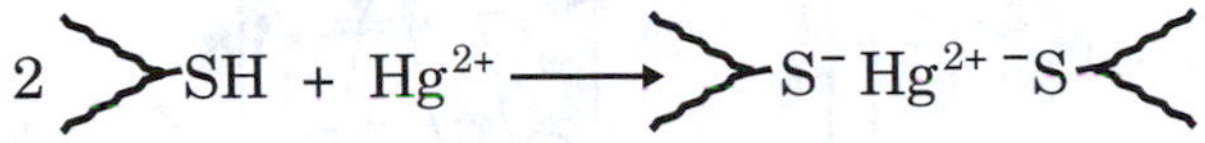

$$2 \,{>}SH + Hg^{2+} \longrightarrow {>}S^-\,Hg^{2+}\,{}^-S{<}$$

Insoluble precipitate

Victims swallowing Hg^{2+} or Pb^{2+} ions are often treated with an antidote of a food rich in proteins, which can combine with mercury or lead ions in the victim's stomach and, hopefully, prevent absorption! Milk and raw egg white are used most often. The insoluble complexes are then immediately removed from the stomach by an emetic.

4. *The xanthoprotein test.* This is a characteristic reaction of proteins that contain phenyl rings.

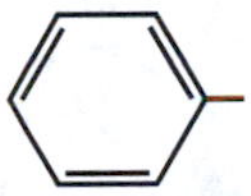

Concentrated nitric acid reacts with the phenyl ring to give a yellow-colored aromatic nitro compound. Addition of alkali at this point will deepen the color to orange.

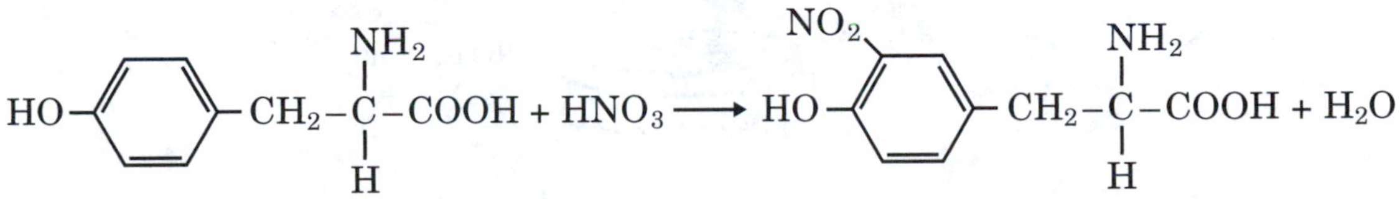

$$HO-C_6H_4-CH_2-CH(NH_2)-COOH + HNO_3 \longrightarrow HO-C_6H_3(NO_2)-CH_2-CH(NH_2)-COOH + H_2O$$

Tyrosine Colored compound

The yellow stains on the skin caused by nitric acid are the result of the xanthoprotein reaction.

OBJECTIVES

1. To isolate the casein from milk under isoelectric conditions.
2. To perform some chemical tests to identify proteins.

PROCEDURE

Isolation of Casein

1. To a 250-mL Erlenmeyer flask, add 50.00 g of milk and heat the flask in a water bath (a 600-mL beaker containing about 200 mL of tap water; see Figure 40.1). Stir the solution constantly with a stirring rod. When

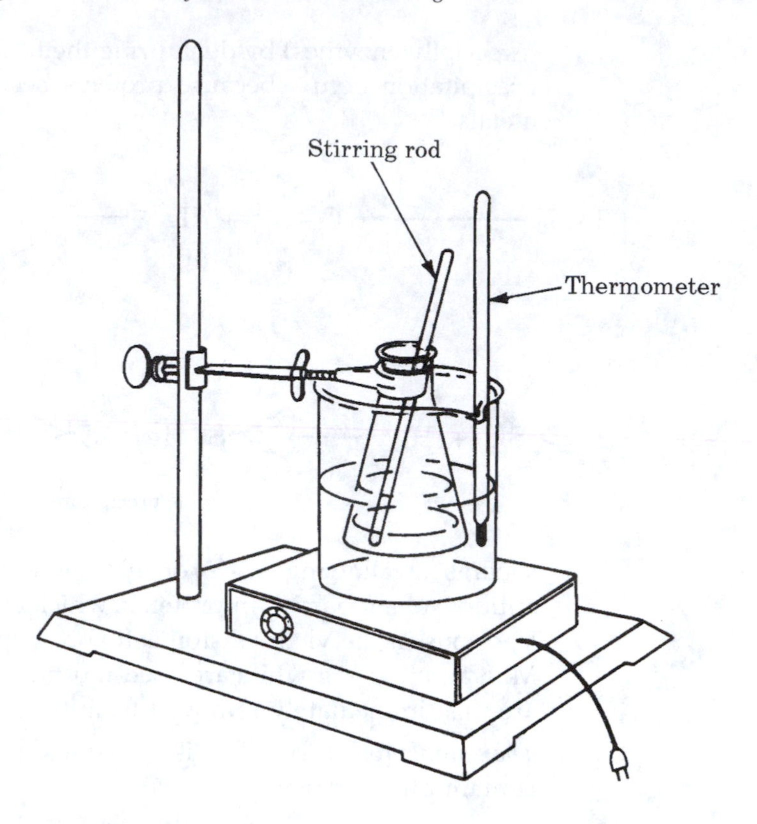

Figure 40.1
Precipitation of casein.

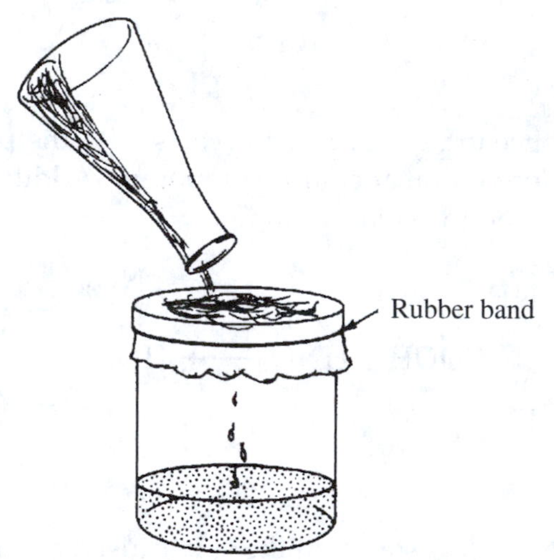

Figure 40.2
Filtration of casein.

the bath temperature has reached about 40°C, remove the flask from the water bath, and add about 10 drops of glacial acetic acid while stirring. Observe the formation of a precipitate.

2. Filter the mixture into a 100-mL beaker by pouring it through a cheese cloth which is fastened with a rubber band over the mouth of the beaker (Figure 40.2). Remove most of the water from the precipitate by squeezing the cloth gently. Discard the filtrate in the beaker. Using a spatula, scrape the precipitate from the cheese cloth into the empty flask.
3. Add 25 mL of 95% ethanol to the flask. After stirring the mixture for 5 min., allow the solid to settle. Carefully decant (pour off) the liquid that contains fats into a beaker. Discard the liquid into a waste container.

4. To the residue, add 25 mL of a 1:1 mixture of diethyl ether–ethanol. After stirring the resulting mixture for 5 min., collect the solid by vacuum filtration (see Figure 33.2, p. 404).

CAUTION

Diethyl ether is highly flammable. Make sure there is no open flame in the lab.

5. Spread the casein on a paper towel and let it dry. Weigh the dried casein and calculate the percentage of casein in the milk. Record it on your Report Sheet.

$$\% \text{ casein} = \frac{\text{weight of solid (casein)}}{50.00 \text{ g of milk}} \times 100$$

Chemical Analysis of Proteins

1. *The biuret test.* Place 15 drops of each of the following solutions in five clean, labeled test tubes. (For all the tests use 100 × 13-mm test tubes.)
   a. 2% glycine
   b. 2% gelatin
   c. 2% albumin
   d. Casein prepared in **Isolation of Casein** (one-quarter of a full spatula) + 15 drops of distilled water
   e. 1% tyrosine

   To each of the test tubes, add 5 drops of 10% NaOH solution and 2 drops of a dilute $CuSO_4$ solution while swirling. The development of a purplish-violet color is evidence of the presence of proteins. Record your results on the Report Sheet.
2. *The ninhydrin test.* Place 15 drops of each of the following solutions in five clean, labeled test tubes.
   a. 2% glycine
   b. 2% gelatin
   c. 2% albumin
   d. Casein prepared in **Isolation of Casein** (one-quarter of a full spatula) + 15 drops of distilled water
   e. 1% tyrosine

   To each of the test tubes, add 5 drops of ninhydrin reagent and heat the test tubes in a boiling-water bath for about 5 min. Record your results on the Report Sheet.
3. *Heavy metal ions test.* Place 2 mL of milk in each of three clean, labeled test tubes. Add a few drops of each of the following metal ions to the corresponding test tubes as indicated below:
   a. Pb^{2+} as $Pb(NO_3)_2$ in test tube no. 1
   b. Hg^{2+} as $Hg(NO_3)_2$ in test tube no. 2
   c. Na^+ as $NaNO_3$ in test tube no. 3

   Record your results on the Report Sheet.

CAUTION

The following test will be performed by your instructor.

4. *The xanthoprotein test.* (Perform the experiment under the hood.) Place 15 drops of each of the following solutions in five clean, labeled test tubes:
   a. 2% glycine
   b. 2% gelatin
   c. 2% albumin
   d. Casein prepared in **Isolation of Casein** (one-quarter of a full spatula) + 15 drops of distilled water
   e. 1% tyrosine

   To each test tube, add 10 drops of concentrated HNO_3 while swirling. Heat the test tubes carefully in a warm water bath. Observe any change in color. Record the results on your Report Sheet.

CHEMICALS AND EQUIPMENT

1. Hot plate
2. Büchner funnel in a no. 7 one-hole rubber stopper
3. 500-mL filter flask
4. Filter paper (Whatman no. 2, 7 cm)
5. Cheese cloth
6. Rubber band
7. Boiling stones
8. 95% ethanol
9. Diethyl ether–ethanol mixture
10. Regular milk
11. Glacial acetic acid
12. Concentrated nitric acid
13. 2% albumin
14. 2% gelatin
15. 2% glycine
16. 5% copper(II) sulfate
17. 5% lead(II) nitrate
18. 5% mercury(II) nitrate
19. Ninhydrin reagent
20. 10% sodium hydroxide
21. 1% tyrosine
22. 5% sodium nitrate

name | section | date

partner | grade

Pre-Lab Questions

1. In the experiment, fat is separated from the precipitated casein by dissolving in ethanol. Why does the fat dissolve in the ethanol?

2. Why does your skin turn yellow when concentrated nitric acid gets on it?

3. Casein does not precipitate from fresh milk, but it does precipitate when milk ''sours'' (curdles). Explain this observation.

4. Why are victims of heavy metal poisoning often given milk and raw egg white as a first-aid treatment? Why is it then necessary to use an emetic to induce vomiting?

name ____________________ section __________ date __________

partner ____________________ grade __________

40

EXPERIMENT 40

Report Sheet

Isolation of casein

1. Weight of milk __________ g

2. Weight of dried casein __________ g

3. Percentage of casein in milk __________ %

Chemical analysis of proteins

Biuret test

Substance	*Color Formed*
2% glycine	
2% gelatin	
2% albumin	
casein + H_2O	
1% tyrosine	

Which of these chemicals gives a positive test with this reagent? __________

Ninhydrin test

Substance	*Color Formed after Heating*
2% glycine	
2% gelatin	
2% albumin	
casein + H_2O	
1% tyrosine	

Which of these chemicals gives a positive test with this reagent? ______________

Heavy metal ions test

Substance	*Precipitates Formed*
$Pb(NO_3)_2$	
$Hg(NO_3)_2$	
$NaNO_3$	

Which of these metal ions gives a positive test with casein in milk? ______________

Xanthoprotein test

Substance	*Color Formed before or after Heating*
2% glycine	
2% gelatin	
2% albumin	
casein + H_2O	
1% tyrosine	

Which of these chemicals gives a positive test with this reagent? ______________

Post-Lab Questions

1. "Little Miss Muffet sat on a tuffet, eating her curds and whey." What was she eating and where did it come from?

2. In order for the ninhydrin test to work, what functional group must be present?

3. According to the equation for the ninhydrin reaction with an amino acid (see **Background** section), an aldehyde and CO_2 are products. What parts of the amino acid give rise to these products?

4. Using the percentage you obtained for casein in milk, how many grams of casein are in a glass of milk (175 g)?

5. Phenylalanine will give a xanthoprotein test with concentrated nitric acid. Predict the structure of the expected yellow-colored compound.

6. Does gelatin contain tyrosine? How do you know?

Properties of Enzymes

BACKGROUND

Each cell in our body operates like a chemical factory. Chemicals are broken down for raw material and energy, and new chemicals are synthesized. Our diet supplies only a few of the many compounds required for the operation of our body. Most are synthesized within the cell by the hundreds of different types of reactions as part of metabolism. What is remarkable is that these reactions would not be able to take place at body temperature, or if they were able to take place at all, they would occur at an extremely slow rate. What allows these reactions to be rapid and efficient are the chemical compounds called *enzymes*.

Enzymes are large protein molecules that behave as *catalysts*. A catalyst is a chemical agent that allows a chemical reaction to go faster without itself undergoing any change—it takes part in the reaction but looks the same after the reaction is done. Because it looks the same at the end of a reaction, it can be used repeatedly in chemical reactions. Think about the difficulty we would experience if food would take weeks to be digested and our muscles and nerves would not work as they do. Certainly our life would be very different.

Like any catalyst, enzymes do not alter the position of an equilibrium. If a reaction would not take place, an enzyme would not make it happen. Enzymes influence rates only; in their presence, rates are faster. They do this by lowering the energy of activation (Figure 41.1). This is the energy necessary for bond-breaking and bond-forming processes to take place. It is the barrier molecules must overcome for a reaction to occur. Enzymes' effectiveness is remarkable: Depending on the reaction and the enzyme, rates can be from 10^9 to 10^{20} times faster than the uncatalyzed reaction under the same set of conditions.

Another aspect of enzymes makes them remarkable. Within each cell there are 2000 to 3000 different enzymes. Depending on the type of cell, there are different sets of enzymes. However, because enzymes are composed of proteins, these large molecules are highly specific: *An enzyme will catalyze one specific reaction with one specific substrate*. For example, the enzyme urease catalyzes only the hydrolysis of urea and not that of any other amides, even though they may be closely related.

$$(NH_2)_2C{=}O + H_2O \xrightarrow{\text{urease}} 2NH_3 + CO_2$$

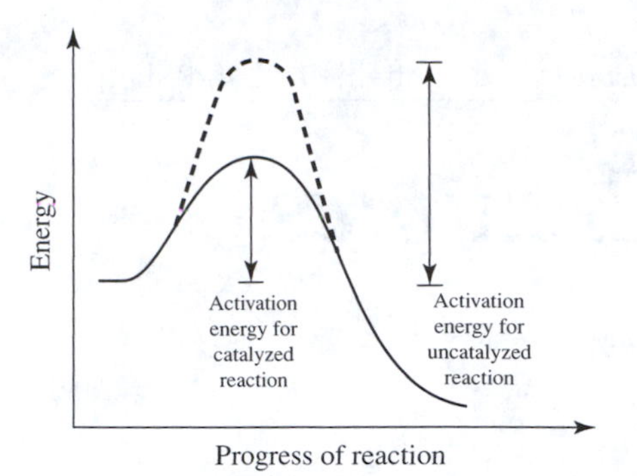

Figure 41.1
Effect of an enzyme on the energy of activation.

Catylase or peroxidase is an enzyme found in most cells and serves to break down peroxides produced from some metabolic processes. You probably have seen its effects if you ever applied hydrogen peroxide to an open cut. The white foam produced is due to the action of the enzyme on the hydrogen peroxide releasing oxygen; the reaction is faster than without the catalyst.

$$2H_2O_2 \xrightarrow{\text{peroxidase}} 2H_2O + O_2$$

For another example, let's consider amylase in some detail. This enzyme is found in human saliva and has a shape that can recognize the polysaccharide amylose (starch). It helps to hydrolyze α-1,4-glycosidic linkages in amylose into the smaller sugar units, including the disaccharide molecule, maltose. It tends to work best in the mouth, where the pH is neutral to slightly alkaline. But when you swallow and the food gets to the stomach, it stops working because the stomach pH of 2 is too acidic for it. However, the food eventually gets to the gut, where the pancreas secretes more amylase and the hydrolysis continues to produce maltose. The enzyme maltase then converts the maltose into glucose, which crosses the intestinal barrier and enters the blood.

$$\text{Amylose} \xrightarrow{\text{amylase}} \text{Oligosaccharides} \xrightarrow{\text{amylase}} \text{Maltose} \xrightarrow{\text{maltase}} \text{Glucose}$$

As these examples show, enzymes are highly specific. In order to do their specific job, they must have the correct shape. Because enzymes are made from proteins, their activity will be easily affected by heat, pH, and heavy metals. Why? One description of how enzymes work is by the ''lock and key'' mechanism—a specific key (enzyme) is required in order to open the lock (substrate). Should anything happen to the shape of the key, it no longer will open the lock. So, twist the key, lose a tooth, or have some foreign substance on the key and it will not work in the lock. Enzymes become altered by temperature changes, pH changes, and metal inhibitors. The conditions must be correct in order to work. Also, even though only small amounts of these catalysts are required, their function is concentration dependent.

Factors Affecting Enzyme Activity

Concentration

In order for an enzyme to show its catalytic effect, the enzyme must combine with the starting substrate. The resulting enzyme–substrate complex comes about when there is a proper fit at the place of the reaction,

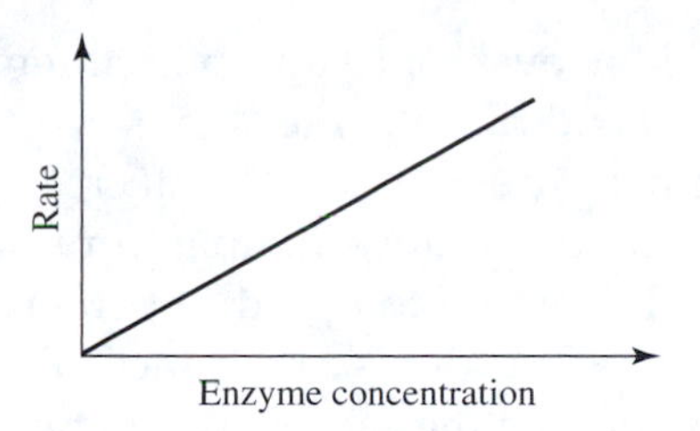

Figure 41.2
The effect of enzyme concentration on the rate of an enzyme-catalyzed reaction: a linear curve.

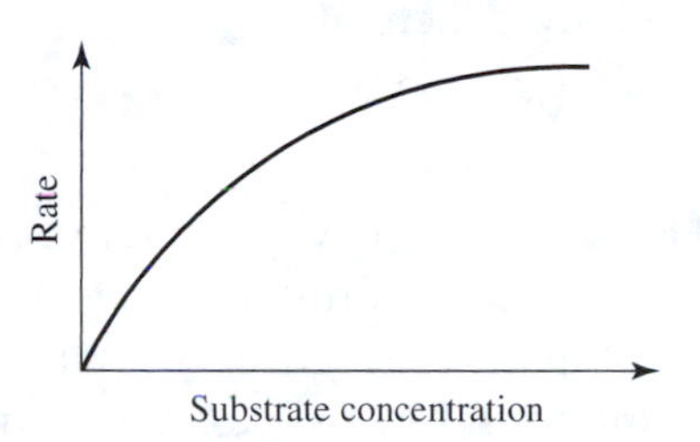

Figure 41.3
The effect of substrate on the rate of an enzyme-catalyzed reaction: a saturation curve.

the *active site*. Products are released and the enzyme is then recycled for another reaction. Schematically, it can be represented by the following:

$$\text{Enzyme} + \text{Substrate} \rightleftharpoons \text{Enzyme–substrate complex} \rightleftharpoons \text{Enzyme} + \text{Products}$$

$$E \quad + \quad S \quad \rightleftharpoons \quad E\text{-}S^{\ddagger} \quad \rightleftharpoons \quad E \quad + \quad P$$

Because the enzyme is a catalyst, its concentration is usually very much smaller than that of the substrate. In this model, if the substrate concentration is held constant, any increase in the enzyme concentration will result in a rate increase. In fact, there is a direct relationship, so that the rate increases linearly (Figure 41.2): double the enzyme concentration and the rate of the reaction increases by a factor of two.

There is another effect of concentration. Suppose the concentration of the enzyme is held constant this time. If substrate concentration is initially low and is then increased, the rate will also increase. However, this initial increase will not be continuous. As enzyme and substrate combine and release products, more and more active sites become occupied. Eventually, as substrate concentration increases, all available active sites become occupied, and further increases in substrate concentration will no longer show a rate increase. The reaction is proceeding at its maximum rate and the rate stays the same; enzymes are cycling as fast as possible and excess substrate cannot find any active sites to which to attach. We have reached saturation (Figure 41.3).

Temperature

In an earlier experiment (see Experiment 16), we saw that the rate of a reaction increased as the temperature increased. This was for an uncatalyzed reaction. With a reaction that is catalyzed by an enzyme, the effect of temperature takes a different course. If the temperature is low, any increase in temperature will cause an increase in the rate. This is what is normally expected by the kinetic-molecular theory—more effective collisions between enzyme molecules and substrate molecules leads to

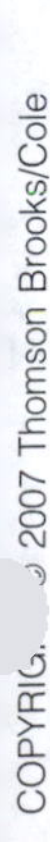

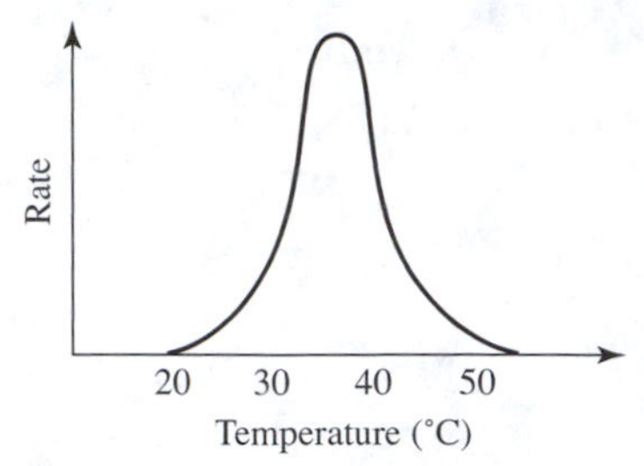

Figure 41.4
The effect of temperature on enzyme activity.

more product. However, once an optimal temperature is reached, the rate is at a maximum and the enzyme shows its greatest efficiency. Any further increase in temperature will lead to a decrease in the rate (Figure 41.4). The decline comes about because temperature affects enzyme conformation. Enzyme and substrate no longer fit together properly, so the rate leading to product falls. Consider some practical application of these effects: (1) deactivation of enzymes at low temperature is the reason refrigeration works in the preservation of food; (2) sterilization in an autoclave works because bacterial proteins and enzymes are denatured at elevated temperatures and are deactivated.

pH

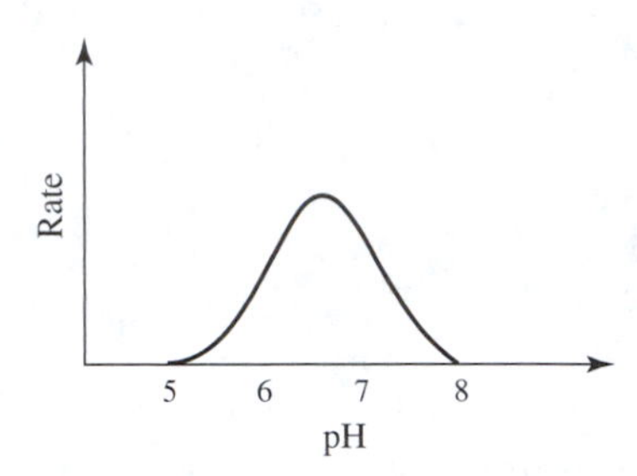

Figure 41.5
The effect of pH on enzyme activity.

Changes in pH also affect enzyme activity. As with temperature, pH can alter the conformation of proteins and, thus, enzymes. Each enzyme operates at an optimal pH (Figure 41.5) and is most active. For example, pepsin, an enzyme found in the stomach, functions at a pH of 2; trypsin, found in the gut, operates efficiently at a pH of 8. Deviation from these optimum values leads to deactivation, often due to denaturation.

Inhibitors

Any substance that reduces or stops an enzyme from behaving as a catalyst is an *inhibitor*. If the substance binds to the active site (the site where the substrate usually combines with the enzyme during reaction), then the substrate is prevented from entering the active site. The inhibitor competes with the substrate for the enzyme. This *competitive inhibitor* causes the enzyme to lose its activity. Some inhibitors bind to some other portion of the enzyme structure and cause a change in the shape of the active site, and the substrate cannot bind. This *noncompetitive inhibitor* also causes enzyme activity to decrease. Poisons, toxins, heavy metals, and some drugs behave either as competitive or noncompetitive inhibitors. The net result is a loss in enzyme activity.

OBJECTIVES

1. To demonstrate the role of enzymes as catalysts.
2. To observe how concentration, temperature, pH, and inhibitors affect enzyme activity.

PROCEDURE

General

You will need three water baths for this experiment. Half-fill three 400-mL beakers. Keep one at 0–5°C with ice water; keep a second at 35–40°C and monitor the temperature of this one carefully; keep the third one at boiling, 100°C.

CAUTION

Wear safety glasses throughout this experiment. Handle hot beakers with beaker tongs.

Catalysis

1. Take three clean, dry test tubes (100 × 13 mm) and label them no. 1, no. 2, and no. 3, respectively. Add 2 mL 3% H_2O_2 (hydrogen peroxide) to each test tube.
2. Place all three test tubes into the water bath held at 35–40°C.
3. To test tube no. 1, add a small, pea-sized piece of raw liver. To test tube no. 2, add a small, pea-sized piece of cooked liver. Add nothing to test tube no. 3.
4. Which solution quickly produces foam? Record your observation on the Report Sheet (1). What accounts for the differences?

Enzyme Concentration

1. Take three clean, dry test tubes (100 × 13 mm) and label them no. 1, no. 2, and no. 3, respectively.
2. Prepare solutions according to the table below.

Test Tube No.	*2% Rennin Solution, mL*	*Distilled Water, mL*	*Milk, mL*
1	4	0	3
2	2	2	3
3	1	3	3

3. Add 1 mL 3 M HCl (hydrochloric acid) to each test tube. Mix by tapping with your finger.
4. At the same time, place the three test tubes into the water bath at 35–40°C. With a stopwatch or seconds hand of a watch, keep track of the time the test tubes are in the bath.
5. Note the order in which the solutions form a precipitate (coagulate). Which test tube solution forms a precipitate soonest? How long did it take? What about the others? Record your observations on the Report Sheet (2).

Temperature

1. Collect 6 mL of your own saliva; saliva contains salivary amylase. Use a 10-mL graduated cylinder to collect your saliva. Until you use the saliva, keep the graduated cylinder in the water bath at 35–40°C.

CAUTION

Use gloves when collecting and handling the saliva, even your own. Pathogens may be present in any body fluid.

2. Place 2 mL of 1% starch solution in each of three clean, dry test tubes (100 × 13 mm) labeled no. 1, no. 2, and no. 3. Into each test tube, add 2 mL of saliva and mix. Place test tube no. 1 into the ice-water bath; place test tube no. 2 into the water bath at 35–40°C; place test tube no. 3 into the boiling-water bath. Allow the test tubes with their contents to stand in their respective baths for at least 30 min. (While these are heating, you can work on the next section.)
3. After heating for 30 min., transfer 3 drops of each solution into separate depressions of a white spot plate. Use clean, separate medicine droppers for the transfers. To each sample, add 2 drops of

iodine solution. (For reference, test starch in a separate depression, and only the iodine solution in another.)

4. Record the color of the solutions on your Report Sheet and estimate the extent of the hydrolysis by the color of the test (3). [Starch can form an intense, brilliant, dark blue or violet-colored complex with iodine. The straight-chain component of starch, the amylose, gives a blue color; the branched component, the amylopectin, yields a purple color. In the presence of iodine the amylose forms helixes inside of which the iodine molecules assemble as long polyiodide chains. The helix-forming branches of amylopectin are much shorter than those of amylose. Therefore, the polyiodide chains are also much shorter in the amylopectin–iodine complex than in the amylose–iodine complex. The result is a different color (purple). When starch is hydrolyzed and broken down to small carbohydrate units (oligosaccharides) the iodine will not give a dark blue (or purple) color; the color will vary in shades of brown. The iodine test is used in this experiment to indicate the degree of completion of the hydrolysis.]
5. Now place test tubes no. 1 and no. 2 into the water bath at 35–40°C. Keep them in this bath for an additional 30 min.
6. After 30 min., transfer 3 drops of each solution into separate depressions of the white spot plate. Use clean, separate medicine droppers for the transfers. To each sample, add 2 drops of iodine solution.
7. Record the color of the solutions on your Report Sheet and estimate the extent of the hydrolysis by the color from the test (3) after the additional 30 min.

pH

1. Take five clean, dry test tubes (150 × 18 mm) and label them no. 1, no. 2, no. 3, no. 4, and no. 5, respectively. Prepare solutions according to the table below:

Test Tube No.	*Buffer Solution, mL*		*Amylase Solution, mL*	*1% Starch*
1	pH 2	2	2	2
2	pH 4	2	2	2
3	pH 7	2	2	2
4	pH 9	2	2	2
5	pH 12	2	2	2

2. Mix. Then at the same time, place the test tubes into the water bath at 35–40°C. (While these are heating, you can work on the next section.)
3. After heating for 15 min., transfer 3 drops of each solution into separate depressions of a white spot plate. Use clean, separate medicine droppers for the transfers. To each sample, add 2 drops of iodine solution. (For reference, test starch in a separate depression and only the iodine solution in another.)
4. Record the color of the solutions on your Report Sheet and estimate the extent of the hydrolysis from the color (4a).

5. After heating for 30 min., transfer 3 drops of each solution into separate depressions of a white spot plate. Use clean, separate medicine droppers for the transfers. To each sample, add 2 drops of iodine solution.
6. Record the color of the solutions on your Report Sheet and estimate the extent of the hydrolysis from the color (4b).

Inhibitor

1. Take a clean, dry test tube (150 × 18 mm) and add to it 2 mL amylase solution (or saliva), 2 mL 1% starch, and 10 drops of 0.1 M $Pb(NO_3)_2$.
2. Mix and place the test tube into the water bath at 35–40°C. Heat for at least 30 min.
3. After heating for 30 min., transfer 3 drops of solution into a depression of a white spot plate. Use a clean medicine dropper for the transfer. To the sample, add 2 drops of iodine solution.
4. Record the color of the solution on your Report Sheet and estimate the extent of the hydrolysis by the color of the test (5).

CHEMICALS AND EQUIPMENT

1. Hot plates
2. Medicine droppers
3. Stopwatch
4. White spot plate
5. Amylase solution
6. Buffer, pH 2
7. Buffer, pH 4
8. Buffer, pH 7
9. Buffer, pH 9
10. Buffer, pH 12
11. 3 M HCl
12. 3% H_2O_2, hydrogen peroxide
13. Liver
14. Milk
15. 2% rennin solution
16. 1% starch solution
17. 0.01 M iodine in KI
18. 0.1 M $Pb(NO_3)_2$

name

section

date

partner

grade

Pre-Lab Questions

1. What is the action of an enzyme?

2. What affects the activity of an enzyme?

3. Reactions take place at what part of the enzyme?

4. Why does an enzyme lose activity at high temperatures?

5. Can an enzyme function at any pH? Why or why not?

6. If a reaction could not take place, would an enzyme (or any catalyst) make the reaction occur? Explain.

name	section	date
partner	grade	

41 EXPERIMENT 41

Report Sheet

Effect of a catalyst

1.

Test Tube	*Observation*	*Conclusion*
1		
2		
3		

Effect of enzyme concentration

2.

Test Tube	*Observation*	*Time*
1		
2		
3		

Effect of temperature

3.

Test Tube	*Observation*	*Conclusion*	*Reheating 30 min. at 35–40°C*	*Conclusion*
1				
2				
3				

Effect of pH

4.

Test Tube	*Observation*	*Conclusion*
a. After 15 min.		
1		
2		
3		
4		
5		
b. After 30 min.		
1		
2		
3		
4		
5		

Effect of an inhibitor

5.

Observation	*Conclusion*

Post-Lab Questions

1. Both carbon monoxide and oxygen bind at the same active site in hemoglobin. However, carbon monoxide binds so strongly that oxygen cannot displace it. What kind of an inhibitor is carbon monoxide?

2. The experiment with rennin simulates the environment in the stomach.
   a. How would the digestion of milk in the stomach be influenced by concentration and pH?

   b. Why was the experiment carried out at 35–40°C?

   c. Based on your observations in the experiment carried out at different temperatures, how effective would rennin be at 0°C and at 100°C? Explain.

3. The enzyme of a bacterium is optimal at 40°C. Would this enzyme be more active or less active at normal body temperature or when a person has a fever?

4. At what pH does amylase work best? Does this simulate the environment in the mouth?

Neurotransmission: An Example of Enzyme Specificity

BACKGROUND

When our brain wants to communicate a message to the skeletal muscles or to the sympathetic ganglia, the nerve cells, or *neurons,* use a *neurotransmitter*. Neurotransmitters are compounds that are chemical messengers that communicate between a neuron and another target cell. That target cell could be a muscle cell, another neuron, or the cell of a gland. The chemical the nerve cell uses as the neurotransmitter is acetylcholine. This messenger is usually stored in vesicles within the neurons.

Upon a signal, when the calcium ion concentration in the neurons increases, the vesicles migrate to the cell membrane. There the two membranes fuse and acetylcholine is dumped into the synapse between the neuron and the muscle cell. Acetylcholine then rapidly crosses the synapse and enters the receptor located on the muscle cell. The presence of the acetylcholine triggers a conformational change in the receptor that allows the passage of Na^+ and K^+ ions across the membrane of the muscle cell. The Na^+ ions flow in and the K^+ ions flow out of the muscle cell through the channels opened within the receptor. This sets up a potential across the membrane. In this manner the chemical signal of the messenger acetylcholine is converted into an electric signal. The electric signal can induce the muscles to contract. All this happens within a few milliseconds. The channel in the receptor closes again.

In special cases, the electricity is stored in columns of cells called electroplax. This happens in the electric eel, Torpedo, which can then discharge the electricity, killing or maiming its enemy.

In order for a new signal to be transmitted and the channel reopened, the acetylcholine must be removed from the receptor. This is accomplished by a specific enzyme, acetylcholine esterase. Acetylcholine esterase is a hydrolase and catalyzes the hydrolysis of the ester bond, yielding choline and acetate:

$$CH_3-\overset{\overset{\displaystyle O}{\|}}{C}-O-CH_2-CH_2-\overset{\overset{\displaystyle CH_3}{|}}{\underset{\underset{\displaystyle CH_3}{|}}{N^+}}-CH_3 + H_2O \xrightarrow{\text{Acetylcholine esterase}} CH_3-\overset{\overset{\displaystyle O}{\|}}{C}-O^- + HO-CH_2-CH_2-\overset{\overset{\displaystyle CH_3}{|}}{\underset{\underset{\displaystyle CH_3}{|}}{N^+}}-CH_3$$

Acetylcholine — Acetate — Choline

The main feature of acetylcholine esterase is its specificity and speed of action. It can hydrolyze an acetylcholine molecule in 40 microseconds. The neurons thus can transmit 1000 signals per second, because the muscle membrane recovers its resting potential within a fraction of a millisecond.

Other esterases have much less specificity and can hydrolyze many different esters. One such enzyme is the carboxyl esterase of liver. Even though the two enzymes hydrolyze the same ester bond, they are completely different in their primary, secondary, and tertiary structures, and even in their active sites. We shall compare the specificity of these two enzymes by measuring their activity against two substrates, (1) acetylcholine and (2) *o*-nitrophenyl acetate.

$$C_6H_4(NO_2)-O-\overset{\overset{\displaystyle O}{\|}}{C}-CH_3 + H_2O \longrightarrow CH_3-\overset{\overset{\displaystyle O}{\|}}{C}-OH + C_6H_4(NO_2)-OH$$

o-Nitrophenyl acetate — Acetic acid — *o*-Nitrophenol

The way we measure the activity is a colorimetric technique. With acetylcholine as a substrate we add a yellow-colored compound, *m*-nitrophenol, to the reaction mixture. *m*-Nitrophenol absorbs light in the 400–440-nm range. As the reaction proceeds, acetic acid and choline are liberated upon hydrolysis. The increase in the intensity of the color of *m*-nitrophenol at 420 nm is in proportion to the choline concentration liberated. Thus the rate of change in absorption is a measure of the enzyme activity. We shall define the activity as the change in absorbance/min./mg enzyme.

The colorimetric technique with *o*-nitrophenyl acetate does not require an indicator. *o*-Nitrophenyl acetate is colorless and does not absorb at 420 nm. On the other hand, the liberated *o*-nitrophenol gives an intense yellow color and the absorbance at 420 nm is proportional to the *o*-nitrophenol liberated. Again, we shall define the activity as the change in absorbance/min./mg enzyme.

OBJECTIVES

1. To study the specificity of two esterases toward different substrates.
2. To learn to measure enzyme activity by colorimetric technique.
3. To learn the use of micropipets.

PROCEDURE

General

1. Before you begin doing the experiments in this laboratory, learn to use micro-delivery pipets, or micropipets. Your instructor will demonstrate to you the use of the micropipets with 1–200-μL pipet tips.
2. You will also be using a spectrophotometer. Turn on the spectrophotometer and let it warm up for a few minutes. Turn the wavelength control knob to read 420 nm. With no sample tube in the sample compartment, adjust the amplifier control knob to read infinite absorbance (or 0.0% transmittance). Insert the sample tube containing 2 mL distilled water into the spectrophotometer. Adjust the reading to 0.0 absorbance (or 100% transmittance). This zeroing must be performed every 15 min., before each enzyme activity run, because some instruments have a tendency to drift.

Acetylcholine Esterase with *m*-Nitrophenol

1. Place 2.0 mL of *m*-nitrophenol solution at pH 8.0 into the sample tube. Add 0.4 mL acetylcholine solution. Mix. Read the absorbance of the solution at 420 nm. Record it on your Report Sheet (1).
2. Add 0.6 mL diluted acetylcholine esterase solution to the sample tube containing acetylcholine. Mix. Note the exact time of addition on your Report Sheet (2). (If you have a stopwatch, the time you start is at zero min. If you are using a regular watch, record the exact hour, minute, and second you start.)
3. Read the absorbance of your sample exactly 30 sec. after the addition and subsequently, exactly 1, 2, 4, 6, and 10 min. after the addition. Record these absorbance readings on your Report Sheet (3).

Carboxyl Esterase with *m*-Nitrophenol

1. With 2 mL distilled water in the sample tube, zero the spectrophotometer as described in the **General** section, step 2.
2. Place 2.0 mL of *m*-nitrophenol solution at pH 8.0 into the sample tube. Add 0.4 mL acetylcholine solution. Mix. Read the absorbance of the solution at 420 nm. Record it on your Report Sheet (4).
3. Add 0.51 mL of 0.15 M NaCl solution at pH 8.0 to the sample. Mix. Add 0.09 mL of the carboxyl esterase solution to the sample tube containing acetylcholine. Note the exact time of addition (with a stopwatch or regular watch, as above) on your Report Sheet (5).
4. Read the absorbance of the sample exactly 30 sec. after the addition. Repeat the readings exactly 1, 2, 4, 6, and 10 min. after the addition. Record these absorbance readings on your Report Sheet (6).

Plot Your Data

Plot your data, Absorbance (y-axis) vs. Time (x-axis), for both enzymes on the graph paper provided on your Report Sheet (7).

Acetylcholine Esterase with *o*-Nitrophenyl Acetate

1. With 2 mL distilled water in your sample tube, zero the spectrophotometer as described in the **General** section, step 2.
2. Take 2.0 mL of *o*-nitrophenyl acetate solution and read its absorbance. Record it on your Report Sheet (8).

3. Add 0.6 mL diluted acetylcholine esterase. Mix. Note the exact time of addition (with a stopwatch or regular watch, as above) on your Report Sheet (9).
4. Read the absorbance of the sample exactly 30 sec. after the addition. Repeat the readings exactly 1, 2, 3, 5, 7, and 10 min. after the addition. Record these absorbance readings on your Report Sheet (10).

Carboxyl Esterase with *o*-Nitrophenyl Acetate

1. With 2 mL distilled water in your sample tube, zero the spectrophotometer as described in the **General** section, step 2.
2. Take 2.0 mL of *o*-nitrophenyl acetate solution and read its absorbance. Record it on your Report Sheet (11).
3. Add 0.58 mL of 0.15 M NaCl solution at pH 8.0. Mix. Add 0.02 mL carboxyl esterase solution. Mix. Note the exact time of addition (with a stopwatch or regular watch, as above) on your Report Sheet (12).
4. Read the absorbance of the sample exactly 30 sec. after the addition. Repeat the readings at exactly 1, 2, 3, 4, and 5 min. Record these absorbance readings on your Report Sheet (13).

Plot Your Data

Plot your data, Absorbance (y-axis) vs. Time (x-axis), for both enzymes on the graph paper provided on your Report Sheet (14).

CHEMICALS AND EQUIPMENT

1. Spectrophotometer
2. Micropipet
3. Pipet tips (1–200 μL)
4. 0.02 M NaOH
5. 0.15 M NaCl at pH 8.0
6. *m*-Nitrophenol solution
7. Acetylcholine solution
8. *o*-Nitrophenyl acetate solution
9. Acetylcholine esterase
10. Carboxyl esterase

name | section | date

partner | grade

EXPERIMENT 42

Pre-Lab Questions

1. What is a neurotransmitter?

2. Acetylcholine is an example of a neurotransmitter. Where is it found?

3. Acetylcholine esterase, 3.0 mg, was dissolved in 10.0 mL solution. From this solution, 0.60 mL was added to the sample tube, bringing the volume to a total of 3.0 mL. How many mg of acetylcholine esterase did you have in the sample tube?

4. The commercial sample of carboxyl esterase contained 15.0 mg of enzyme in 0.30 mL of suspension. In your experiment, you add 0.090 mL of this suspension to the sample tube that contains a total volume of 3.0 mL. How many mg of carboxyl esterase enzyme do you have in the sample tube?

name | section | date

partner | grade

42

EXPERIMENT 42

Report Sheet

Acetylcholine esterase with *m*-nitrophenol

1. Absorbance of the acetylcholine in *m*-nitrophenol solution ______

2. Time of addition of acetylcholine esterase solution ______

3. Absorbance of the mixture

after 30 sec. ______ after 4 min. ______

after 1 min. ______ after 6 min. ______

after 2 min. ______ after 10 min. ______

Carboxyl esterase with *m*-nitrophenol

4. Absorbance of the acetylcholine in *m*-nitrophenol solution ______

5. Time of addition of carboxyl esterase solution ______

6. Absorbance of the mixture

after 30 sec. ______ after 4 min. ______

after 1 min. ______ after 6 min. ______

after 2 min. ______ after 10 min. ______

7. Plot your data: Absorbance vs. Time for both enzyme solutions

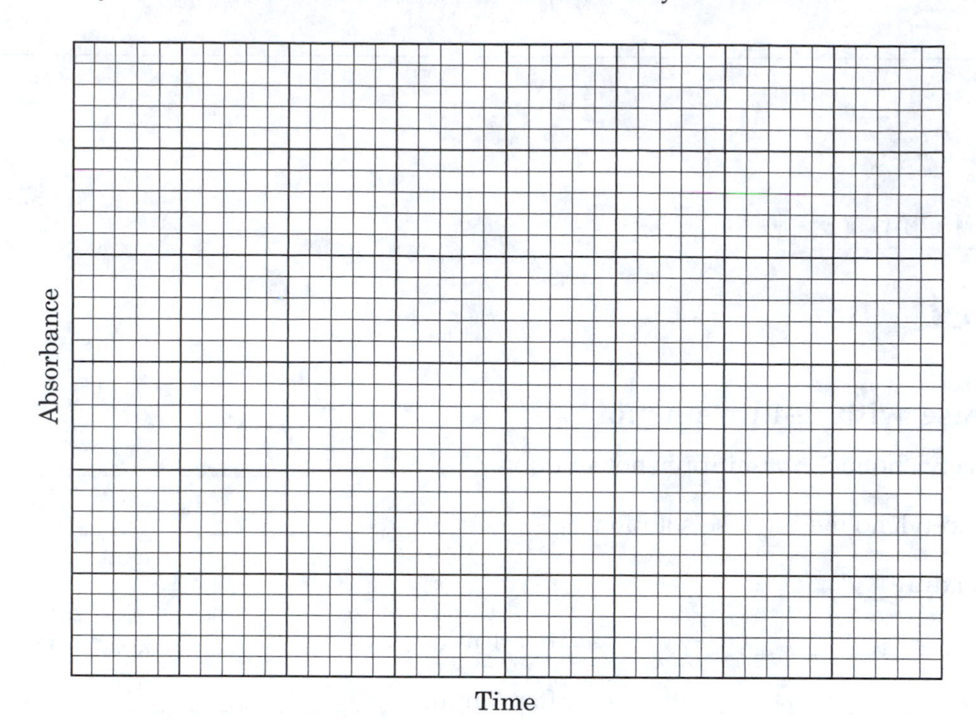

Acetylcholine esterase with *o*-nitrophenyl acetate

8. Absorbance of the *o*-nitrophenyl acetate solution ________________

9. Time of addition of acetylcholine esterase solution ________________

10. Absorbance of the mixture

after 30 sec. ________________ after 5 min. ________________

after 1 min. ________________ after 7 min. ________________

after 2 min. ________________ after 10 min. ________________

after 3 min. ________________

Carboxyl esterase with *o*-nitrophenyl acetate

11. Absorbance of *o*-nitrophenyl acetate solution ______________

12. Time of addition of carboxyl esterase solution ______________

13. Absorbance of the mixture

after 30 sec. ______________ after 3 min. ______________

after 1 min. ______________ after 4 min. ______________

after 2 min. ______________ after 5 min. ______________

14. Plot your data: Absorbance vs. Time for both enzymes

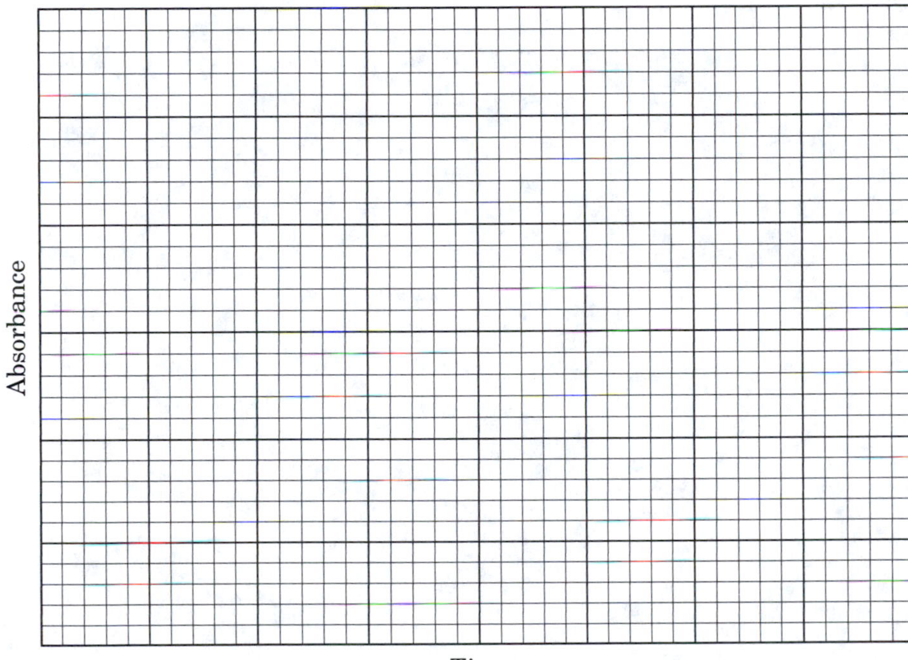

Post-Lab Questions

1. Calculate the enzyme activities with acetylcholine as the substrate for both enzymes. Using the change in absorbencies between 30 sec. and 2 min., calculate the rate of reaction as change (Δ) in absorbance/min. for

   a. acetylcholine esterase

   b. carboxyl esterase

   c. You have calculated the amount of acetylcholine esterase that you had in the sample tube (**Pre-Lab Question** no. 3). Use this number to calculate the activity of this enzyme toward acetylcholine in Δ absorbance/min./mg enzyme.

   d. You have calculated the amount of carboxyl esterase that you had in the sample tube (**Pre-Lab Question** no. 4). Use this number to calculate the activity of this enzyme toward acetylcholine in Δ absorbance/min./mg enzyme.

   e. Which enzyme catalyzes the hydrolysis of acetylcholine faster? How many times faster?

2. Calculate the enzyme activities of *o*-nitrophenyl acetate as a substrate for both enzymes. Using the change in absorbance between 30 sec. and 2 min. in the reaction, calculate the rate of reaction toward *o*-nitrophenyl acetate in Δ absorbance/min. for

   a. acetylcholine esterase

   b. carboxyl esterase

   c. You have calculated the amount of acetylcholine esterase in your sample tube (**Pre-Lab Question** no. 3). Use this number to calculate the activity of this enzyme toward *o*-nitrophenyl acetate in Δ absorbance/min./mg enzyme.

   d. You have calculated the amount of carboxyl esterase in your sample tube (**Pre-Lab Question** no. 4). Use this number to calculate the activity of this enzyme toward *o*-nitrophenyl acetate in Δ absorbance/min./mg enzyme.

   e. Which enzyme catalyzes the hydrolysis of *o*-nitrophenyl acetate faster? How many times faster?

Isolation and Identification of DNA from Onion

BACKGROUND

Hereditary traits are transmitted by genes. Genes are parts of giant deoxyribonucleic acid (DNA) molecules. The DNA is found coiled around basic protein molecules called histones. The combination of the DNA unit with the histone is a nucleosome. In lower organisms, such as bacteria and yeast, both DNA and RNA (ribonucleic acid) occur in the cytoplasm, while in higher organisms, most of the DNA is inside the nucleus and the RNA is outside the nucleus, in other organelles, and in the cytoplasm.

In this experiment, we will isolate DNA molecules from onion cells. The first task is to break up the cells. This is achieved by a combination of different techniques and agents. Grinding up the cells with sand disrupts them and the cytoplasm of many onion cells is spilled out. However, this is not a complete process. The addition of a detergent accomplishes two functions: 1) it helps to solubilize cell membranes and thereby further weakens the cell structures, and 2) it helps to inactivate the nucleic acid–degrading enzymes, nucleases, that are present. Without this inhibition, the nucleases would degrade the nucleic acids to their constituent nucelotides.

Once the nucleic acids are in solution, they must be separated from the other constituents of the cell. These are proteins and small water-soluble molecules. The addition of NaCl solution minimizes the interaction between the positively charged histone (protein) molecules and the negatively charged DNA.

The addition of ethanol precipitates the large molecules (DNA, RNA, and proteins) and leaves the small molecules in solution. DNA, being the largest fibrous molecule, forms thread-like precipitates that can be spooled off onto a rod. The protein and RNA form a gelatinous precipitate that cannot be picked up by winding them on a glass rod. Thus, the spooling separates DNA from RNA and proteins. This is a crude separation and the DNA thus isolated may contain some contaminant proteins. In order that you should be able to follow the step-by-step procedure, a flow chart of the DNA isolation process is provided here (Figure 43.1).

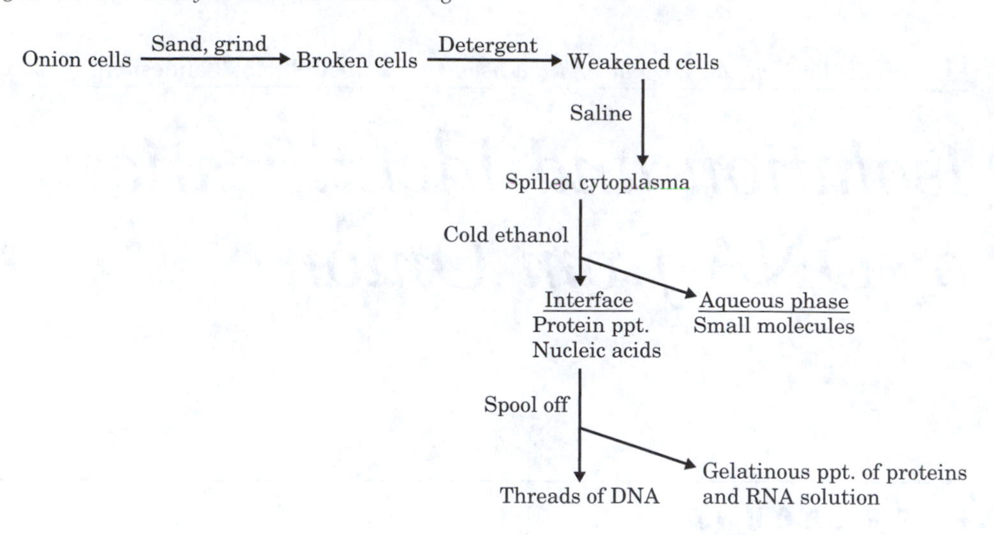

Figure 43.1
DNA isolation process.

After the isolation of DNA, we will probe its identity by the diphenylamine test. The blue color of this test is specific for deoxyribose. The appearance of blue color can be used to identify the deoxyribose-containing DNA molecule. The diphenylamine reagent contains small amounts of acetaldehyde. This compound is an eye irritant, although not in the concentration that exists in the reagent. Still, some stockrooms prefer not to handle acetaldehyde and prepare a less selective diphenylamine test without acetaldehyde in which the yellow-green color of deoxyribose appears very fast, while other sugars, glucose, ribose, etc., give faint coloration that evolves slowly.

OBJECTIVES

1. To demonstrate the separation of DNA molecules from other cell constituents.
2. To prove their identity.

PROCEDURE

1. Cool a mortar in ice water. Add 20 to 30 g of diced yellow onion and 20 g of acid-washed sand. Grind the onion and the sand vigorously with a pestle for 5 min. to disrupt the cells. A yellowish mush will form.
2. Using a hot plate, warm 100 mL distilled water to 60°C. Add 10 g NaCl and add 10 mL of detergent. Stir to dissolve.
3. Add the ground onion and sand mush to 30 mL of the saline-detergent solution. Using a glass rod, stir the solution for 5 min., all the while maintaining the temperature at approximately 60°C.
4. Support a funnel with a ring clamp on a ring stand above a 150-mL beaker. Add cheese cloth to the funnel and decant the cell suspension onto the cheese cloth. Let the yellowish liquid filter into the beaker, while gently squeezing the cheese cloth to obtain the maximum yield, leaving the cellular debris and the sand behind. Cool the solution to room temperature.

5. To the viscous DNA-containing aqueous solution, slowly add twice its volume of cold absolute ethanol, taking care that when you add the alcohol, it flows along the side of the beaker, settling on top of the aqueous solution. *The next step is critical.* Insert a flame-sterilized glass rod gently into the solution reaching **just below the interface of the DNA-alcohol solution.** The DNA forms a thread-like precipitate at the interface. *Rotate (do not stir) the glass rod* in one direction (clockwise or counterclockwise). The rotation spools all the DNA precipitate onto the glass rod. Transfer the spooled DNA on the rod into a test tube containing 95% ethanol.
6. Discard the alcohol/detergent solution remaining in the beaker into specially labeled waste jars. Do not pour the solution down the sink.
7. Remove the rod and the spooled DNA from the test tube. Dry the DNA with a clean filter paper. Describe the appearance of the crude DNA (1). Dissolve the isolated crude DNA in 2 mL citrate buffer (0.15 M NaCl, 0.015 M sodium citrate).
   a. Set up 4 dry and clean test tubes. Into the test tubes add 2 mL each of the following:

Test Tube	*Solution*
1	1% glucose
2	1% ribose
3	1% deoxyribose
4	crude DNA solution

   Add 5 mL diphenylamine reagent to each test tube.

CAUTION

Diphenylamine reagent contains glacial acetic acid and concentrated sulfuric acid. Handle with care. Use gloves.

   Mix the contents of the test tubes. Heat the test tubes in a boiling-water bath for 10 min. Note the color (2).

   b. As an alternative procedure, one can use 1% diphenylamine dissolved in concentrated sulfuric acid. Set up the same 4 test tubes as above under **a.** but add only 1 mL of solutions to each.

   Keep the test tubes in a test tube rack pointed **away from your face. Using gloves,** add slowly, **dropwise,** from a pipet, a maximum of 1 mL of 1% diphenylamine solution to the first test tube. Stop adding diphenylamine reagent when coloration appears. Holding the top of the test tube between your fingers, swirl it slowly after each addition. The addition of concentrated sulfuric acid to the aqueous solution generates heat that helps the development of the color. **Do not touch the bottom of the test tubes. They are hot.** Note the color and estimate in seconds how fast it develops.

   Repeat the procedure with the three other test tubes.

CHEMICALS AND EQUIPMENT

1. Yellow onion
2. Sand
3. Mortar and pestle
4. Ivory liquid dishwashing detergent
5. NaCl
6. Absolute ethanol
7. Citrate buffer
8. Glucose solution
9. Ribose solution
10. Deoxyribose solution
11. Diphenylamine reagent
12. 95% ethanol
13. Cheese cloth

name | section | date

partner | grade

Pre-Lab Questions

1. How are hereditary traits transmitted?

2. Where are genes found?

3. What are the nucleic acids found in cells?

4. Consult your textbook and draw the structures of ribose and deoxyribose. What is the difference between the two molecules?

5. How can you test in order to determine the presence of deoxyribose?

6. The **Procedure** calls for *vigorous* grinding of the onion cells with sand. Would you be able to isolate DNA without this step? What is the purpose of the grinding process?

name | section | date

partner | grade

Report Sheet

1. Describe the appearance of the crude DNA preparation.

2. Diphenylamine test

Solution	*Color*
1% glucose	
1% ribose	
1% deoxyribose	
crude DNA sample	

Did the diphenylamine test confirm the identity of DNA?

Post-Lab Questions

1. What are histones? What is a nucleosome? Where is DNA located in the nucleosome?

2. In the experiment, how did you minimize the interaction between the DNA and histones (proteins)?

3. What compounds remain in the ethanol solution after the DNA is removed by spooling onto a glass rod?

4. Could you do without using the detergent in the procedure? Explain your answer.

5. Is the diphenylamine reagent able to distinguish between ribose and deoxyribose? Could it also distinguish between DNA and RNA?

6. DNA, RNA, and proteins all are large molecules. However, DNA can be isolated by the simple "spooling" procedure. Why does it work for DNA and not the other molecules?

Viscosity and Secondary Structure of DNA

BACKGROUND

In 1953, James Watson and Francis Crick proposed a three-dimensional structure of DNA that is a cornerstone in the history of biochemistry and molecular biology. The double helix they proposed for the secondary structure of DNA gained immediate acceptance, partly because it explained all known facts about DNA, and partly because it provided a beautiful model for DNA replication.

In the DNA double helix, two polynucleotide chains run in opposite directions. This means that at each end of the double helix there is one 5′-OH and one 3′-OH terminal. The sugar phosphate backbone is on the outside, and the bases point inward. These bases are paired so that for each adenine (A) on one chain a thymine (T) is aligned opposite it on the other chain. Each cytosine (C) on one chain has a guanine (G) aligned with it on the other chain. The AT and GC base pairs form hydrogen bonds with each other. The AT pair has two hydrogen bonds; the GC pair has three hydrogen bonds (Figure 44.1).

Most of the DNA in nature has the double helical secondary structure. The hydrogen bonds between the base pairs provide the stability of the double helix. Under certain conditions the hydrogen bonds are broken. During the replication process itself, this happens and parts of the double helix unfold. Under other conditions, the whole molecule unfolds, becomes single stranded, and assumes a random coil conformation. This can happen in denaturation processes aided by heat, extreme acidic or basic conditions, etc. Such a transformation is often referred to as helix-to-coil transition. There are a number of techniques that can monitor such a transition. One of the most sensitive is the measurement of viscosity of DNA solutions.

Viscosity is the resistance to flow of a liquid. Honey has a high viscosity and gasoline a low viscosity, at room temperature. In a liquid flow, the molecules must slide past each other. The resistance to flow comes from the interaction between the molecules as they slide past each other. The stronger this interaction, i.e., hydrogen bonds vs. London

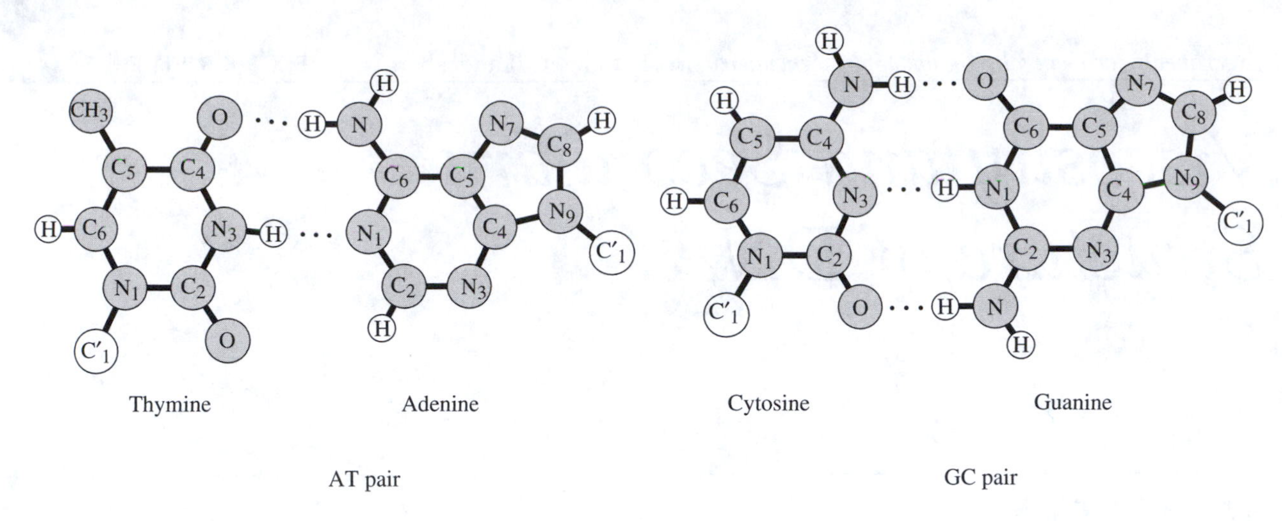

Figure 44.1
Hydrogen bonding between base pairs.

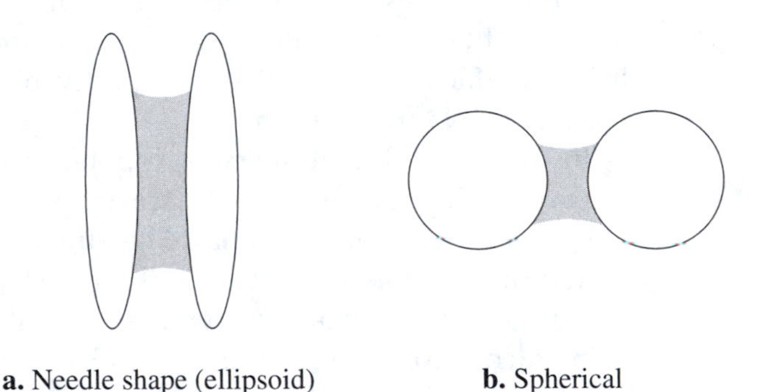

Figure 44.2
Surface area of interaction between molecules of different shapes.

dispersion forces, the greater the resistance and the higher the viscosity. Even more than the nature of the intermolecular interaction, the size and the shape of the molecules influence their viscosity. A large molecule has greater surface over which it interacts with other molecules than does a small molecule. Therefore, its viscosity is greater than that of a small molecule. If two molecules have the same size and the same interaction forces but have different shapes, their viscosity will be different. For example, needle-shaped molecules, when aligned parallel by the flow of liquid, have greater surfaces of interaction than spherical molecules of the same molecular weight (Figure 44.2). The needle-shaped molecule will have a higher viscosity than the spherical molecule. The DNA double helix is a rigid structure held together by hydrogen bonds. Its long axis along the helix exceeds by far its short axis perpendicular to it. Thus the DNA double helix has large surface area and consequently high viscosity. When the hydrogen bonds are broken and the DNA molecule becomes single stranded, it assumes a random coil shape, which has much lower surface area and lower viscosity. Thus a helix-to-coil transition is accompanied by a drop in viscosity.

In practice, we can measure viscosity by the efflux time of a liquid in a viscometer (Figure 44.3). The capillary viscometer is made of two bulbs connected by a tube in which the liquid must flow through a capillary tube. The capillary tube provides a laminary flow in which concentric layers of

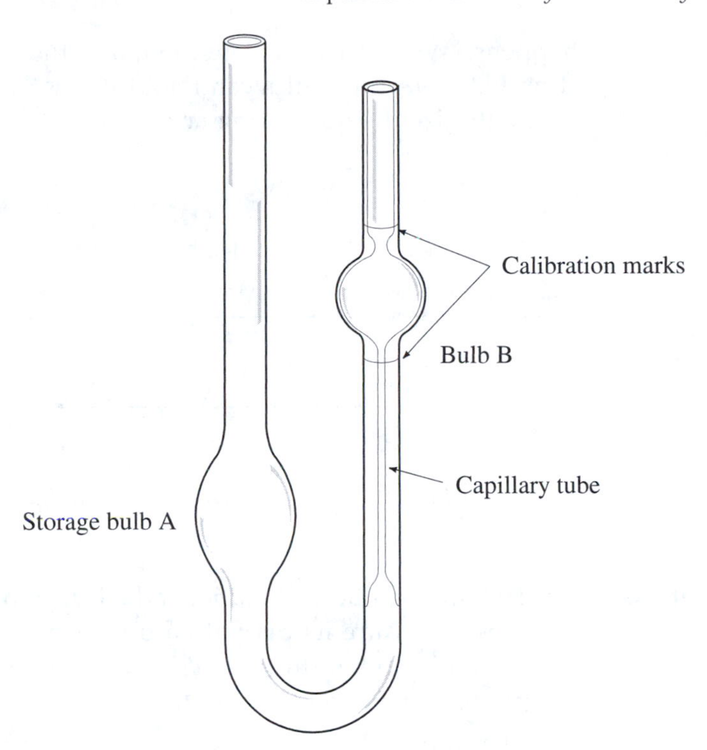

Figure 44.3
Ostwald capillary viscometer.

the liquid slide past each other. Originally, the liquid is placed in the storage bulb (A). By applying suction above the capillary, the liquid is sucked up past the upper calibration mark. With a stopwatch in hand, the suction is released and the liquid is allowed to flow under the force of gravity. The timing starts when the meniscus of the liquid hits the upper calibration mark. The timing ends when the meniscus of the liquid hits the lower calibration mark of the viscometer. The time elapsed between these two marks is the efflux time.

With dilute solutions, such as the DNA in this experiment, the viscosity of the solution is compared to the viscosity of the solvent. The efflux time of the solvent, aqueous buffer, is t_o and that of the solution is t_s. The relative viscosity of the solution is

$$\eta_{\mathbf{rel}} = \mathbf{t_s/t_o}$$

The viscosity of a solution also depends on the concentration; the higher the concentration, the higher the viscosity. In order to make the measurement independent of concentration, a new viscometric parameter is used, which is called intrinsic viscosity, $[\eta]$. This number is

$$[\eta] = (\mathbf{log}\ \eta_{\mathbf{rel}})/\mathbf{c}$$

which is almost a constant for a particular solute (DNA in our case) in very dilute solutions.

In this experiment, we follow the change in the viscosity of a DNA solution when we change the pH of the solution from the very acidic (pH 2.0) to very basic (pH 12.0). At extreme pH values, we expect that the hydrogen bonds will break and the double helix will become single-stranded random coils. A change in the viscosity will tell at what pH this happens. We shall also determine whether two acid-denatured single-stranded DNA molecules can refold themselves into a double helix when we neutralize the denaturing acid.

OBJECTIVES

1. To demonstrate helix-to-coil-to-helix transitions.
2. To learn how to measure viscosity.

PROCEDURE

Because of the cost of viscometers the students may work in groups of 5–6. Each group will need two viscometers.

Viscosity of DNA Solutions

1. To 3 mL of a buffer solution, add 1 drop of 1.0 M HCl using a Pasteur pipet. Measure its pH with a universal pH paper. If the pH is above 2.5, add another drop of 1 M HCl. Measure the pH again. Record the pH on your Report Sheet (1).
2. Clamp one clean and dry viscometer on a stand. Pipet 3 mL of your acidified buffer solution into bulb A of your viscometer. Using a suction bulb of a Spectroline pipet filler (see Experiment 2), raise the level of the liquid in the viscometer above the upper calibration mark. Release the suction by removing the suction bulb and time the efflux time between the two calibration marks. Record this as t_o on your Report Sheet (2). Remove all the liquid from your viscometer by pouring the liquid out from the wide arm. Then apply pressure with the suction bulb on the capillary arm of the viscometer and blow out any remaining liquid into the storage bulb (A); pour out this residual liquid.
3. Take 3 mL of the prepared DNA solution. Add the same amount of 1 M HCl as above (1 or 2 drops). Mix it thoroughly by shaking the solution. Test the pH of the solution with a universal pH paper and record the pH (3) and the DNA concentration of the prepared solution on your Report Sheet (4).
4. Pour the acidified DNA solution into the wide arm (bulb A) of your viscometer. Using a suction bulb, raise the level of your liquid above the upper calibration mark. Release the suction by removing the suction bulb and measure and record the efflux time of the acidified DNA solution on your Report Sheet (5).
5. Add the same amount (1 or 2 drops) as above of neutralizing 1 M NaOH solution to the liquid in the wide arm of your viscometer. With the suction bulb on the capillary arm blow a few air bubbles through the solution to mix the ingredients. Repeat the measurement of the efflux time and record it on your Report Sheet (6). For the next 100 min. or so, repeat the measurement of the efflux time every 20 min. and record the results on your Report Sheet (7–11).

pH Dependence of the Viscosity of DNA Solutions

6. While the efflux time measurements in viscometer no. 1 are repeated every 20 min., another dry and clean viscometer will be used for establishing the pH dependence of the viscosity of DNA solutions.

First, measure the pH of the buffer solution with a universal pH paper. Record it on your Report Sheet (12). Second, transfer 3 mL of the buffer into viscometer no. 2 and measure and record its efflux time on your Report Sheet (13). Empty the viscometer as instructed in step 2 above. Test the pH of the DNA solution with a universal pH paper (14) and transfer 3 mL into the viscometer. Measure its efflux time and record it on your Report Sheet (15). Empty your viscometer.

7. Repeat the procedure described in step 6, but this time, with the aid of a Pasteur pipet, add one drop of 0.1 M HCl to the 3-mL buffer solution as well as to the 3-mL DNA solution. Measure the pH and the efflux times of both buffer and DNA solutions and record them (16–19) on your Report Sheet. *Make sure that you empty the viscometer after each viscosity measurement.*
8. Repeat the procedure described in step 6, but this time add one drop of 0.1 M NaOH solution to both the 3-mL buffer and 3-mL DNA solutions. Measure their pH and efflux times and record them on your Report Sheet (20–23).
9. Repeat the procedure described in step 6, but this time add 2 drops of 1 M NaOH to both buffer and DNA solutions (3 mL of each solution). Measure and record their pH and efflux times on your Report Sheet (24–27).

Optional

10. If time allows, you may repeat the procedure at other pH values; for example, by adding two drops of 1 M HCl (28–31), or two drops of 0.1 M HCl (32–35), or two drops of 0.1 M NaOH (36–39) to the separate samples of buffer and DNA solutions.

CHEMICALS AND EQUIPMENT

1. Viscometers, 3-mL capacity
2. Stopwatch or watch with a second hand
3. Stand with utility clamp
4. Pasteur pipets
5. Buffer at pH 7.0
6. Prepared DNA solution
7. 1 M HCl
8. 0.1 M HCl
9. 1 M NaOH
10. 0.1 M NaOH
11. Spectroline pipet fillers

name   section   date

partner   grade

EXPERIMENT 44

Pre-Lab Questions

1. What is *viscosity*?

2. How is viscosity measured?

3. What provides the stability that is found in the double helix of DNA?

4. What happens to native DNA in extremely acidic or basic conditions?

5. Why does DNA in a double helix have a greater surface interaction than the same DNA as a single-stranded random coil?

6. Viscosity increases when the surface interaction between the molecules sliding past each other increases. In the following pairs, predict which will have a greater viscosity and justify your choice.

a. water or CH_3Cl

b. tRNA (molecular weight 3×10^4) or DNA (molecular weight 1×10^8)

name _______________ section _______ date _______

partner _______________ grade _______

44 EXPERIMENT 44

Report Sheet

Viscosity of DNA solutions

1. pH of acidified buffer ____________
2. Efflux time of acidified buffer (t_o) ____________ sec.
3. pH of acidified DNA solution ____________
4. Concentration of DNA solution ____________
5. Efflux time of acidified DNA solution ____________ sec.
6. Efflux time of neutralized DNA solution at time of neutralization ____________ sec.
7. 20 min. later ____________ sec.
8. 40 min. later ____________ sec.
9. 60 min. later ____________ sec.
10. 80 min. later ____________ sec.
11. 100 min. later ____________ sec.

pH dependence of the viscosity of DNA solutions

12. pH of neutral buffer ____________
13. Efflux time of neutral buffer ____________ sec.
14. pH of DNA solution in neutral buffer ____________
15. Efflux time of DNA in neutral buffer ____________ sec.

After addition of 1 drop of 0.1 M HCl

16. pH of buffer ______________

17. Efflux time of buffer ______________ sec.

18. pH of DNA solution ______________

19. Efflux time of DNA solution ______________ sec.

After addition of 1 drop of 0.1 M NaOH

20. pH of buffer ______________

21. Efflux time of buffer ______________ sec.

22. pH of DNA solution ______________

23. Efflux time of DNA solution ______________ sec.

After addition of 2 drops of 1 M NaOH

24. pH of buffer ______________

25. Efflux time of buffer ______________ sec.

26. pH of DNA solution ______________

27. Efflux time of DNA solution ______________ sec.

Optional

After addition of 2 drops of 1 M HCl

28. pH of buffer ______________

29. Efflux time of buffer ______________ sec.

30. pH of DNA solution ______________

31. Efflux time of DNA solution ______________ sec.

After addition of 2 drops of 0.1 M HCl

32. pH of buffer ______________

33. Efflux time of buffer ______________ sec.

34. pH of DNA solution ______________

35. Efflux time of DNA solution ______________ sec.

After addition of 2 drops of 0.1 M NaOH

36. pH of buffer ______________

37. Efflux time of buffer ______________ sec.

38. pH of DNA solution ______________

39. Efflux time of DNA solution ______________ sec.

Tabulate your data on the pH dependence of relative viscosity.

	pH		η_{rel}
(3)	______________	(5)/(2)	______________
(14)	______________	(15)/(13)	______________
(18)	______________	(19)/(17)	______________
(22)	______________	(23)/(21)	______________
(26)	______________	(27)/(25)	______________
(30)	______________	(31)/(29)	______________
(34)	______________	(35)/(33)	______________
(38)	______________	(39)/(37)	______________

Post-Lab Questions

1. Plot your tabulated data—relative viscosity on the y-axis and pH on the x-axis.

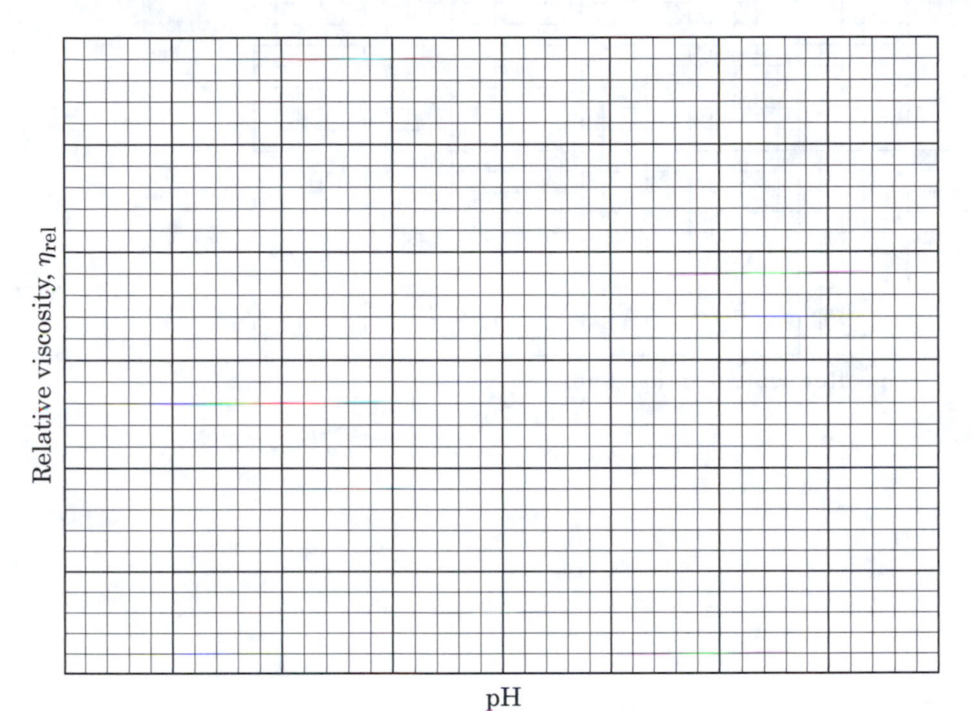

2. At what pH values did you observe helix-to-coil transitions?

3. Plot your data on the refolding of DNA double helix (5)–(11). Plot the time on the x-axis and the efflux times on the y-axis.

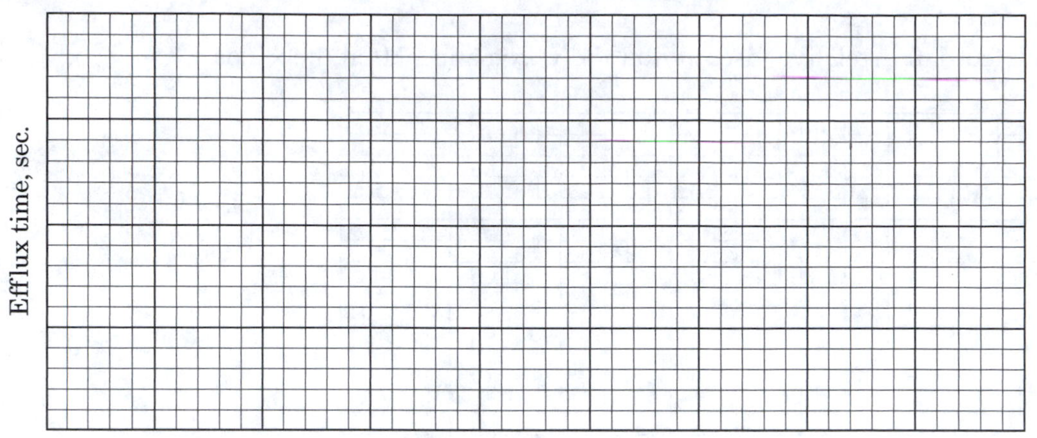

4. Was there any indication that, upon neutralization of the denaturing acid, the DNA did refold into a double helix? Explain.

5. Compare the efflux time of the neutral DNA (15) to that of the denatured DNA 100 min. after neutralization (11). What does the difference between these two efflux times tell you regarding the refolding process?

6. Calculate the intrinsic viscosity of your DNA at

a. neutral pH $= 2.3 \times \{\log[(15)/(13)]\}/(4) =$

b. acidic pH $= 2.3 \times \{\log[(5)/(2)]\}/(4) =$

c. basic pH $= 2.3 \times \{\log[(27)/(25)]\}/(4) =$

d. neutralized pH 100 min. after neutralization $= 2.3 \times \{\log[(11)/(13)]\}/(4) =$

7. A high intrinsic viscosity implies a double helix; a low intrinsic viscosity means a random coil. What do you think is the shape of the DNA after acid denaturation and subsequent neutralization? [See **6d.**] Explain your answer.

Kinetics of Urease-Catalyzed Decomposition of Urea

BACKGROUND

Enzymes speed up the rates of reactions by forming an enzyme–substrate complex (see Experiment 41). The reactants can undergo the reaction on the surface of the enzyme, rather than finding each other by collision. Thus the enzyme lowers the energy of activation of the reaction.

Urea decomposes according to the following equation:

$$H_2N-\overset{\overset{\displaystyle O}{\|}}{C}-NH_2 + H_2O \rightleftharpoons CO_2 + 2NH_3 \qquad (1)$$

This reaction is catalyzed by a highly specific enzyme, urease. Urease is present in a number of bacteria and plants. The most common source of the enzyme is jack bean or soybean. Urease was the first enzyme that was crystallized. James B. Sumner, in 1926, proved unequivocally that enzymes are protein molecules.

Urease is an —SH group thiol–containing enzyme. The cysteine residues of the protein molecule must be in the reduced —SH form in order for the enzyme to be active. Oxidation of these groups will form —S—S—, disulfide bridges, and the enzyme loses its activity. Reducing agents such as cysteine or glutathione can reactivate the enzyme.

Heavy metals such as Ag^+, Hg^{2+}, or Pb^{2+}, which form complexes with the —SH groups, also deactivate the enzyme. For example, the poison phenylmercuric acetate is a potent inhibitor of urease.

$$\underset{\text{Active}}{\text{enzyme}-SH} + \underset{\text{Phenylmercuric acetate}}{CH_3-\overset{\overset{\displaystyle O}{\|}}{C}-O^-Hg^+C_6H_5} \rightleftharpoons \underset{\text{Acetic acid}}{CH_3\overset{\overset{\displaystyle O}{\|}}{C}-OH} + \underset{\text{Inactive}}{\text{enzyme}-S-Hg-C_6H_5} \qquad (2)$$

In this experiment, we study the kinetics of urea decomposition. As shown in equation (1), the products of the reaction are carbon dioxide,

CO_2, and ammonia, NH_3. Ammonia, being a base, can be titrated with an acid, HCl, and in this way we can determine the amount of NH_3 that is produced.

$$NH_3(aq) + HCl(aq) \rightleftharpoons NH_4Cl(aq) \qquad (3)$$

Example

A 5-mL aliquot of the reaction mixture is taken before the reaction starts. We use this as a blank. We titrate this with 0.05 M HCl to an end point. The amount of acid used was 1.5 mL. This blank then must be subtracted from all subsequent titration values. Next, we take a 5-mL sample of the reaction mixture after the reaction has proceeded for 10 min. We titrate this with 0.05 M HCl and, let's assume, get a value of 5.0 mL HCl. Therefore, 5.0 − 1.5 = 3.5 mL of 0.05 M HCl was used to neutralize the NH_3 produced in a 10-min. reaction time. This means that

$$(3.5 \cancel{mL} \times 0.05 \text{ moles HCl})/1000 \cancel{mL} = 1.75 \times 10^{-4} \text{ moles HCl}$$

was used up. According to reaction (3), one mole of HCl neutralizes 1 mole of NH_3; therefore, the titration indicates that in our 5-mL sample, 1.75×10^{-4} moles of NH_3 was produced in 10 min. Equation (1) also shows that for each mole of urea decomposed, 2 moles of NH_3 are formed. Therefore, in 10 min.

$$(1 \text{ mole urea} \times 1.75 \times 10^{-4} \cancel{\text{moles } NH_3})/2 \cancel{\text{moles } NH_3}$$
$$= 0.87 \times 10^{-4} \text{ moles urea or } 8.7 \times 10^{-5} \text{ moles of urea}$$

were decomposed. Thus the rate was 8.7×10^{-6} moles of urea per min. This is the result we obtained using a 5-mL sample in which 1 mg of urease was dissolved. This rate of reaction corresponds to 8.7×10^{-6} moles urea/mg enzyme-min.

A *unit of activity* of urease is defined as the micromoles (1×10^{-6} moles) of urea decomposed in 1 min. Thus the enzyme in the preceding example had an activity of 8.7 units per mg enzyme.

In this experiment we also study the rate of the urease-catalyzed decomposition in the presence of an inhibitor. We use a dilute solution of phenylmercuric acetate to inhibit, but not completely deactivate, urease.

CAUTION

Mercury compounds are poisons. Take extra care to avoid getting the mercuric salt solution in your mouth or swallowing it.

Many of the enzymes in our body are also —SH-containing enzymes, and these will be deactivated if we ingest such compounds. As a result of mercury poisoning, many body functions will be inhibited.

OBJECTIVES

1. To demonstrate how to measure the rate of an enzyme-catalyzed reaction.
2. To investigate the effect of an inhibitor on the rate of reaction.
3. To calculate urease activity.

PROCEDURE

Enzyme Kinetics in the Absence of Inhibitor

1. Prepare a 37°C water bath in a 250-mL beaker. Maintain this temperature by occasionally adding hot water to the bath. To a 100-mL Erlenmeyer flask, add 20 mL of 0.05 M Tris buffer and 20 mL of 0.3 M urea in a Tris buffer. Mix the two solutions, and place the corked Erlenmeyer flask into the water bath for 5 min. This is your *reaction vessel.*
2. Set up a buret filled with 0.05 M HCl. Place into a 100-mL Erlenmeyer flask 3 to 4 drops of a 1% $HgCl_2$ solution. This will serve to stop the reaction, once the sample is pipeted into the titration flask. Add a few drops of methyl red indicator. This Erlenmeyer flask will be referred to as the *titration vessel.*
3. Take the *reaction vessel* from the water bath. Add 10 mL of urease solution to your reaction vessel. The urease solution contains a specified amount of enzyme (e.g., 20 mg enzyme in 10 mL of solution). Note the time of adding the enzyme solution as zero reaction time. Immediately pipet a 5-mL aliquot of the urea mixture into your *titration vessel.* Stopper the reaction vessel, and put it back into the 37°C bath.
4. Titrate the contents of the titration vessel with 0.05 M HCl to an end point. The end point is reached when the color changes from yellow to pink and stays that way for 10 sec. Record the amount of acid used. This is your blank.
5. Wash and rinse your titration vessel after each titration and reuse it for subsequent titrations.
6. Take a 5-mL aliquot from the reaction vessel **every 10 min**. Pipet these aliquots into the cleaned titration vessel into which methyl red indicator and $HgCl_2$ inhibitor were already placed, similar to the procedure in step 2 that you used in your first titration (blank). Record the time you placed the aliquots into the titration vessels and titrated them with HCl to an end point. Record the amount of HCl used in your titration. Use five samples over a period of 50 min.

Enzyme Kinetics in the Presence of Inhibitor

CAUTION

Be careful with the phenylmercuric acetate solution. Do not get it in your mouth or eyes.

1. Use the same water bath as in the first experiment. Maintain the temperature at 37°C. To a new 100-mL reaction vessel, add 19 mL of 0.05 M Tris buffer, 20 mL of 0.3 M urea solution, and 1 mL of

phenylmercuric acetate (1×10^{-3} M). Mix the contents, and place the reaction vessel into the water bath for 5 min.

2. Ready the *titration vessel* as before by adding a few drops of $HgCl_2$ and methyl red indicator. To the *reaction vessel,* add 10 mL of urease solution. Note the time of addition as zero reaction time. Mix the contents of the *reaction vessel*. Transfer immediately a 5-mL aliquot into the *titration vessel*. This will serve as your blank.
3. Titrate it as before. Record the result. **Every 10 min**. take a 5-mL aliquot for titration. The duration of this experiment should be 40 min.

CHEMICALS AND EQUIPMENT

1. Tris buffer
2. 0.3 M urea
3. 0.05 M HCl
4. 1×10^{-3} M phenylmercuric acetate
5. 1% $HgCl_2$
6. Methyl red indicator
7. Urease solution
8. 50-mL buret
9. 10-mL graduated pipets
10. 5-mL volumetric pipets
11. 10-mL volumetric pipets
12. Buret holder
13. Spectroline pipet filler

name

partner

section

date

grade

45

EXPERIMENT 45

Pre-Lab Questions

1. What kind of an enzyme is urease?

2. What chemicals will deactivate the enzyme?

3. In order for the urease to be active, what functional group must be present at the active site?

4. How do we follow the decomposition of urea?

5. What is the purpose of the *blank*?

name | section | date

partner | grade

45 EXPERIMENT 45

Report Sheet

Enzyme kinetics in the absence of inhibitor

Reaction Time (min.)	*Buret Readings before Titration (A)*	*Buret Readings after Titration (B)*	*mL Acid Titrated (B) – (A)*	*mL 0.05 M HCl Used up in the Reaction (B) – (A) – Blank*
0 (blank)				
10				
20				
30				
40				
50				

Enzyme kinetics in the presence of inhibitor

Reaction Time (min.)	*Buret Readings before Titration (A)*	*Buret Readings after Titration (B)*	*mL Acid Titrated (B) – (A)*	*mL 0.05 M HCl Used up in the Reaction (B) – (A) – Blank*
0 (blank)				
10				
20				
30				
40				

1. Present the preceding data in graphical form by plotting reaction time (column 1) on the x-axis and the mL of 0.05 M HCl used (column 5) on the y-axis for both reactions.

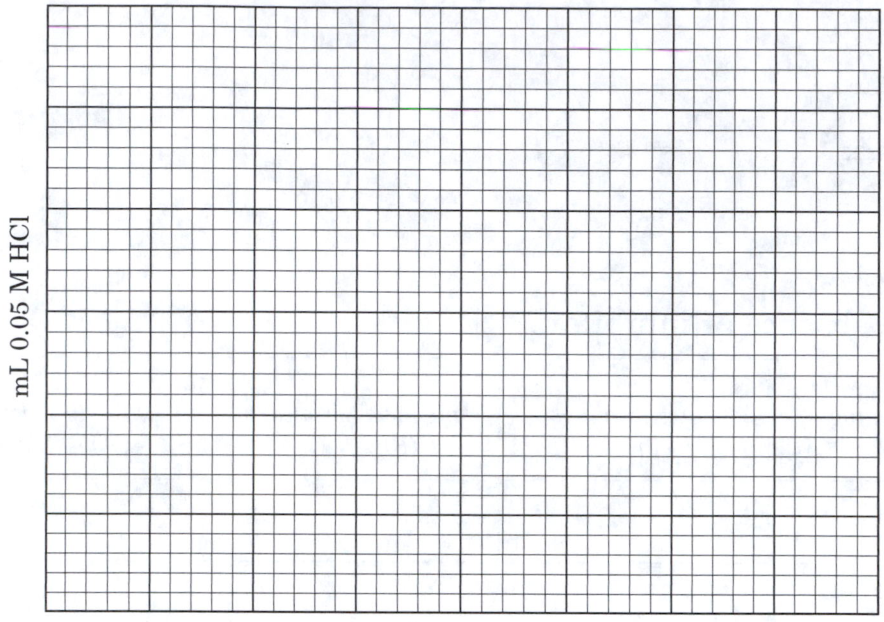

2. Calculate the urease activity only for the reaction without the inhibitor. Use the titration data from the first 10 min. of reaction (initial slope).

Urease activity

$$= \frac{\text{X } \cancel{\text{mL HCl}} \text{ consumed} \times 0.05 \cancel{\text{ moles HCl}} \times 1 \cancel{\text{ mole } NH_3} \times 1 \text{ mole urea} \times 50 \cancel{\text{ mL sol.}}}{10 \text{ min.} \times 5 \cancel{\text{ mL sol.}} \times 1000 \cancel{\text{ mL HCl}} \times 1 \cancel{\text{ mole HCl}} \times 2 \cancel{\text{ moles } NH_3} \times 20 \text{ mg urease}}$$

= Z units activity/mg enzyme

Post-Lab Questions

1. What would be the urease activity if you used the slope between 40 and 50 min. instead of the initial slope from your diagram?

2. Your instructor will provide the activity of urease as it was specified by the manufacturer. Compare this activity with the one you calculated. Can you account for the difference? (Enzymes usually lose their activity in long storage.)

3. What was the purpose of adding $HgCl_2$ to the titration vessel? Is there a substitute for $HgCl_2$ if this chemical was not available?

4. A student carried out the titration of a 5.0-mL sample, but did not add any $HgCl_2$ to the titration vessel. Would the amount of HCl used be higher, lower, or the same as what you needed to reach the end point? Explain your answer.

5. Why could you use 0.05 M HNO_3 in place of the 0.05 M HCl used in this experiment? Would it affect your calculations?

6. Compare the initial rates (the first 10 min.) of the enzyme reactions with and without inhibitor. Is there a difference in rates? If so, what is the relative rate difference?

7. What role did the urease play in the decomposition of urea?

Isocitrate Dehydrogenase: An Enzyme of the Citric Acid Cycle

BACKGROUND

The citric acid cycle is the first unit of the common metabolic pathway through which most of our food is oxidized to yield energy. In the citric acid cycle, the partially fragmented food products are broken down further. The carbons of the C_2 fragments are oxidized to CO_2, released as such, and expelled in the respiration. The hydrogens and the electrons of the C_2 fragments are transferred to the coenzyme, nicotinamide adenine dinucleotide, NAD^+, or to flavin adenine dinucleotide, FAD, which in turn become NADH + H^+ or $FADH_2$, respectively. These enter the second part of the common pathway, oxidative phosphorylation, and yield water and energy in the form of ATP.

The first enzyme of the citric acid cycle to catalyze both the release of one carbon dioxide and the reduction of NAD^+ is isocitrate dehydrogenase. The overall reaction of this step is as follows:

$$\begin{array}{c} COO^- \\ | \\ CH_2 \\ | \\ CH-COO^- \\ | \\ HO-CH \\ | \\ COO^- \end{array} + NAD^+ \xrightarrow{\text{enzyme}} \begin{array}{c} COO^- \\ | \\ CH_2 \\ | \\ CH_2 \\ | \\ C=O \\ | \\ COO^- \end{array} + NADH + CO_2$$

Isocitrate α-Ketoglutarate

The reduction of the NAD^+ itself is given by the equation:

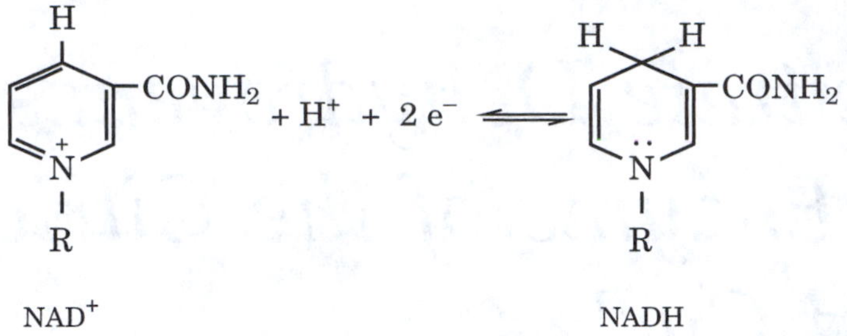

$$\underset{NAD^+}{\text{(H, CONH}_2\text{, } \overset{+}{N}\text{–R)}} + H^+ + 2\,e^- \rightleftharpoons \underset{NADH}{\text{(H, H, CONH}_2\text{, } \ddot{N}\text{–R)}}$$

The enzyme has been isolated from many tissues, the best sources being a heart muscle or yeast. The isocitrate dehydrogenase requires the presence of cofactors Mg^{2+} or Mn^{2+}. As an allosteric enzyme, it is regulated by a number of modulators. ADP, adenosine diphosphate, is a positive modulator and therefore stimulates enzyme activity. The enzyme has an optimum pH of 7.0. As is the case with all enzymes of the citric acid cycle, isocitrate dehydrogenase is found in the mitochondria.

In the present experiment, you will determine the activity of isocitrate dehydrogenase extracted from pork heart muscle. The commercial preparation comes in powder form and it uses $NADP^+$ rather than NAD^+ as a coenzyme. The basis of the measurement of the enzyme activity is the absorption spectrum of NADPH. This reduced coenzyme has an absorption maximum at 340 nm. Therefore, an increase in the absorbance at 340 nm indicates an increase in NADPH concentration, hence the progress of the reaction. We define the unit of isocitrate dehydrogenase activity as one that causes an increase of 0.01 absorbance per min. at 340 nm.

Example

A 10-mL solution containing isocitrate, isocitrate dehydrogenase, and $NADP^+$ exhibits a 0.04-unit change in the absorbance in 2 min.; the enzyme activity is

$$\frac{0.04\ \cancel{\text{abs.}}}{2\ \cancel{\text{min.}} \times 10\ \text{mL}} \times \frac{1\ \text{unit}}{0.01\ \cancel{\text{abs.}}/1\ \cancel{\text{min.}}} = 0.2\ \text{units/mL}$$

The 10-mL test solution contains 1 mL of isocitrate dehydrogenase solution with a concentration of 1 mg powder/1 mL of enzyme solution; the activity will be

$$\frac{0.2\ \text{units}}{1\ \cancel{\text{mL test soln.}}} \times \frac{10\ \cancel{\text{mL test solution}}}{1\ \cancel{\text{mL enzyme soln.}}} \times \frac{1\ \cancel{\text{mL enzyme solution}}}{1\ \text{mg enzyme powder}} =$$

$$2\ \text{units/mg enzyme powder}$$

OBJECTIVES

To measure the activity of an enzyme of the citric acid cycle, isocitrate dehydrogenase, and the effect of enzyme concentration on the rate of reaction.

PROCEDURE

General

1. In using the graduated pipets, it is important that you adhere to the following procedure before each use: (1) rinse with distilled water; (2) rinse with the solution you expect to use and discard the rinse; (3) then fill the pipet with solution.
2. **Do not pipet directly from the reagent bottle.** Avoid contamination by pouring the small amount you expect to use into a small test tube. Pipet from the test tube.
3. Turn *on* the spectrophotometer and let it warm up for a few minutes. Turn the wavelength control knob to read 340 nm. With *no sample in the sample compartment*, adjust the amplifier control knob so that infinite absorbance (or 0% transmittance) is read.
4. *"Cocktail" preparation.* Prepare a mixture of reactants, a "cocktail," in the following manner: In a 10-mL test tube, mix 2.0 mL phosphate buffer, 1.0 mL 0.1 M $MgCl_2$ solution, 1.0 mL 15 mM isocitrate solution, and 5 mL distilled water.

Preparation of Solutions

1. Solutions are prepared according to Table 46.1 and are detailed below. It is important to measure the volumes accurately with graduated pipets.
2. *Blank preparation.* To prepare a "blank" for the spectrophotometric reading, take a sample tube and add to it 1.0 mL of reagent cocktail (prepared as above), 0.2 mL $NADP^+$ solution, and 1.0 mL of distilled water. Mix the resulting solution by shaking the sample tube. *Be careful to pipet exactly 0.2 mL* $NADP^+$.
3. *Zeroing the spectrophotometer.* Insert the sample tube with the blank solution into the spectrophotometer. Adjust the reading to 0.00 absorbance (or 100% transmittance). This zeroing operation must be performed every 10 min., or before each enzyme activity run, because some instruments have a tendency to drift. The instrument is now ready to measure enzyme activity. Save this sample tube and solution for zeroing operations below.

Table 46.1

	Samples				
Reagent	*Blank, mL*	*1, mL*	*2, mL*	*3, mL*	*4, mL*
Cocktail	1.0	1.0	1.0	1.0	1.0
Distilled water	1.0	0.7	0.7	0.8	0.6
$NADP^+$ solution	0.2	0.2	0.2	0.2	0.2
Enzyme	0.0	0.3	0.3	0.2	0.4
Total volume	2.2	2.2	2.2	2.2	2.2

Enzyme Activity Measurements

1. *Sample 1.* Take a sample tube and add 1.0 mL of reagent cocktail and 0.7 mL of distilled water. Next, add 0.2 mL $NADP^+$ solution. *Be careful to pipet exactly 0.2 mL* $NADP^+$. Mix the contents of the sample tube. Readjust the spectrophotometer with the blank (prepared above) to read 0.00 absorbance (or 100% transmission). Remove the blank and save it for future adjustments. *In the next step, the timing is very important.* Take a watch (or a stopwatch), and *at a set time* (for example, 2 hr. 15 min. 00 sec. for the watch, or 0.0 for the stopwatch), add *exactly* 0.3 mL enzyme solution to the sample tube. Mix it thoroughly and quickly by shaking the tube. Insert the sample tube into the spectrophotometer and take a first reading 1 min. after the mixing time (i.e., 2 hr. 16 min. 00 sec., or 60 sec.). Record the absorbance on your Report Sheet in column 1 (for sample 1). Thereafter, take a reading from the spectrophotometer every 30 sec. and record the readings for a total of 6 min. on your Report Sheet in column 1.
2. *Sample 2.* Repeat the experiment exactly as above for sample 1: Prepare the solution in the same way; readjust the instrument with the blank; add the enzyme; then read the absorbance after 1 min. and thereafter every 30 sec. for 6 min. Record the spectrophotometric readings on your Report Sheet in column 2.
3. *Sample 3.* Prepare a new solution in a clean, dry sample tube with the following reagents: 1.0 mL reagent cocktail, 0.8 mL distilled water, and 0.2 mL $NADP^+$ solution. *Be careful to pipet exactly 0.2 mL* $NADP^+$. Mix it thoroughly. Readjust the spectrophotometer with the blank to 0.00 absorbance (or 100% transmission). *At a set time,* add *exactly* 0.2 mL of enzyme solution. Mix the sample tube and insert it into the spectrophotometer. Take your first reading 1 min. after mixing and every 30 sec., for 6 min., thereafter. Record the absorbance readings on your Report Sheet in column 3.
4. *Sample 4.* Prepare a new solution in a clean, dry sample tube with the following reagents: 1.0 mL reagent cocktail, 0.6 mL distilled water, and 0.2 mL $NADP^+$ solution. *Again, be careful to pipet exactly 0.2 mL* $NADP^+$. Mix it thoroughly. Readjust the spectrophotometer with the blank to 0.00 absorbance (or 100% transmission). *At a set time,* add *exactly* 0.4 mL enzyme solution to the sample tube. Mix it thoroughly. Insert the sample tube into the spectrophotometer. Take a first reading 1 min. after mixing and every 30 sec., for 6 min., thereafter. Record the absorbance readings on your Report Sheet in column 4.

Plotting Data

1. Plot the numerical data you recorded in the four columns on the graph paper provided (1a). Note that somewhere *between 3 and 5 min.*, your graphs are linear.
2. Obtain slopes of these linear portions (the change in absorbance divided by the change in time over this range). Record them on your Report Sheet (1b).
3. Calculate the activities of the enzyme, first as units per mL of sample solution (2a), and second as units per mg of enzyme powder (2b).

CHEMICALS AND EQUIPMENT

1. Phosphate buffer, pH 7.0
2. 0.1 M $MgCl_2$ solution
3. 6.0 mM $NADP^+$ solution
4. 15.0 mM isocitrate solution
5. Isocitrate dehydrogenase (0.2 mg powder/mL solution)
6. Spectrophotometers with 5 cuvettes each
7. 1-mL graduated pipets

name

section

date

partner

grade

Pre-Lab Questions

1. What is the importance of the citric acid cycle?

2. What is the origin of the CO_2 formed in the reaction?

3. Consult your text for the structures of citrate and isocitrate. Explain the differences between these two structures. Why was the isomerization a necessary step?

4. How is isocitrate dehydrogenase activity measured?

5. Consult your textbook for the structures of NAD^+ and $NADP^+$ and explain the difference between them. What is the role of these coenzymes?

name | *section* | *date*

partner | *grade*

46 EXPERIMENT 46

Report Sheet

Time (sec.) after Mixing	Absorbance of Sample			
	Column 1	*2*	*3*	*4*
60				
90				
120				
150				
180				
210				
240				
270				
300				
330				
360				

1a. Plot your data: Absorbance versus time.

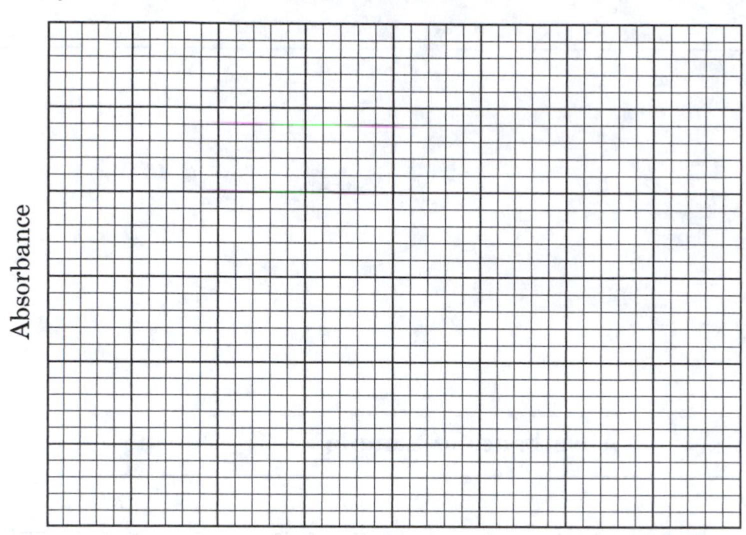

1b.

Sample	*Slope*
1	
2	
3	
4	

2. Calculate the enzyme activity:

a. Units of enzyme activity/mL reaction mixture: The slope of the plot is usually a straight line. If so, read the value of change in absorbance per min. Divide it by 0.01. This gives you the number of enzyme activity units/reaction mixture. (One unit of enzyme activity is 0.01 absorbance/min.) Your reaction mixture had a volume of 2.2 mL. Thus, dividing by 2.2 will give you the activities in units/mL reaction mixture.

	(1)	*(2)*	*(3)*	*(4)*
Units/reaction mixture				
Units/mL reaction mixture				

b. Calculate the isocitrate dehydrogenase activity per mg powder extract. For example, your enzyme solution contained 0.2 mg powder extract/mL solution. If you added 0.2 mL of enzyme solution, it contained 0.04 mg powder extract. Dividing the units/reaction mixture (obtained above) by the number of mg of powder extract added gives you the units/mg powder extract.

	(1)	*(2)*	*(3)*	*(4)*
Units of enzyme activity/mg powder extract				

Post-Lab Questions

1. In the procedure, for the preparation of the "cocktail," $MgCl_2$ was added to the reaction mixture. What was the purpose of this reagent? If this reagent was omitted, would the activity of the enzyme be different from the result obtained? If so, in what way?

2. Suppose that the collection of absorbance data was started after 100 sec. had elapsed. Would the calculated enzyme activity be (a) the same, (b) greater, or (c) smaller than if the experiment had been performed as directed in the procedure? Explain your answer.

3. What do your results show for the different enzyme concentrations used in this experiment? Is there an observable trend or do you get the same enzyme activity (units/mg powder)?

4. What kind of result would you expect if acid was introduced into the solution and the pH became 2?

5. You ran two experiments with the same enzyme concentration in (1) and (2). Calculate the average activity for this concentration of enzyme in units/mg powder. Does the reproducibility fall within ±5%? Account for any large error.

Quantitative Analysis of Vitamin C Contained in Foods

BACKGROUND

Ascorbic acid is commonly known as vitamin C. It was one of the first vitamins that played a role in establishing the relationship between a disease and its prevention by proper diet. The disease *scurvy* has been known for ages, and a vivid description of it was given by Jacques Cartier, a 16th-century explorer of the American continent: "Some did lose their strength and could not stand on their feet. . . . Others . . . had their skin spotted with spots of blood . . . their mouth became stinking, their gums so rotten that all the flesh did fall off." Prevention of scurvy can be obtained by eating fresh vegetables and fruits. The active ingredient in fruits and vegetables that helps to prevent scurvy is ascorbic acid. It is a powerful biological antioxidant (reducing agent). It helps to keep the iron in the enzyme, prolyl hydroxylase, in the reduced form and, thereby, it helps to maintain the enzyme activity. Prolyl hydroxylase is essential for the synthesis of normal collagen. In scurvy, the abnormal collagen causes skin lesions and broken blood vessels.

Vitamin C cannot be synthesized in the human body and must be obtained from the diet (e.g., citrus fruits, broccoli, turnip greens, sweet peppers, tomatoes) or by taking synthetic vitamin C (e.g., vitamin C tablets, "high-C" drinks, and other vitamin C–fortified commercial foods). The minimum recommended adult daily requirement of vitamin C to prevent scurvy is 60 mg. Some people, among them the late Linus Pauling, twice Nobel Laureate, suggested that very large daily doses (250 to 10,000 mg) of vitamin C could help prevent the common cold, or at least lessen the symptoms for many individuals. No reliable medical data support this claim. At present, the human quantitative requirement for vitamin C is still controversial and requires further research.

In this experiment, the amount of vitamin C is determined quantitatively by titrating the test solution with a water-soluble form of iodine, I_3^-:

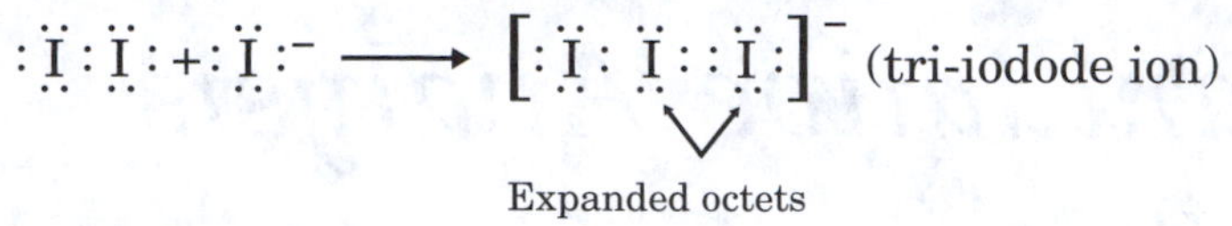

$$:\ddot{I}:\ddot{I}: + :\ddot{I}:^- \longrightarrow [:\ddot{I}:\ddot{I}::\ddot{I}:]^- \text{ (tri-iodode ion)}$$

Expanded octets

Vitamin C is oxidized by I_2 (as I_3^-) according to the following chemical reaction:

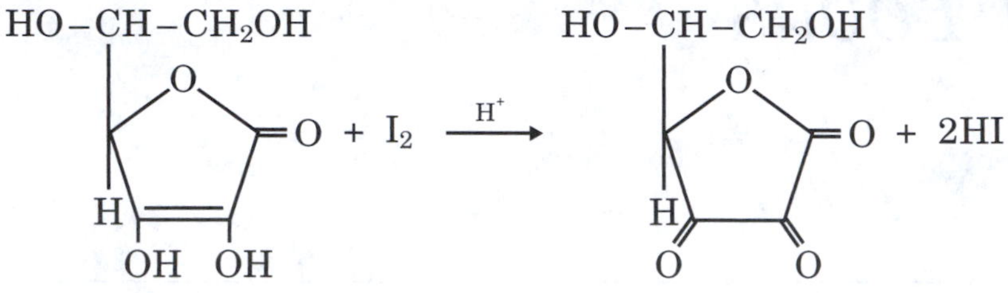

HO–CH–CH_2OH (Vitamin C, MW 176) + I_2 (MW 254) $\xrightarrow{H^+}$ HO–CH–CH_2OH (Oxidized product, MW 174) + 2HI

Vitamin C
(MW 176)

(MW 254)

Oxidized product
(MW 174)

As vitamin C is oxidized by iodine, I_2 becomes reduced to I^-. When the end point is reached (no vitamin C is left), the excess of I_2 will react with a starch indicator to form a starch–iodine complex, which is blackish-blue in color.

$$I_2 + \textbf{starch} \rightarrow \textbf{iodine–starch complex (blackish-blue)}$$

It is worthwhile to know that although vitamin C is very stable when dry, it is readily oxidized by air (oxygen) when in solution; therefore, a solution of vitamin C should not be exposed to air for long periods. The amount of vitamin C can be calculated by using the following conversion factor:

$$\textbf{1 mL of } I_2 \textbf{ (0.01 M) = 1.76 mg vitamin C}$$

OBJECTIVE

To determine the amount of vitamin C that is present in certain commercial food products by the titration method.

PROCEDURE

Prior to this laboratory, there are techniques that you should review. You will find the relevant techniques in the following experiments: (a) Use of the Spectroline pipet filler is in Experiment 2; (b) buret use and titration techniques are in Experiment 19.

1. Pour about 60 mL of a fruit drink that you wish to analyze into a clean, dry 100-mL beaker. The fruit drink should be light colored, apple, orange, or grapefruit, but not dark colored, such as grape. Record the kind of drink on the Report Sheet (1).
2. If the fruit drink is cloudy or contains suspended particles, it can be clarified by the following procedure: Add Celite (about 0.5 g), used as a

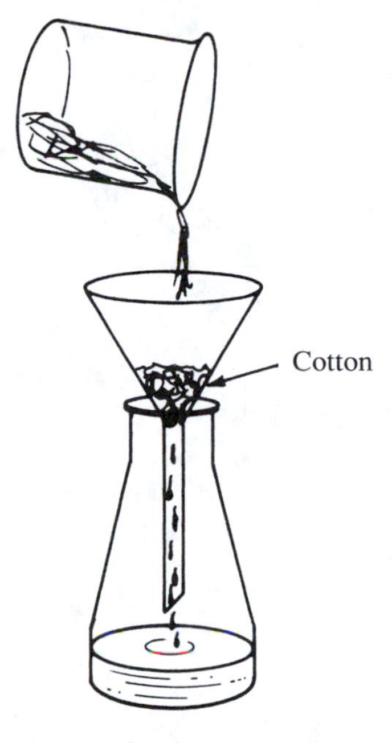

Figure 47.1
Clarification of fruit drinks.

filter aid, to the fruit drink. After swirling it thoroughly, filter the solution through a glass funnel, bedded with a large piece of cotton. Collect the filtrate in a 50-mL Erlenmeyer flask (Figure 47.1).

3. Using a 10-mL volumetric pipet and a Spectroline pipet filler, transfer 10.00 mL of the fruit drink into a 125-mL Erlenmeyer flask (sample 1). Then add 20 mL of distilled water, 5 drops of 3 M HCl (as a catalyst), and 10 drops of 2% starch solution to the flask.
4. Clamp a clean, dry 50-mL buret onto the buret stand. Rinse the buret twice with 5-mL portions of iodine solution. Let the rinses run through the tip of the buret and discard them. Fill the buret slightly above the zero mark with a standardized iodine solution. (A dry funnel may be used for easy transfer.) Air bubbles should be removed by turning the stopcock several times to force the air bubbles out of the tip. Record the molarity of standardized iodine solution (2). Record the initial reading of standardized iodine solution to the nearest 0.02 mL (3a).
5. Place the flask that contains vitamin C sample 1 under the buret and add the iodine solution dropwise, while swirling, until the indicator just changes to dark blue. This color should persist for at least 20 sec. Record the final buret reading (3b).
6. Calculate the total volume of iodine solution required for the titration (3c), the weight of vitamin C in the sample (4), and the percent (w/v) of vitamin C in the drink (5).
7. Repeat this titration procedure with two additional samples of the same fruit drink: (a) sample 2, a 20-mL portion; (b) sample 3, a 30-mL portion. Record the volumes of iodine solution that are required for each titration in the appropriate spaces on the Report Sheet. Carry out the calculations as above.

CHEMICALS AND EQUIPMENT

1. 50-mL buret
2. Buret clamp
3. Spectroline pipet filler
4. 10-mL volumetric pipet
5. 50-mL Erlenmeyer flask
6. Cotton
7. Filter aid
8. Hi-C apple drink
9. Hi-C orange drink
10. Hi-C grapefruit drink
11. 0.01 M iodine in potassium iodide
12. 3 M HCl
13. 2% starch solution

name | section | date

partner | grade

47 EXPERIMENT 47

Pre-Lab Questions

1. Is there any chemical difference between the vitamin C found in fruits and vegetables and the vitamin C synthesized in the laboratory?

2. How is vitamin C obtained by the human body? Is this a vitamin that the body can make through a biosynthetic pathway?

3. What is the biological role of vitamin C in the body?

4. What enzyme is kept in its active form by the presence of vitamin C in the diet?

5. Vitamin C is essential for the prevention of what disease?

6. How does the starch indicator work to indicate the end point in the titration experiment?

name ____________________ section __________ date __________

partner ____________________ grade __________

47 EXPERIMENT 47

Report Sheet

1. The kind of fruit drink ____________________

2. Molarity of iodine solution ____________________

3. Titration results

	Sample 1 (10.0 mL)	*Sample 2 (20.0 mL)*	*Sample 3 (30.0 mL)*
a. Initial buret reading	______ mL	______ mL	______ mL
b. Final buret reading	______ mL	______ mL	______ mL
c. Total volume of iodine solution used: (b − a)	______ mL	______ mL	______ mL
4. The weight of vitamin C in the fruit drink sample: [(3c) × 1.76 mg/mL]	______ mg	______ mg	______ mg
5. Concentration of vitamin C in the fruit drink (mg/100 mL): [(4)/volume of drink] × 100	______	______	______

6. Average concentration of vitamin C in the fruit drink ______ mg/100 mL

Post-Lab Questions

1. Using your average concentration of vitamin C, calculate the volume of fruit drink that would satisfy your minimum daily requirement of vitamin C. Show your work.

2. Vitamin C is ascorbic acid. If you examine the structure given in the **Background** section, there is no –COOH group present. The group present is a lactone (a cyclic ester). Draw the structure of ascorbic acid, circle the lactone, and draw the structure that results from hydrolysis of the lactone.

3. Refer to the oxidized structure of vitamin C in the **Background** section. What functional group forms as a result of the oxidation?

4. When the vitamin C gets oxidized, what happens to the iodine?

5. In the titration, 5 drops of HCl are added. What is the purpose of the acid?

6. The iodine used in the experiment was in the form of I_3^-, not I_2. Why did the experiment require this form of iodine?

Analysis of Vitamin A in Margarine

BACKGROUND

Vitamin A, or retinol, is one of the major fat-soluble vitamins. It is present in many foods; the best natural sources are liver, butter, margarine, egg yolk, carrots, spinach, and sweet potatoes. Vitamin A is the precursor of retinal, the essential component of the visual pigment rhodopsin. Examination of the structure of these molecules shows that they are *conjugated*, that is, the molecules contain double bonds that alternate with single bonds. A characteristic of a conjugated system is that it readily absorbs energy in the ultraviolet or visible region of the electromagnetic spectrum (light). For example, lycopene is red and β-carotene is yellow-orange (see Experiment 23); these molecules absorb in the visible region of light.

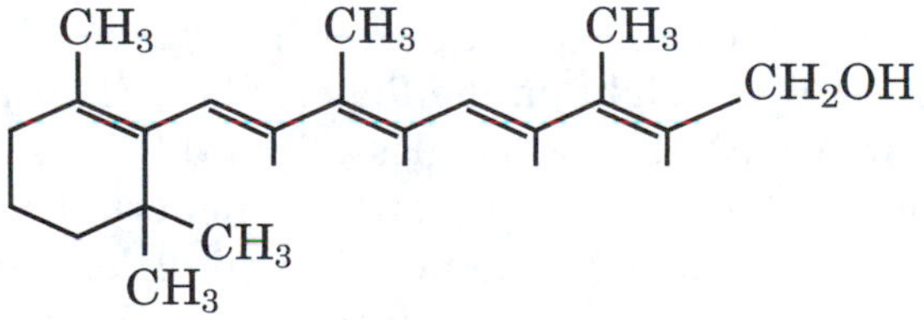

Vitamin A (All-*trans*-retinol)

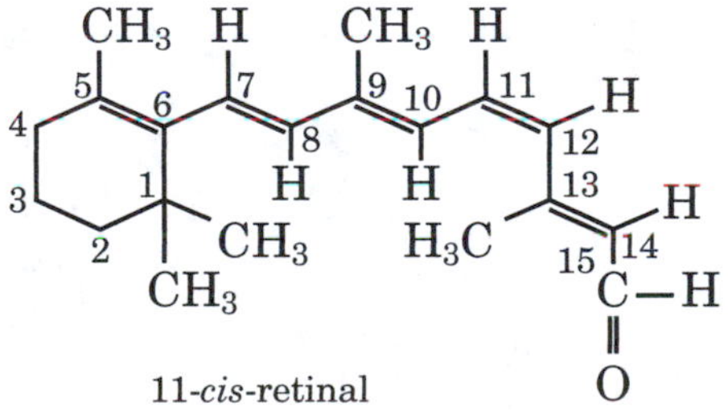

11-*cis*-retinal

When a photon of light penetrates the eye, it is absorbed by the 11-*cis*-retinal. The absorption of light converts the 11-*cis*-retinal to all-*trans*-retinal:

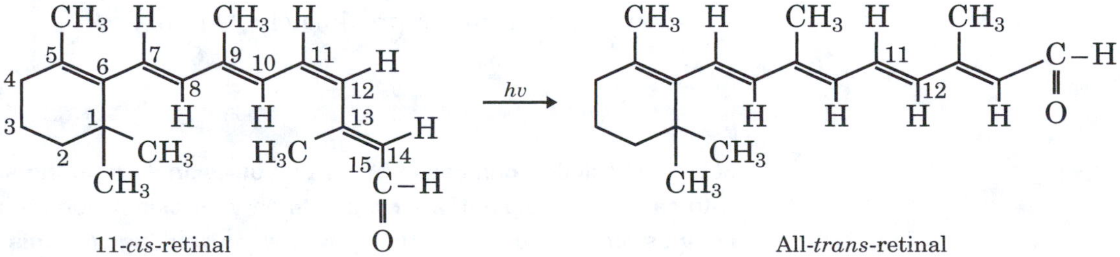

11-*cis*-retinal $\xrightarrow{h\nu}$ All-*trans*-retinal

This isomerization converts the energy of a photon into an atomic motion, which in turn is converted into an electrical signal. The electrical signal generated in the retina of the eye is transmitted through the optic nerve into the brain's visual cortex.

Even though part of the all-*trans*-retinal is regenerated in the dark to 11-*cis*-retinal, for good vision—especially night vision—a constant supply of vitamin A is needed. The recommended daily allowance of vitamin A is 750 μg. Deficiency in vitamin A results in night blindness and keratinization of epithelium. The latter compromises the integrity of healthy skin. In young animals, vitamin A is also required for growth. On the other hand, large doses of vitamin A, sometimes recommended in faddish diets, can be harmful. A daily dose above 1500 μg can be toxic.

OBJECTIVE

To analyze the vitamin A content of margarine by the spectrophotometric method.

PROCEDURE

The analysis of vitamin A requires a multistep process. To help you follow the step-by-step procedure, a flow chart is provided in Figure 48.1.

1. Margarine is largely fat. In order to separate vitamin A from the fat in margarine, first the sample must be saponified. This converts the fat to water-soluble products, glycerol and potassium salts of fatty acids. Vitamin A can be extracted by diethyl ether from the products of the saponification process. To start, weigh a watch glass to the nearest 0.1 g. Report this weight on the Report Sheet (1). Add approximately 10 g of margarine to the watch glass. Record the weight of watch glass plus sample to the nearest 0.1 g on your Report Sheet (2). Transfer the sample from the watch glass into a 250-mL Erlenmeyer flask with the aid of a glass rod, and wash it in with 75 mL of 95% ethanol. Add 25 mL of 50% KOH solution. Cover the Erlenmeyer flask *loosely* with a cork and put it on an electric hot plate. Bring it gradually to a boil. Maintain the boiling for 5 min. with an occasional swirling of the flask using tongs. The stirring should aid the complete dispersal of the sample. Remove the Erlenmeyer from the hot plate and let it cool to room temperature (approximately 20 min.).

CAUTION

50% KOH solution can cause burns on your skin. Handle the solution with care; do not spill it. If a drop gets on your skin, wash it immediately with copious amounts of water. Use gloves when working with this solution.

2. While the sample is cooling, prepare a chromatographic column. Take a 25-mL buret. Add a small piece of glass wool. With the aid of a glass rod, push it down near the stopcock. Add 15–16 mL of petroleum ether to the buret. Open the stopcock slowly, and allow the solvent to fill the tip of the buret. Close the stopcock. You should have 12–13 mL of

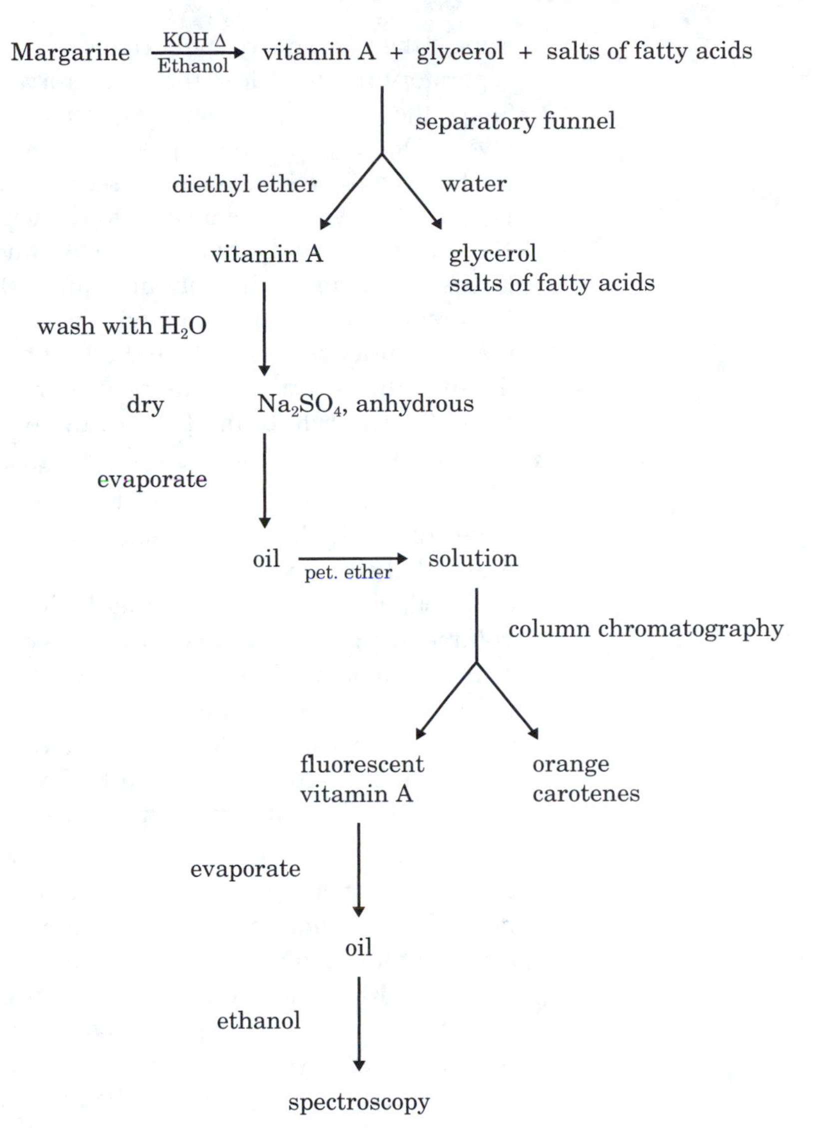

Figure 48.1
Flow chart for vitamin A analysis.

petroleum ether above the glass wool. Weigh about 20 g of alkaline aluminum oxide (alumina) in a 100-mL beaker. Place a small funnel on top of your buret. Pour the alumina slowly, in small increments, into the buret. Allow it to settle to form a 20-cm column. Drain the solvent but do not allow the column to run dry. **Always have at least 0.5 mL of clear solvent on top of the column.** If the alumina adheres to the walls of the buret, wash it down with more solvent.

CAUTION

Petroleum ether and diethyl ether are very volatile and flammable. Make certain that there are no open flames, not even a hot electrical plate, in the vicinity of the operation.

3. Transfer the solution (from your reaction in step 1) from the Erlenmeyer flask to a 500-mL separatory funnel. Rinse the flask with 30 mL of distilled water and add the rinsing to the separatory funnel (see Experiment 31 for a review of separatory funnel techniques).

Repeat the rinsing two more times. Add 100 ml of diethyl ether to the separatory funnel. Close the separatory funnel with the glass stopper. Shake the separatory funnel vigorously. Allow it to separate into two layers. Drain the bottom aqueous layer into an Erlenmeyer flask. Add the top (diethyl ether) layer to a second clean 250-mL Erlenmeyer flask. Pour back the aqueous layer into the separatory funnel. Add another 100-mL portion of diethyl ether. Shake and allow it to separate into two layers. Drain again the bottom (aqueous) layer and discard. Combine the first diethyl ether extract with the residual diethyl ether extract in the separatory funnel. Add 100 mL of distilled water to the combined diethyl ether extracts in the separatory funnel. Agitate it gently and allow the water to drain. Discard the washing.

4. Transfer the diethyl ether extracts into a clean 400-mL beaker. Add 3–5 g of anhydrous Na_2SO_4 and stir it gently for 5 min. to remove traces of water (the anhydrous salt binds water as water of hydration). Decant the diethyl ether extract into a clean 400-mL beaker. Add a boiling stone or a boiling stick. Evaporate the diethyl ether solvent to about 25 mL volume by placing the beaker **in the hood** on a steam bath. Transfer the sample to a 50-mL beaker and continue to evaporate on the steam bath until an oily residue forms. Remove the beaker from the steam bath. Cool it in an ice bath for 1 min. Add 5 mL of petroleum ether and transfer the liquid (without the boiling stone) to a 10-mL volumetric flask. Add sufficient petroleum ether to bring it to volume.
5. Add 5 mL of extracts in petroleum ether to your chromatographic column. By opening the stopcock, drain the sample into your column, but **take care not to let the column run dry**. (Always have about 0.5 mL of liquid on top of the column.) Continue to add solvent to the top of your column. Collect the eluents in a beaker. First you will see the orange-colored carotenes moving down the column. With the aid of a UV lamp, you can also observe a fluorescent band following the carotenes. This fluorescent band contains your vitamin A. Allow all the orange color band to move to the bottom of your column and into the collecting beaker. When the fluorescent band reaches the bottom of the column, close the stopcock. By adding petroleum ether on the top of the column continuously, elute the fluorescent band from the column into a 25-mL graduated cylinder. Continue the elution until all the fluorescent band has been drained into the graduated cylinder. Close the stopcock, and record the volume of the eluate in the graduated cylinder on your Report Sheet (4). Add the vitamin A in the petroleum ether eluate to a dry and clean 50-mL beaker. Evaporate the solvent in the hood on a steam bath. The evaporation is complete when an oily residue appears in the beaker. Add 5 mL of absolute ethanol to the beaker. Transfer the sample into a 10-mL volumetric flask and bring it to volume by adding absolute ethanol.
6. Place your sample in a 1-cm-length quartz spectroscopic cell. The control (blank) spectroscopic cell should contain absolute ethanol. Read the absorbance of your sample against the blank, according to the instructions of your spectrophotometer, at 325 nm. Record the absorption at 325 nm on your Report Sheet (5).
7. Calculate the amount of margarine that yielded the vitamin A in the petroleum ether eluate. Remember that you added only half (5 mL) of

the extract to the column. Report this value on your Report Sheet (6). Calculate the grams of margarine that would have yielded the vitamin A in 1 mL absolute ethanol by dividing (6)/10 mL. Record it on your Report Sheet (7). Calculate the vitamin A in a pound of margarine by using the following formula:

$$\mu g \text{ vitamin A/lb of margarine} = \text{Absorption} \times 5.5 \times [454/(7)].$$

Record your value on the Report Sheet (8).

CHEMICALS AND EQUIPMENT

1. Separatory funnel (500 mL)
2. Buret (25 mL)
3. UV lamp
4. Spectrophotometer (near UV); quartz cells
5. Margarine
6. Petroleum ether (30–60°C)
7. 95% ethanol
8. Absolute ethanol
9. Diethyl ether
10. Glass wool
11. Alkaline aluminum oxide (alumina)
12. 50% potassium hydroxide
13. Sodium sulfate, anhydrous
14. Boiling stones

name

section

date

partner

grade

EXPERIMENT 48

Pre-Lab Questions

1. The structure of β-carotene is given below. What is the difference between β-carotene and vitamin A? How many units of vitamin A can you get from β-carotene?

β-carotene

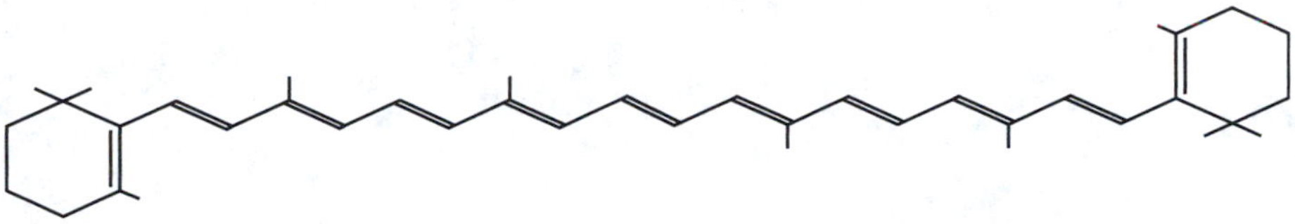

2. Why can you characterize retinol and retinal as conjugated systems?

3. Going from 11-*cis*-retinal to all-*trans*-retinal is what kind of a reaction? What brings this conversion about?

4. Fats in this experiment are hydrolyzed by saponification with base. Why do you hydrolyze with KOH and not with an acid like HCl?

5. In the **Procedure** section, you are cautioned to use petroleum ether and diethyl ether in the hood and not to have any flames in the laboratory. Why must you exercise these precautions?

name ______ section ______ date ______

partner ______ grade ______

48 EXPERIMENT 48

Report Sheet

1. Weight of watch glass ______ g
2. Weight of watch glass + margarine ______ g
3. Weight of margarine: (2) − (1) ______ g
4. Volume of petroleum ether eluate ______ mL
5. Absorption at 325 nm ______
6. Grams margarine in 1 mL of petroleum ether eluate: 2 × [(3)/(4)] ______ g
7. Grams of margarine in 1 mL of absolute ethanol: (6)/10 mL ______ g
8. μg vitamin A/lb margarine: (5) × 5.5 × [454/(7)] ______

Post-Lab Questions

1. Consider the structures of vitamin A and β-carotene. Why does vitamin A absorb in the ultraviolet range whereas β-carotene absorbs in the visible range?

2. You are able to extract the vitamin A from the saponified margarine with diethyl ether. Why can you use this solvent?

3. In the isolation procedure you removed traces of water with anhydrous sodium sulfate. Why can you use a salt, and how does this salt "dry" the organic solvent?

4. The passage of an organic material through a chromatography column with alumina depends on the polarity of the compounds being separated. Comment on the polarities of vitamin A and the carotenes from your observations in this experiment. How do you account for the differences?

5. The label on a commercial margarine sample states that 1 g of it contains 15% of the recommended daily allowance for vitamin A. How does your result compare with this stated amount?

List of Apparatus and Equipment in Student's Locker

AMOUNT AND DESCRIPTION

(1) Beaker, 50 mL
(1) Beaker, 100 mL
(1) Beaker, 250 mL
(1) Beaker, 400 mL
(1) Beaker, 600 mL
(1) Bunsen burner
(3) Clamps, extension
(3) Clamps, holder
(1) Cylinder, graduated by 0.1 mL, 10 mL
(1) Cylinder, graduated by 1 mL, 100 mL
(1) Dropper, medicine with rubber bulb
(1) Evaporating dish
(2) Flasks, Erlenmeyer, 50 mL
(2) Flasks, Erlenmeyer, 125 mL
(1) Flask, Erlenmeyer, 250 mL
(1) Flask, Erlenmeyer, 500 mL
(1) File, triangular
(1) Forceps
(1) Funnel, short stem
(1) Gauze, wire
(1) Spatula, stainless steel
(1) Sponge
(1) Striker (or box of matches)
(12) Test tubes, 150 × 18 mm
(12) Test tubes, 100 × 13 mm
(1) Test tube brush
(1) Test tube holder

(1) Test tube rack
(1) Thermometer, 150°C
(1) Tongs, crucible
(1) Wash bottle, plastic
(1) Watch glass, 3 in.
(1) Watch glass, 5 in.

List of Common Equipment and Materials in the Laboratory

Each laboratory should be equipped with hoods and safety-related items such as fire extinguisher, fire blankets, safety shower, and eye wash fountain. The equipment and materials listed here for 25 students should be made available in each laboratory.

Acid tray
Aspirators (splashgun type) on sink faucet
Balances, single pan, triple beam (or Centogram) or top-loading (electronic)
Barometer
Clamps, extension
Clamps, thermometer
Clamps, utility
Containers for solid chemical waste disposal
Containers for liquid organic waste disposal
Corks
Cotton
Detergent for washing glassware
Drying oven
Filter paper
Glass rods, 4 and 6 mm OD
Glass tubing, 6 and 8 mm OD
Glass wool
Glycerol (glycerine) in dropper bottles
Hot plates
Ice maker
Kimwipes®
Paper towel dispensers
Pasteur pipets
Rings, support, iron, 76 mm OD

Ring stands
Rubber tubing, pressure
Rubber tubing, latex (0.25 in. OD)
Tongs, beaker
Water, deionized or distilled
Weighing dishes, polystyrene, disposable, 73 × 73 × 25 mm
Weighing paper

Special Equipment and Chemicals

In the instructions below, every time a solution is to be made up in "water" you must use *distilled water*.

1 EXPERIMENT 1

Laboratory Techniques: Using the Laboratory Gas Burner; Making Laboratory Measurements

SPECIAL EQUIPMENT

(25)	Bunsen or Tirrill burners
(25)	50-mL graduated beakers
(25)	50-mL graduated Erlenmeyer flasks
(25)	100-mL graduated cylinders
(25)	Metersticks or rulers, with both English and metric scales
(25)	Hot plates
(25)	Single pan, triple beam balances (Centogram)
(5)	Platform triple beam balances
(5)	Top-loading balances
(25)	Quarters

2 EXPERIMENT 2

Density Determination

SPECIAL EQUIPMENT

(5)	250-mL beakers (labeled for unknown metals)
(25)	Solid wood blocks, rectangular or cubic
(25)	Spectroline pipet fillers
(25)	10-mL volumetric pipets

CHEMICALS

Unknown Metals

(100 g)	Aluminum, pellets or rod
(100 g)	Iron, pellets
(100 g)	Lead, shot
(100 g)	Tin, pellets or cut strips
(100 g)	Zinc, pellets

Unknown Liquids

(250 mL)	Ethanol
(250 mL)	Ethylene glycol (commercial permanent antifreeze may be used instead)
(250 mL)	Hexane
(250 mL)	Milk, homogenized, regular
(250 mL)	Olive oil
(250 mL)	Sea water: this can be a 3.5% NaCl solution by weight. Dissolve 8.75 g of NaCl in enough water to make 250 mL.

3 EXPERIMENT 3

Separation of the Components of a Mixture

SPECIAL EQUIPMENT

(2)	Top-loading balances (weigh to 0.001 g)
(15)	Single pan, triple beam (Centogram) balances (weigh to 0.01 g) as an alternative
(25)	Evaporating dishes, porcelain, 6 cm dia.
(25)	Rubber policeman
(1 box)	Filter paper, 15 cm, fast flow
(25)	Mortar and pestle

CHEMICALS

(30 g) Unknown mixture: mix 3.0 g naphthalene (10%), 15 g sodium chloride (50%), 12 g sea sand (40%)

(1 jar) Boiling stones (silicon carbide chips if available)

EXPERIMENT 4

Resolution of a Mixture by Distillation

SPECIAL EQUIPMENT

(25) Distillation kits with 19/22 standard taper joints: 100-mL round-bottom flasks (2); distilling head; thermometer adapter; 110°C thermometer; condenser; vacuum adapter

(25) Nickel wires

CHEMICALS

(1 jar) Boiling stones

(1 jar) Silicone grease

(2 L) 5% sodium chloride solution: dissolve 100 g NaCl in enough water to make 2 L

(100 mL) 0.5 M silver nitrate: dissolve 8.5 g $AgNO_3$ in enough water to make 100 mL solution

(50 mL) Concentrated nitric acid, HNO_3

5

EXPERIMENT 5

The Empirical Formula of a Compound: The Law of Constant Composition

SPECIAL EQUIPMENT

(1 box) Filter papers (Whatman no. 2), 7.0 cm

(25) Hot plates

(25) Rubber policemen

(25) Vacuum filtration setup: (2) 250-mL filter flasks; rubber stopper (no. 6, 1 hole) with glass tubing inserted (10-cm length × 7 mm OD); (2) vacuum tubing, 2-ft. lengths; Büchner funnel (65 mm OD) in a no. 6 1-hole rubber stopper

CHEMICALS

(25) Aluminum wires, no. 18 (45 cm; approx. 1.5 g)

(500 mL) Acetone (use dropper bottles)

(500 mL)	6 M aqueous ammonia, 6 M $NH_3(aq)$: add 200 mL concentrated NH_3 (28%) into a 500-mL volumetric flask and add enough water to bring to the mark. Place the prepared solution in dropper bottles. **Prepare in the hood using a face shield, rubber apron, and rubber gloves.**
(200 g)	Copper(II) chloride, $CuCl_2$

6 EXPERIMENT 6

Determination of the Formula of a Metal Oxide

SPECIAL EQUIPMENT

(25)	Porcelain crucibles and covers
(25)	Clay triangles
(25)	Crucible tongs
(25)	Eye droppers

CHEMICALS

(25)	Magnesium ribbon, 12-cm strip
(500 mL)	6 M HCl: take 250 mL 12 M HCl and add to enough ice-cold water to make 500 mL of solution. **Wear a face mask, rubber gloves, and a rubber apron during the preparation. Do in the hood.**

EXPERIMENT 7

Classes of Chemical Reactions

SPECIAL EQUIPMENT

(1 box)	Wood splints

CHEMICALS

(25 pieces)	Aluminum foil (2 × 0.5 in. each)
(50 pieces)	Aluminum wire (1 cm each)
(25 pieces)	Copper foil (2 × 0.5 in. each)
(25)	Pre-1982 copper penny (optional)
(20 g)	Ammonium carbonate, $(NH_4)_2CO_3$
(20 g)	Potassium iodide, KI
(40 g)	Potassium iodate, KIO_3
(20 g)	Calcium turnings, Ca

(20 g)	Magnesium ribbon, Mg
(20 g)	Mossy zinc, Zn
(20 g)	Lead shot, Pb

All the following solutions should be placed in dropper bottles:

(100 mL)	3 M hydrochloric acid, 3 M HCl: 25 mL 12 M HCl diluted with ice-cold water to 100 mL
(100 mL)	6 M hydrochloric acid, 6 M HCl: 50 mL 12 M HCl diluted with ice-cold water to 100 mL
(100 mL)	3 M sulfuric acid, 3 M H_2SO_4: 16.7 mL 18 M H_2SO_4 is slowly added to 60 mL ice-cold water; stir slowly and dilute with water to 100 mL
(100 mL)	3 M sodium hydroxide, 3 M NaOH: dissolve 12 g NaOH per 100 mL solution

In preparing the above solutions, rubber gloves, a rubber apron, and a face shield should be worn. Do all preparations in the hood.

(100 mL)	0.1 M silver nitrate, 0.1 M $AgNO_3$: dissolve 1.70 g $AgNO_3$ in enough water to make 100 mL
(100 mL)	0.1 M sodium chloride, 0.1 M NaCl: dissolve 0.58 g NaCl in enough water to make 100 mL
(100 mL)	0.1 M sodium nitrate, 0.1 M $NaNO_3$: dissolve 0.85 g $NaNO_3$ in enough water to make 100 mL
(100 mL)	0.1 M sodium carbonate, 0.1 M Na_2CO_3: dissolve 1.24 g $Na_2CO_3 \cdot H_2O$ in enough water to make 100 mL
(100 mL)	0.1 M potassium nitrate, 0.1 M KNO_3: dissolve 1.01 g KNO_3 in enough water to make 100 mL
(100 mL)	0.1 M potassium chromate, 0.1 M K_2CrO_4: dissolve 1.94 g K_2CrO_4 in enough water to make 100 mL
(100 mL)	0.1 M barium chloride, 0.1 M $BaCl_2$: dissolve 2.08 g $BaCl_2$ in enough water to make 100 mL
(100 mL)	0.1 M copper(II) nitrate, 0.1 M $Cu(NO_3)_2$: dissolve 1.88 g $Cu(NO_3)_2$ in enough water to make 100 mL
(100 mL)	0.1 M copper(II) chloride, 0.1 M $CuCl_2$: dissolve 1.70 g $CuCl_2 \cdot 2H_2O$ in enough water to make 100 mL
(100 mL)	0.1 M lead(II) nitrate, 0.1 M $Pb(NO_3)_2$: dissolve 3.31 g $Pb(NO_3)_2$ in enough water to make 100 mL
(100 mL)	0.1 M iron(III) nitrate, 0.1 M $Fe(NO_3)_3$: dissolve 4.04 g $Fe(NO_3)_3 \cdot 9H_2O$ in enough water to make 100 mL

8 **EXPERIMENT 8**

Chemical Properties of Consumer Products

SPECIAL EQUIPMENT

(1 roll)	Copper wire
(2 vials)	Litmus paper, blue
(2 vials)	Litmus paper, red

CHEMICALS

All solutions should be placed in dropper bottles. In preparing all acid and base solutions, observe personal safety practices. Use a face shield, rubber gloves, and a rubber apron. Do preparations in the hood.

(100 mL) Commercial ammonia solution, NH_3 (2.8% ammonia solution can be substituted: 10 mL 28% NH_3 solution diluted to 100 mL with water)
(100 mL) Commercial bleach containing sodium hypochlorite, NaOCl
(100 g) Commercial baking soda, $NaHCO_3$
(100 g) Commercial detergent containing sodium phosphate, Na_3PO_4
(100 g) Garden fertilizer containing ammonium phosphate, $(NH_4)_3PO_4$
(100 g) Epsom salt, $MgSO_4 \cdot 7H_2O$
(100 g) Table salt, sodium chloride, NaCl
(100 g) Ammonium chloride, NH_4Cl
(100 g) Potassium iodide, KI
(100 mL) Ammonium molybdate reagent: preparation
solution 1: dissolve 100 g of ammonium molybdate, $(NH_4)_6Mo_7O_{24}$, in 400 mL water; add slowly, with stirring, 80 mL concentrated NH_3 (15 M NH_3).
solution 2: add 400 mL concentrated HNO_3 (16 M HNO_3), slowly, with stirring, to 600 mL ice-cold water.
Mix *solution 1* and *solution 2* in the proportion of 1:2.
(100 mL) 1 M barium chloride, 1 M $BaCl_2$: dissolve 24.42 g $BaCl_2 \cdot 2H_2O$ in enough water to make 100 mL
(100 mL) 5% barium hydroxide, 5% $Ba(OH)_2$: dissolve 5.0 g $Ba(OH)_2 \cdot 8H_2O$ in enough water to make 100 mL
(200 mL) 3 M nitric acid, 3 M HNO_3: 50 mL concentrated HNO_3 (12 M HNO_3) diluted to 200 mL solution with ice-cold water
(200 mL) 6 M nitric acid, 6 M HNO_3: 100 mL concentrated HNO_3 (12 M HNO_3) diluted to 200 mL solution with ice-cold water
(100 mL) 0.1 M silver nitrate, 0.1 M $AgNO_3$: dissolve 1.70 g $AgNO_3$ in enough water to make 100 mL
(100 mL) 6 M sodium hydroxide, 6 M NaOH: dissolve 24 g NaOH in enough water to make 100 mL
(100 mL) 1 M sodium phosphate, 1 M Na_3PO_4: dissolve 38 g $Na_3PO_4 \cdot 12H_2O$ in enough water to make 100 mL
(100 mL) 6 M sulfuric acid, 6 M H_2SO_4: pour 33.3 mL concentrated H_2SO_4 (18 M H_2SO_4) into 50 mL ice-cold water. Stir slowly. Dilute to 100 mL volume.
(100 mL) 0.01% *p*-nitrobenzene-azoresorcinol: dissolve 0.01 g *p*-nitrobenzene-azoresorcinol in 100 mL 0.025 M NaOH
(100 mL) Hexane, $CH_3CH_2CH_2CH_2CH_2CH_3$

9 EXPERIMENT 9

Calorimetry: The Determination of the Specific Heat of a Metal

SPECIAL EQUIPMENT

(25)	Wire loops for stirring
(25)	Rubber rings (cut from latex tubing)
(50)	Styrofoam cups (8 oz.)
(25)	Lids for styrofoam cups
(25)	Thermometers, 110°C
(25)	Stopwatches
(25)	Thermometer clamps
(25)	Test tubes, 150 × 25 mm

CHEMICALS

(1 kg)	Lead shot, no. 8
(1 kg)	Aluminum metal, turnings or wire
(1 kg)	Iron metal
(1 kg)	Tin metal, granular (mossy)
(1 kg)	Zinc shot

10 EXPERIMENT 10

Boyle's Law: The Pressure–Volume Relationship of Gases

SPECIAL EQUIPMENT

(25)	Boyle's Law apparatus: the apparatus can be constructed as follows. A piece of glass tubing, 30 cm in length, 3 mm OD, is sealed at one end. A Pasteur disposable pipet is drawn out to form a capillary; the capillary needs to be only small enough to be inserted into the 3-mm OD glass tubing and long enough to reach half the length of the tubing (approx. 15 cm). Mercury is transferred with the pipet into the tubing; enough mercury is placed in the tube to give a 10-cm long column. The tube is attached to a 1-ft. ruler by means of rubber bands. The ruler should read with both English (to nearest 1/16 in.) and metric (to nearest mm) scales.
(25)	30°-60°-90° plastic triangles

11

EXPERIMENT 11

Charles's Law: The Volume–Temperature Relationship of Gases

SPECIAL EQUIPMENT

(25)	Bunsen burner (or hot plate)
(50)	250-mL Erlenmeyer flask
(50)	800-mL beaker
(50)	Clamp
(25)	Glass tubing (5- to 8-cm length; 7 mm OD)
(25)	Marking pencil
(25)	One-hole rubber stopper (size no. 6)
(25)	Premade stopper assembly for 250-mL Erlenmeyer flask (optional alternative)
(25)	Ring stand
(25)	Ring support
(25)	Rubber tubing, latex (2-ft. length)
(25)	Thermometer, 110°C
(25)	Wire gauze

CHEMICALS

(1 jar)	Boiling stones

12

EXPERIMENT 12

Properties of Gases: Determination of the Molar Mass of a Volatile Liquid

SPECIAL EQUIPMENT

(25)	Aluminum foil, 2.5 × 2.5 in.
(25)	Aluminum foil, 3 × 3 in.
(1 roll)	Copper wire
(25)	Beaker tongs
(25)	Crucible tongs
(25)	Rubber bands
(25)	Hot plates
(25)	Lead sinkers

CHEMICALS

(1 jar)	Boiling stones

The following liquids should be placed in dropper bottles.

(100 mL)	Pentane
(100 mL)	Acetone
(100 mL)	Methanol (methyl alcohol)
(100 mL)	Hexane
(100 mL)	Ethanol (ethyl alcohol)
(100 mL)	2-Propanol (isopropyl alcohol)

13 EXPERIMENT 13

Physical Properties of Chemicals: Melting Point, Sublimation, and Boiling Point

SPECIAL EQUIPMENT

(1 roll)	Aluminum foil
(1 bottle)	Boiling stones
(1)	Commercial melting point apparatus (if available)
(25)	Glass tubing, 20-cm segments
(25)	Hot plates
(100)	Melting-point capillary tubes
(50)	Rubber rings (cut 0.25-in. rubber tubing into narrow segments)
(25)	Thermometer clamps
(25)	Thiele tube melting-point apparatus

CHEMICALS

(20 g)	Acetamide
(20 g)	Acetanilide
(20 g)	Adipic acid
(20 g)	Benzophenone
(20 g)	Benzoic acid
(20 g)	*p*-Dichlorobenzene
(20 g)	Naphthalene, pure
(50 g)	Naphthalene, impure: mix 47.5 g (95%) naphthalene and 2.5 g (5%) charcoal powder
(20 g)	Stearic acid

The following liquids should be placed in dropper bottles.

(200 mL)	Acetone
(200 mL)	Cyclohexane
(200 mL)	Ethyl acetate
(200 mL)	Hexane
(200 mL)	2-Propanol (isopropyl alcohol)
(200 mL)	Methanol (methyl alcohol)
(200 mL)	1-Propanol

14 EXPERIMENT 14

Solubility and Solutions

SPECIAL EQUIPMENT

(12)	Electrical conductivity apparatus (one for each pair of students)
(50)	Beakers, 30 mL
(25)	Beaker tongs
(25)	Hot plates
(25)	Test tubes, 150 × 25 mm

CHEMICALS

(10 g)	Granulated table sugar, sucrose
(10 g)	Table salt, NaCl
(10 g)	Naphthalene
(10 g)	Iodine
(500 mL)	Ethanol (ethyl alcohol)
(500 g)	Potassium nitrate, KNO_3
(500 mL)	Acetone
(500 mL)	Petroleum ether (b.p. 30–60°C)
(500 mL)	1 M NaCl: dissolve 29.22 g of NaCl in water and bring to 500 mL volume
(500 mL)	0.1 M NaCl: take 50 mL of 1 M NaCl and add enough water to make 500 mL
(500 mL)	1 M sucrose: dissolve 171.15 g sucrose in water and bring to 500 mL volume
(500 mL)	0.1 M sucrose: take 50 mL of 1 M sucrose and add enough water to make 500 mL
(500 mL)	1 M HCl: add 41.7 mL concentrated HCl (12 M HCl) to 200 mL of ice-cold water; add water to bring to 500 mL volume **Use a face shield, rubber gloves, and a rubber apron during the preparation. Do in the hood.**
(500 mL)	0.1 M HCl: add 50 mL of 1 M HCl to enough water to make 500 mL **(use the same precautions as in the above preparation)**
(500 mL)	Glacial acetic acid
(500 mL)	0.1 M acetic acid: take 3 mL glacial acetic acid and add water to bring to 500 mL volume

15 EXPERIMENT 15

Water of Hydration

SPECIAL EQUIPMENT

(25)	Crucibles and covers
(25)	Clay triangles
(25)	Crucible tongs
(25)	Ring stands

CHEMICALS

(25 g)	Calcium chloride, anhydrous, $CaCl_2$ (beads)
(100 g)	Copper(II) sulfate pentahydrate, $CuSO_4 \cdot 5H_2O$

16 EXPERIMENT 16

Factors Affecting Reaction Rates

SPECIAL EQUIPMENT

(5)	Mortars
(5)	Pestles
(25)	10-mL graduated pipets
(25)	5-mL volumetric pipets

CHEMICALS

Solutions should be put into dropper bottles. In preparing the solutions, wear a face shield, rubber gloves, and a rubber apron. Do in the hood.

(100 mL)	3 M H_2SO_4; dissolve 16.7 mL concentrated H_2SO_4 (18 M H_2SO_4) in 60 mL ice-cold water. Stir gently and bring to 100 mL volume.
(500 mL)	6 M HCl: add 250 mL concentrated HCl (12 M HCl) to 200 mL ice-cold water. Mix and bring it to 500 mL volume.
(100 mL)	2 M H_3PO_4: add 13.3 mL concentrated H_3PO_4 (15 M H_3PO_4) to 50 mL ice-cold water. Mix and bring it to 100 mL volume.
(100 mL)	6 M HNO_3: add 50.0 mL concentrated HNO_3 (12 M HNO_3) to 50 mL ice-cold water. Mix and bring it to 100 mL volume.
(100 mL)	6 M acetic acid: add 34.4 mL glacial acetic acid (99–100%) to 50 mL water. Mix and bring it to 100 mL volume.
(500 mL)	0.1 M KIO_3: **Caution!** *This solution must be fresh. Prepare it on the day of the experiment.* Dissolve 10.7 g KIO_3 in enough water to make 500 mL volume.
(250 mL)	4% starch indicator: add 10 g soluble starch to 50 mL cold water. Stir it to make a paste. Bring 200 mL water to a boil in a 500-mL beaker. Pour the starch paste into the boiling water. Stir and cool to room temperature.
(500 mL)	0.01 M $NaHSO_3$: dissolve 0.52 g $NaHSO_3$ in 100 mL water. Add slowly 2 mL concentrated sulfuric acid (18 M H_2SO_4). Stir and bring it to 500 mL volume.
(250 mL)	3% hydrogen peroxide: take 25 mL concentrated H_2O_2 (30%) and bring it to 250 mL volume with water.
(150)	Mg ribbons, 1 cm long
(25)	Zn ribbons, 1 cm long
(25)	Cu ribbons, 1 cm long
(25 g)	MnO_2

17 EXPERIMENT 17

The Law of Chemical Equilibrium and Le Chatelier's Principle

SPECIAL EQUIPMENT

(2 rolls)	Litmus paper, blue
(2 rolls)	Litmus paper, red

CHEMICALS

(50 mL)	0.1 M copper(II) sulfate: dissolve 0.80 g $CuSO_4$ (or 1.25 g $CuSO_4 \cdot 5H_2O$) in enough water to make 50 mL
(50 mL)	1 M ammonia: dilute 3.3 mL concentrated NH_3 (28%) with water to 50 mL volume. **In the preparation wear a face shield, rubber gloves, and a rubber apron. Do in the hood.**
(25 mL)	Concentrated HCl (12 M HCl)
(100 mL)	1 M hydrochloric acid: add 8.5 mL concentrated HCl (12 M HCl) to 50 mL ice water; add enough water to bring to 100 mL. **In the preparation wear a face shield, rubber gloves, and a rubber apron. Do in the hood.**
(150 mL)	0.1 M phosphate buffer: dissolve 1.74 g K_2HPO_4 in enough water to make 100 mL. Also dissolve 1.36 g KH_2PO_4 in enough water to make 100 mL. Mix 100 mL K_2HPO_4 with 50 mL of KH_2PO_4 solution.
(100 mL)	0.1 M potassium thiocyanate: dissolve 0.97 g KSCN in enough water to make 100 mL
(100 mL)	0.1 M iron(III) chloride: dissolve 2.7 g $FeCl_3 \cdot 6H_2O$ (or 1.6 g $FeCl_3$) in enough water to make 100 mL
(100 mL)	Saturated saline solution: add 290 g NaCl to warm (60°C) water. Stir until dissolved. Cool to room temperature.
(50 mL)	1.0 M cobalt chloride: dissolve 11.9 g $CoCl_2 \cdot 6H_2O$ in enough water to make 50 mL

18 EXPERIMENT 18

pH and Buffer Solutions

SPECIAL EQUIPMENT

(5)	pH meters
(12 rolls)	pHydron paper (pH range 0 to 12)
(5 boxes)	Kimwipes
(5)	Wash bottles

(100)	10-mL graduated pipets
(25)	Spot plates
(25)	10-mL beakers

CHEMICALS

(250 mL)	0.1 M acetic acid, 0.1 M CH_3COOH: dissolve 1.4 mL glacial acetic acid in enough water to make 250 mL
(500 mL)	0.1 M sodium acetate, 0.1 M CH_3COONa: dissolve 6.8 g $CH_3COONa \cdot 3H_2O$ in enough water to make 500 mL
(1 L)	0.1 M hydrochloric acid, 0.1 M HCl: add 8.5 mL concentrated HCl (12 M HCl) to 100 mL ice-cold water with stirring; dilute with water to 1 L. **Prepare in the hood; wear a face shield, rubber gloves, and a rubber apron.**
(1 L)	0.1 M sodium bicarbonate, 0.1 M $NaHCO_3$: dissolve 8.2 g $NaHCO_3$ in enough water to make 1 L
(500 mL)	Saturated carbonic acid, H_2CO_3: use a bottle of Club Soda or Seltzer water; these solutions are approximately 0.1 M carbonic acid.

The following solutions should be placed in dropper bottles.

(100 mL)	0.1 M HCl prepared above
(100 mL)	0.1 M ammonia, 0.1 M NH_3: dilute 0.7 mL concentrated NH_3 (28%) with enough water to make 100 mL
(100 mL)	0.1 M sodium hydroxide, 0.1 M NaOH: dissolve 0.4 g NaOH in enough water to make 100 mL

19 EXPERIMENT 19

Analysis of Vinegar by Titration

SPECIAL EQUIPMENT

(25)	25-mL (or 50-mL) burets
(25)	Buret clamps
(25)	Ring stands
(25)	5-mL volumetric pipets
(25)	Small funnels
(25)	Spectroline pipet fillers

CHEMICALS

(500 mL)	Vinegar
(2 L)	0.2 M NaOH standardized solution: dissolve 16.8 g NaOH in enough water to make 2 L. Standardize the solution as follows: place approximately 1 g potassium hydrogen phthalate, $KC_8H_5O_4$, in a tared weighing bottle. Weigh it to the nearest 0.001 g. Dissolve it in 20 mL water. Add a few drops of phenolphthalein indicator and titrate with the NaOH solution

prepared above. The molarity of NaOH is calculated as follows: M = mass of phthalate/(0.2043 × mL NaOH used in titration). Write the calculated molarity of the NaOH on the bottle of the standardized NaOH solution.

(100 mL) Phenolphthalein indicator: dissolve 0.1 g phenolphthalein in 60 mL 95% ethanol and bring it to 100 mL volume with water

20

EXPERIMENT 20

Analysis of Antacid Tablets

SPECIAL EQUIPMENT

(25) 25-mL (or 50-mL) burets
(25) 100-mL burets
(25) Buret clamps
(25) Ring stands
(5) Balances to read to 0.001 g

CHEMICALS

(5 bottles) Commercial antacids such as Alka-Seltzer, Gelusil, Maalox, Rolaids, Di-Gel, Tums, etc. Have at least two different kinds available.

(1 L) 0.2 M NaOH, sodium hydroxide, standardized: dissolve 8.4 g NaOH in 1 L water. Standardize to 3 significant figures as follows: accurately weigh to the nearest 0.001 g approximately 1 g potassium hydrogen phthlate, $KC_8H_5O_4$, MW = 204.3 g/mole, and dissolve it in 20 mL water. Add a few drops of phenolphthalein and titrate the potassium hydrogen phthalate with the prepared NaOH solution. The molarity (M) of the NaOH solution is calculated as follows: M = mass of phthalate/(0.2043 × mL NaOH used in the titration). Write the calculated molarity on the bottle of the standardized NaOH solution.

(1 L) 0.2 M HCl, hydrochloric acid: add 16.7 mL concentrated HCl (12 M HCl) to 100 mL ice-cold water; dilute with water to 1 L volume. **(Prepare in the hood; wear a face shield, rubber gloves, and a rubber apron.)** Standardize to 3 significant figures the acid solution by titration against the standardized 0.2 M NaOH solution. Write the calculated molarity on the bottle of the standardized HCl solution.

(100 mL) Thymol blue indicator: dissolve 0.1 g thymol blue in 50 mL 95% ethanol and dilute with water to 100 mL volume. Put in a dropper bottle.

(100 mL) Phenolphthalein indicator: dissolve 0.1 g phenolphthalein in 60 mL 95% ethanol and bring to 100 mL volume with water. Put in a dropper bottle.

21 EXPERIMENT 21

Structure in Organic Compounds: Use of Molecular Models. I

SPECIAL EQUIPMENT

(Color of spheres may vary depending on the set; substitute as necessary.)

(50) Black spheres—4 holes
(300) Yellow spheres—1 hole
(50) Colored spheres (e.g., green)—1 hole
(25) Blue spheres—2 holes
(400) Sticks
(25) Protractors
(75) Springs (optional)

22 EXPERIMENT 22

Stereochemistry: Use of Molecular Models. II

SPECIAL EQUIPMENT

Commercial molecular model kits vary in style, size, material composition, and the color of the components. The set that works best in this exercise is the *Molecular Model Set for Organic Chemistry* (ISBN: 0-205-08136-3) available from Prentice Hall (online at www.prenhall.com). Wood ball-and-stick models work as well. For 25 students, 25 of these sets should be provided. If you wish to make up your own kit, you would need the following for 25 students:

(25) Cyclohexane model kits: each consisting of the following components:
8 carbons—black, 4 holes
18 hydrogens—white, 1 hole
2 substituents—red, 1 hole
24 connectors—bonds
(25) Chiral model kits: each consisting of the following components:
8 carbons—black, 4 holes
32 substituents—8 red, 1 hole; 8 white, 1 hole; 8 blue, 1 hole; 8 green, 1 hole
28 connectors—bonds
(5) Small hand mirrors

23 EXPERIMENT 23

Column and Paper Chromatography: Separation of Plant Pigments

SPECIAL EQUIPMENT

(50)	Melting-point capillaries open at both ends
(25)	25-mL burets
(1 jar)	Glass wool
(25)	Filter papers (Whatman no. 1), 20 × 10 cm
(3)	Heat lamp (optional)
(25)	Ruler with both English and metric scales
(1)	Stapler
(15)	Hot plates with or without water bath

CHEMICALS

(1 lb)	Tomato paste
(500 g)	Aluminum oxide (alumina)
(500 mL)	95% ethanol
(500 mL)	Petroleum ether, b.p. 30–60°C
(500 mL)	Eluting solvent: mix 450 mL petroleum ether with 10 mL toluene and 40 mL acetone
(10 mL)	0.5% β-carotene solution: dissolve 50 mg in 10 mL petroleum ether. Wrap the vial in aluminum foil to protect from light and keep in refrigerator until used.
(150 mL)	Saturated bromine water: mix 5.5 g bromine with 150 mL water. **Prepare in hood; wear a face shield, rubber gloves, and a rubber apron.**
(500 mg)	Iodine crystals

24 EXPERIMENT 24

Identification of Hydrocarbons

SPECIAL EQUIPMENT

(2 vials)	Litmus paper, blue
(250)	100 × 13 mm test tubes

CHEMICALS

(1 g)	Anhydrous aluminum chloride, $AlCl_3$

The following solutions should be placed in dropper bottles.

(100 mL)	Concentrated H_2SO_4 (18 M H_2SO_4)
(100 mL)	Cyclohexene
(100 mL)	Hexane
(100 mL)	Ligroin (b.p. 90–110°C)
(100 mL)	Toluene
(100 mL)	1% Br_2 in cyclohexane **(wear a face shield, rubber gloves, and a rubber apron; prepare under hood):** mix 1.0 mL Br_2 with enough cyclohexane to make 100 mL. *Prepare fresh solutions prior to use; keep in a dark-brown dropper bottle; do not store.*
(100 mL)	1% aqueous $KMnO_4$: dissolve 1.0 g potassium permanganate in 50 mL distilled water by gently heating for 1 hr.; cool and filter; dilute to 100 mL. Store in a dark-brown dropper bottle.
(100 mL)	Unknown A = hexane
(100 mL)	Unknown B = cyclohexene
(100 mL)	Unknown C = toluene

25 EXPERIMENT 25

Identification of Alcohols and Phenols

SPECIAL EQUIPMENT

(125)	Corks (for test tubes 100 × 13 mm)
(125)	Corks (for test tubes 150 × 18 mm)
(25)	Hot plate
(5 rolls)	Indicator paper (pH 1–12)

CHEMICALS

The following solutions should be placed in dropper bottles.

(100 mL)	Acetone (reagent grade)
(100 mL)	1-Butanol
(100 mL)	2-Butanol
(100 mL)	2-Methyl-2-propanol (*t*-butyl alcohol)
(100 mL)	1,2-Dimethoxyethane
(200 mL)	20% aqueous phenol: dissolve 80 g of phenol in 20 mL distilled water; dilute to 400 mL.
(100 mL)	Lucas reagent **(prepare under hood; wear a face shield, rubber gloves, and a rubber apron):** cool 100 mL of concentrated HCl (12 M HCl) in an ice bath; with stirring, add 150 g anhydrous $ZnCl_2$ to the cold acid
(150 mL)	Chromic acid solution **(prepare under hood; wear a face shield, rubber gloves, and a rubber apron):** dissolve 20 g potassium dichromate, $K_2Cr_2O_7$, in 100 mL concentrated sulfuric acid (18 M H_2SO_4). Carefully add this solution to enough ice-cold water to bring to 1 L.
(100 mL)	2.5% iron(III) chloride solution: dissolve 2.5 g anhydrous $FeCl_3$ in 50 mL water; dilute to 100 mL
(100 mL)	Iodine in KI solution: mix 20 g of KI and 10 g of I_2 in 100 mL water

(250 mL) 10% sodium hydroxide, 10% NaOH: dissolve 25.00 g NaOH in 100 mL water. Dilute to 250 mL with water.
(100 mL) Unknown A = 1-butanol
(100 mL) Unknown B = 2-butanol
(100 mL) Unknown C = 2-methyl-2-propanol (*t*-butyl alcohol)
(100 mL) Unknown D = 20% aqueous phenol

26

EXPERIMENT 26

Identification of Aldehydes and Ketones

SPECIAL EQUIPMENT

(250) Corks (to fit 100 × 13 mm test tube)
(125) Corks (to fit 150 × 18 mm test tube)
(1 box) Filter paper (students will need to cut to size)
(25) Hirsch funnels
(25) Hot plates
(25) Neoprene adapters (no. 2)
(25) Rubber stopper assemblies: a no. 6 one-hole stopper fitted with glass tubing (15 cm in length × 7 mm OD)
(25) 50-mL side-arm filter flasks
(25) 250-mL side-arm filter flasks
(50) Vacuum tubing, heavy-walled (2-ft. lengths)

CHEMICALS

(50 g) Hydroxylamine hydrochloride
(100 g) Sodium acetate

The following solutions should be placed in dropper bottles.
(100 mL) Acetone (reagent grade)
(100 mL) Benzaldehyde (freshly distilled)
(100 mL) *Bis*(2-ethoxymethyl) ether
(100 mL) Cyclohexanone
(100 mL) 1,2-Dimethoxyethane
(500 mL) Ethanol (absolute)
(500 mL) Ethanol (95%)
(100 mL) Isovaleraldehyde
(500 mL) Methanol
(100 mL) Pyridine
(150 mL) Chromic acid reagent: dissolve 20 g potassium dichromate, $K_2Cr_2O_7$, in 100 mL concentrated sulfuric acid (18 M H_2SO_4). Carefully add this solution to enough ice-cold water to bring to 1 L. **Wear a face shield, rubber gloves, and a rubber apron during the preparation. Do in the hood.**

Tollens' reagent (Place in dropper bottles.)
(100 mL) Solution A: dissolve 9.0 g silver nitrate in 90 mL of water; dilute to 100 mL

(100 mL) Solution B: 10 g NaOH dissolved in enough water to make 100 mL

The following reagents should be placed in dropper bottles.

(100 mL) 10% ammonia water: 35.7 mL of concentrated (28%) NH_3 diluted to 100 mL

(100 mL) 10% sodium hydroxide, 10% NaOH: dissolve 10.00 g NaOH in enough water to make 100 mL

(500 mL) Iodine-KI solution: mix 100 g of KI and 50 g of iodine in enough distilled to make 500 mL

(100 mL) 2,4-dinitrophenylhydrazine reagent: dissolve 3.0 g of 2,4-dinitrophenylhydrazine in 15 mL concentrated H_2SO_4 (18 M H_2SO_4). In a beaker, mix together 10 mL water and 75 mL 95% ethanol. With vigorous stirring, slowly add the 2,4-dinitrophenylhydrazine solution to the aqueous ethanol mixture. After thorough mixing, filter by gravity through a fluted filter paper. **Wear a face shield, rubber gloves, and a rubber apron during the preparation. Do in the hood.**

(100 mL) Semicarbazide reagent: dissolve 22.2 g of semicarbazide hydrochloride in 100 mL of water

(100 mL) Unknown A = isovaleraldehyde

(100 mL) Unknown B = benzaldehyde

(100 mL) Unknown C = cyclohexanone

(100 mL) Unknown D = acetone

Additional compounds for use as unknowns:

Aldehydes

(100 mL) 2-Butenal (crotonaldehyde)

(100 mL) Octanal (caprylaldehyde)

(100 mL) Pentanal (valeraldehyde)

Ketones

(100 mL) Acetophenone

(100 mL) Cyclopentanone

(100 mL) 2-Pentanone

(100 mL) 3-Pentanone

27 EXPERIMENT 27

Properties of Carboxylic Acids and Esters

SPECIAL EQUIPMENT

(5 rolls) pH paper (range 1–12)

(100) Disposable Pasteur pipets

(5 vials) Litmus paper, blue

(25) Hot plates

CHEMICALS

(10 g) Salicylic acid

(10 g) Benzoic acid

The following solutions are to be placed in dropper bottles.

(75 mL) Acetic acid
(50 mL) Formic acid
(25 mL) Benzyl alcohol
(50 mL) Ethanol (ethyl alcohol)
(25 mL) 2-Methyl-1-propanol (isobutyl alcohol)
(25 mL) 3-Methyl-1-butanol (isopentyl alcohol)
(50 mL) Methanol (methyl alcohol)
(25 mL) Methyl salicylate
(250 mL) 6 M hydrochloric acid, 6 M HCl: take 125 mL of concentrated HCl (12 M HCl) and add to 50 mL of ice-cold water; dilute with enough water to make 250 mL. **Wear a face shield, rubber gloves, and a rubber apron during the preparation. Do in the hood.**
(100 mL) 3 M hydrochloric acid, 3 M HCL: take 50 mL 6 M HCl and bring to 100 mL; **follow the same precautions as above.**
(300 mL) 6 M sodium hydroxide, 6 M NaOH: dissolve 72.00 g NaOH in enough water to bring to 300 mL; **follow the same precautions as above.**
(150 mL) 2 M sodium hydroxide, 2 M NaOH: take 50 mL 6 M NaOH and bring to 150 mL; **follow the same precautions as above.**
(25 mL) Concentrated sulfuric acid (18 M H_2SO_4).

28 **EXPERIMENT 28**

Properties of Amines and Amides

SPECIAL EQUIPMENT

(2 rolls) pH paper (range 0–12)
(25) Hot plates

CHEMICALS

(20 g) Acetamide

The following chemicals and solutions should be placed in dropper bottles.

(25 mL) Triethylamine
(25 mL) Aniline
(25 mL) N,N-Dimethylaniline
(100 mL) Diethyl ether (ether)
(100 mL) 6 M ammonia solution, 6 M NH_3: add 40 mL concentrated NH_3 (28%) to 50 mL water; then add enough water to bring to 100 mL. **Wear a face shield, rubber gloves, and a rubber apron when preparing. Do in the hood.**
(100 mL) 6 M hydrochloric acid, 6 M HCl: add 50 mL concentrated HCl (12 M HCl) to 40 mL ice-cold water; then add enough water to bring to 100 mL. **Wear a face shield, rubber gloves, and a rubber apron when preparing. Do in the hood.**

(50 mL) Concentrated hydrochloric acid (12 M HCl)

(250 mL) 6 M sulfuric acid, 6 M H_2SO_4: pour 83.4 mL concentrated H_2SO_4 (18 M H_2SO_4) into 125 mL ice-cold water. Stir slowly. Then add enough water to bring to 250 mL. **Wear a face shield, rubber gloves, and a rubber apron when preparing. Do in the hood.**

(250 mL) 6 M sodium hydroxide, 6 M NaOH: dissolve 60.00 g NaOH in 100 mL water. Then add enough water to bring to 250 mL. **Wear a face shield, rubber gloves, and a rubber apron when preparing. Do in the hood.**

29

EXPERIMENT 29

Polymerization Reactions

SPECIAL EQUIPMENT

(25) Hot plates
(25) Cylindrical paper rolls or sticks
(25) Bent wire approximately 10 cm long
(25) 10-mL pipets or syringes
(25) Spectroline pipet fillers
(25) Test tubes, 150 × 18 mm, Pyrex
(25) Beaker tongs
(2 Boxes) Microscope slides
(2 Boxes) Filter paper, 18.5 cm

CHEMICALS

(1 kg) Sea sand

The following chemicals and solutions should be placed in dropper bottles.

(75 mL) Styrene
(250 mL) Xylene
(10 mL) *t*-butyl peroxide benzoate (also called *t*-butyl benzoyl peroxide); store at 4°C.
(75 mL) 20% sodium hydroxide: dissolve 15.00 g NaOH in enough water to make 75 mL
(300 mL) 5% adipoyl chloride: dissolve 15.00 g adipoyl chloride in enough cyclohexane to make 300 mL
(300 mL) 5% hexamethylene diamine: dissolve 15.00 g hexamethylene diamine in enough water to make 300 mL
(200 mL) 80% formic acid: add 40 mL water to 160 mL formic acid
(500 mL) 50% ethanol: mix 250 mL absolute ethanol with enough water to make 500 mL

30 EXPERIMENT 30

Preparation of Acetylsalicylic Acid (Aspirin)

SPECIAL EQUIPMENT

(25)	Büchner funnels (65 mm OD)
(25)	Filtervac or no. 2 neoprene adapters
(1 box)	Filter paper (5.5 cm, Whatman no. 2)
(25)	250-mL filter flasks
(25)	Hot plates

CHEMICALS

(1 jar)	Boiling stones
(25)	Commercial aspirin tablets
(100 mL)	Concentrated sulfuric acid (18 M H_2SO_4) (in a dropper bottle)
(100 mL)	1% iron(III) chloride: dissolve 1 g $FeCl_3 \cdot 6H_2O$ in enough distilled water to make 100 mL (in a dropper bottle)
(100 mL)	Acetic anhydride, freshly opened bottle
(300 mL)	Ethyl acetate
(100 g)	Salicylic acid

31 EXPERIMENT 31

Isolation of Caffeine from Tea Leaves

SPECIAL EQUIPMENT

(25)	Cold finger condensers (115 mm long × 15 mm OD)
(1 box)	Filter paper; 7.0 cm, fast flow (Whatman no. 1)
(25)	Hot plates
(50)	Latex tubing, 2-ft. lengths
(1 vial)	Melting-point capillaries
(25)	No. 2 neoprene adapters
(25)	Rubber stopper (no. 6, 1-hole) with glass tubing inserted (10 cm length × 7 mm OD)
(25)	125-mL separatory funnels
(25)	25-mL side-arm filter flasks
(25)	250-mL side-arm filter flasks
(25)	Small sample vials
(1)	Stapler
(50)	Vacuum tubing, 2-ft. lengths
(1 box)	Weighing paper

CHEMICALS

(1 jar)	Boiling stones
(500 mL)	Dichloromethane, CH_2Cl_2
(250 g)	Sodium sulfate, anhydrous, Na_2SO_4
(50 g)	Sodium carbonate, anhydrous, Na_2CO_3
(50)	Tea bags

32 EXPERIMENT 32

Carbohydrates

SPECIAL EQUIPMENT

(50)	Medicine droppers
(125)	Microtest tubes or 25 depressions white spot plates
(2 rolls)	Litmus paper, red

CHEMICALS

(20 g)	Boiling stones
(400 mL)	Fehling's reagent (solutions A and B, from Fisher Scientific Co.)
(200 mL)	3 M NaOH: dissolve 24.00 g NaOH in 100 mL water and then add enough water to bring to 200 mL. Store in a dropper bottle.
(200 mL)	2% starch solution: place 4 g soluble starch in a beaker. With vigorous stirring, add 10 mL water to form a thin paste. Boil 190 mL water in another beaker. Add the starch paste to the boiling water and stir until the solution becomes clear. Store in a dropper bottle.
(200 mL)	2% sucrose: dissolve 4 g sucrose in 200 mL water. Store in a dropper bottle.
(50 mL)	3 M sulfuric acid: add 8.5 mL concentrated H_2SO_4 (18 M H_2SO_4) to 30 mL ice-cold water; **pour the sulfuric acid slowly along the walls of the beaker, so that it will settle on the bottom without much mixing;** stir slowly in order not to generate too much heat; when fully mixed, bring the volume to 50 mL. **Wear a face shield, rubber gloves, and a rubber apron when preparing. Do in the hood.** Store in a dropper bottle.
(100 mL)	2% fructose: dissolve 2 g fructose in enough water to make 100 mL. Store in a dropper bottle.
(100 mL)	2% glucose: dissolve 2 g glucose in enough water to make 100 mL. Store in a dropper bottle.
(100 mL)	2% lactose: dissolve 2 g lactose in enough water to make 100 mL. Store in a dropper bottle.

(100 mL) 0.01 M iodine in KI: dissolve 1.2 g KI in 80 mL water. Add 0.25 g I_2. Stir until the iodine dissolves. Dilute the solution with water to 100 mL. Store in a dark dropper bottle.

33 EXPERIMENT 33

Fermentation of a Carbohydrate: Ethanol from Sucrose

SPECIAL EQUIPMENT

(50) Extension clamps and holders
(50) Ring stands
(25) Heating mantles (or Thermowells)
(25) Variacs
(1 jar) Boiling stones
(1 jar) Celite (Filter Aid)
(1 jar) Silicone grease
(3 pads) Steel wool, coarse (1 package contains 12 pads)
(25) Kits for fractional distillation with 19/22 standard taper joints: 250-mL round-bottom flask; condensers (2); distilling head; thermometer adapter; 110°C thermometer; vacuum adapter; latex tubing (2 lengths, 2 ft. each)

CHEMICALS

(1 L) Calcium hydroxide solution, $Ca(OH)_2$, saturated: mix 75 g of calcium hydroxide with 1 L of water and let stand for several days
(25 g) Disodium hydrogen phosphate, Na_2HPO_4
(50 g) Potassium phosphate, K_3PO_4
(600 g) Sucrose
(50 mL) Mineral oil, in a dropper bottle
(75 g) Dried baker's yeast (Fleischmann's All Natural Yeast comes in individual packets of 7 g each)

EXPERIMENT 34

Preparation and Properties of a Soap

SPECIAL EQUIPMENT

(25) Büchner funnels (85 mm OD)
(25) No. 7 one-hole rubber stoppers
(1 box) Filter paper (7.0 cm, Whatman no. 2)
(1 roll) pHydrion paper (pH range 0–12)

CHEMICALS

(1 jar)	Boiling stones
(1 L)	95% ethanol
(1 L)	Saturated sodium chloride (sat. NaCl): dissolve 360 g NaCl in 1 L water
(1 L)	25% sodium hydroxide (25% NaOH): dissolve 250 g NaOH in enough water to make 1 L
(1 L)	Vegetable oil
(100 mL)	5% iron(III) chloride (5% $FeCl_3$): dissolve 5 g $FeCl_3 \cdot 6H_2O$ in enough water to make 100 mL. Store in a dropper bottle.
(100 mL)	5% calcium chloride (5% $CaCl_2$): dissolve 5 g $CaCl_2 \cdot H_2O$ in enough water to make 100 mL. Store in a dropper bottle.
(100 mL)	Mineral oil. Store in a dropper bottle.
(100 mL)	5% magnesium chloride (5% $MgCl_2$): dissolve 5 g $MgCl_2$ in enough water to make 100 mL. Store in a dropper bottle.

35 EXPERIMENT 35

Preparation of a Hand Cream

SPECIAL EQUIPMENT

(25)	Bunsen burners or hot plates

CHEMICALS

(100 mL)	Triethanolamine
(40 mL)	Propylene glycol (1,2-propanediol)
(500 g)	Stearic acid
(40 g)	Methyl stearate (ethyl stearate may be substituted)
(400 g)	Lanolin
(400 g)	Mineral oil

36 EXPERIMENT 36

Extraction and Identification of Fatty Acids from Corn Oil

SPECIAL EQUIPMENT

(12)	Water baths
(2)	Heat lamps or hair dryers
(25)	15 × 6.5 cm silica gel TLC plates
(25)	Rulers, metric scale
(25 pairs)	Polyethylene surgical gloves

(150)	Capillary tubes, open on both ends
(25)	Glass-stoppered test tubes
(1 jar)	Glass wool

CHEMICALS

(50 g)	Corn oil
(5 mL)	Methyl palmitate solution: dissolve 25 mg methyl palmitate in 5 mL petroleum ether
(5 mL)	Methyl oleate solution: dissolve 25 mg methyl oleate in 5 mL petroleum ether
(5 mL)	Methyl linoleate solution: dissolve 25 mg methyl linoleate in 5 mL petroleum ether
(100 mL)	0.5 M KOH: dissolve 2.81 g KOH in 25 mL water and add enough 95% ethanol to make 100 mL
(500 g)	Sodium sulfate, Na_2SO_4, anhydrous, granular
(100 mL)	Concentrated hydrochloric acid (12 M HCl)
(1 L)	Petroleum ether (b.p. 30–60°C)
(300 mL)	Methanol:perchloric acid mixture: mix 285 mL methanol with 15 mL $HClO_4 \cdot 2H_2O$ (73% perchloric acid)
(400 mL)	Hexane:diethyl ether mixture: mix 320 mL hexane with 80 mL diethyl ether
(10 g)	Iodine crystals, I_2

37 EXPERIMENT 37

Analysis of Lipids

SPECIAL EQUIPMENT

(25)	Hot plates
(25)	Cheese cloth squares, 3 × 3 in.

CHEMICALS

(3 g)	Cholesterol (ash free) 95–98% pure from Sigma-Aldrich Co.
(3 g)	Lecithin (prepared from dried egg yolk) 60% pure from Sigma-Aldrich Co.
(10 g)	Glycerol
(10 g)	Corn oil
(10 g)	Butter
(1)	Egg yolk obtained from one fresh egg before the lab period. Stir and mix.
(10 g)	Sucrose
(250 mL)	Molybdate solution: dissolve 0.8 g $(NH_4)_6Mo_7O_{24} \cdot 4H_2O$ in 30 mL water. Put in an ice bath. Pour slowly 20 mL concentrated sulfuric acid (18 M H_2SO_4) into the solution and stir slowly.

	After cooling to room temperature bring the volume to 250 mL. **Wear a face shield, rubber gloves, and a rubber apron during the preparation. Do in the hood.**
(50 mL)	0.1 M ascorbic acid solution: dissolve 0.88 g ascorbic acid (vitamin C) in water and bring it to 50 mL volume. This must be prepared fresh every week and stored at 4°C.
(250 mL)	6 M sodium hydroxide, 6 M NaOH: dissolve 60.00 g NaOH in water and bring the volume to 250 mL
(250 mL)	6 M nitric acid, 6 M HNO_3: into a 250-mL volumetric flask containing 100 mL ice-cold water, pipet 125 mL concentrated nitric acid (12 M HNO_3); add enough water to bring to 250 mL. **Wear a face shield, rubber gloves, and a rubber apron during the preparation. Do in the hood.**
(200 mL)	Chloroform
(75 mL)	Acetic anhydride
(50 mL)	Concentrated sulfuric acid (18 M H_2SO_4)
(75 g)	Potassium hydrogen sulfate, $KHSO_4$

38 EXPERIMENT 38

Separation of Amino Acids by Paper Chromatography

SPECIAL EQUIPMENT

(1)	Drying oven, 105–110°C
(2)	Heat lamps or hair dryers
(25)	Whatman no. 1 chromatographic paper: 15 × 8 cm; 12 × 4 cm
(25)	Rulers, metric scale
(25 pairs)	Polyethylene surgical gloves
(150)	Capillary tubes, open on both ends
(1 roll)	Aluminum foil
(2)	Wide-mouth jars
(25 each)	Felt-tip pens: red, blue, black

CHEMICALS

(25 mL)	0.12% aspartic acid solution: dissolve 30 mg aspartic acid in 25 mL distilled water
(25 mL)	0.12% phenylalanine solution: dissolve 30 mg phenylalanine in 25 mL distilled water
(25 mL)	0.12% leucine solution: dissolve 30 mg leucine in 25 mL distilled water
(25 mL)	Aspartame solution: dissolve 150 mg Equal sweetener powder in 25 mL distilled water
(50 mL)	3 M HCl solution: place 10 mL ice-cold distilled water into a 50-mL volumetric flask. Add slowly 12.5 mL of concentrated

HCl (12 M HCl) and bring it to volume with distilled water. **Wear a face shield, rubber gloves, and a rubber apron when preparing. Do in the hood.**

(1 L) Solvent mixture: mix 600 mL 1-butanol with 150 mL acetic acid and 250 mL distilled water. **Solvent mixture *must* be freshly prepared the day it is to be used.**

(1 can) Ninhydrin spray reagent (0.2% ninhydrin in ethanol or acetone). Do not use any reagent older than 6 months.

(1 can) Diet Coca-Cola

(4 packets) Equal or NutraSweet, sweeteners

(10 g) Iodine crystals, I_2

39 EXPERIMENT 39

Acid–Base Properties of Amino Acids

SPECIAL EQUIPMENT

(10) pH meters or

(5 rolls) pHydrion short-range papers, from each range: pH: 0.0 to 3.0; 3.0 to 5.5; 5.2 to 6.6; 6.0 to 8.0; 8.0 to 9.5; and 9.0 to 12.0

(25) 20-mL pipets

(25) 50-mL burets

(25) Spectroline pipet fillers

(25) Pasteur pipets

CHEMICALS

(500 mL) 0.25 M NaOH: dissolve 5.00 g NaOH in 100 mL water and then add enough water to bring to 500 mL

(750 mL) 0.1 M alanine solution: dissolve 6.68 g *l*-alanine in 500 mL water; add sufficient 1 M HCl to bring the pH to 1.5. Add enough water to bring to 750 mL.

or

Do as above but use either 5.63 g glycine or 9.84 g *l*-leucine or 12.39 g *l*-phenylalanine or 8.79 g *l*-valine.

EXPERIMENT 40

Isolation and Identification of Casein

SPECIAL EQUIPMENT

(25) Hot plates

(25) 600-mL beakers

(25) Büchner funnels (O.D. 85 mm) in no. 7 1-hole rubber stopper

(7 boxes)	Whatman no. 2 filter paper, 7 cm
(25)	Rubber bands
(25)	Cheese cloths (6 × 6 in.)

CHEMICALS

(1 jar)	Boiling stones
(1 L)	95% ethanol
(1 L)	Diethyl ether:ethanol mixture (1:1)
(0.5 gal)	Regular milk
(500 mL)	Glacial acetic acid

The following solutions should be placed in dropper bottles:

(100 mL)	Concentrated nitric acid (12 M HNO_3)
(100 mL)	2% albumin suspension: dissolve 2 g albumin in enough water to make 100 mL
(100 mL)	2% gelatin: dissolve 2 g gelatin in enough water to make 100 mL
(100 mL)	2% glycine: dissolve 2 g glycine in enough water to make 100 mL
(100 mL)	5% copper(II) sulfate: dissolve 5 g $CuSO_4$ (or 7.85 g $CuSO_4 \cdot 5H_2O$) in enough water to make 100 mL
(100 mL)	5% lead(II) nitrate: dissolve 5 g $Pb(NO_3)_2$ in enough water to make 100 mL
(100 mL)	5% mercury(II) nitrate: dissolve 5 g $Hg(NO_3)_2$ in enough water to make 100 mL
(100 mL)	Ninhydrin reagent: dissolve 3 g ninhydrin in enough acetone to make 100 mL. Do not use a reagent older than 6 months.
(100 mL)	10% sodium hydroxide: dissolve 10 g NaOH in enough water to make 100 mL
(100 mL)	1% tyrosine: dissolve 1 g tyrosine in enough water to make 100 mL
(100 mL)	5% sodium nitrate: dissolve 5 g $NaNO_3$ in enough water to make 100 mL

41 EXPERIMENT 41

Properties of Enzymes

SPECIAL EQUIPMENT

(25)	Hot plates
(100)	Medicine droppers
(10)	Stopwatches
(25)	White spot plates, 25 depressions

CHEMICALS

(20 g)	Liver, raw, cut into small, pea-sized chunks
(20 g)	Liver, cooked, cut into small, pea-sized chunks
(500 mL)	Milk

The following solutions should be placed into dropper bottles.

(300 mL) 1% α-amylase solution: use α-amylase from barley malt with a specific activity of 100,000 units (available from Sigma-Aldrich Co.). Dissolve 3 g of α-amylase in enough water to make 300 mL. *Prepare fresh the day of the experiment.*

(200 mL) Buffer, pH 2 (available commercially as a standardized pH buffer solution)

(200 mL) Buffer, pH 4 (available commercially as a standardized pH buffer solution)

(200 mL) Buffer, pH 7 (available commercially as a standardized pH buffer solution)

(200 mL) Buffer, pH 9 (available commercially as a standardized pH buffer solution)

(200 mL) Buffer, pH 12 (available commercially as a standardized pH buffer solution)

(100 mL) 3 M HCl, 3 M hydrochloric acid: 25 mL concentrated HCl (12 M HCl) diluted with enough ice-cold water to make 100 mL. **Wear a face shield, rubber gloves, and a rubber apron when preparing. Do in the hood.**

(200 mL) 3% H_2O_2, 3% hydrogen peroxide (available from any pharmacy)

(200 mL) 2% rennin solution: use Junket® Rennet tablets. Rennet is available in the pudding section of the supermarket. It may be ordered directly from Redco Foods, Inc., P. O. Box 879, Windsor, CT 06095 (1-800-556-6674). Five tablets are approx. 4 g; pulverize to a powder with a mortar and pestle. Add to 50 mL of water and stir to dissolve (the solution will be cloudy). Add enough water to make 200 mL. *Prepare fresh the day of the experiment.*

(200 mL) 1% starch solution: place 2 g soluble starch in a beaker. With vigorous stirring, add 10 mL water to form a thin paste. Boil 190 mL water in another beaker. Add the starch paste to the boiling water and stir until the solution becomes clear.

(100 mL) 0.01 M iodine in KI: dissolve 1.2 g potassium iodide, KI, in 80 mL water. Add 0.25 g iodine, I_2. Stir until the iodine dissolves. Add enough water to make 100 mL. Store in a dark dropper bottle.

(100 mL) 0.1 M $Pb(NO_3)_2$, 0.1 M lead(II) nitrate: dissolve 3.31 g $Pb(NO_3)_2$ in enough water to make 100 mL

42 **EXPERIMENT 42**

Neurotransmission: An Example of Enzyme Specificity

SPECIAL EQUIPMENT

(5) Spectrophotometers

(5) Micropipets (1–200 μL)

(20) Natural pipet tips (1–200 μL) (USA Scientific Co. # 1111-0810)

CHEMICALS

(1 L)	0.01 M acetic acid: take 0.6 mL glacial acetic acid and add enough water to bring to 1 L
(500 mL)	0.02 M NaOH: dissolve 0.40 g NaOH in enough water to make 500 mL of solution
(1 L)	0.150 M NaCl solution: dissolve 8.766 g NaCl in 800 mL water. Adjust the pH to 8.0 using 0.02 M NaOH. Add enough water to bring to 1.00 L. **The pH of the solution must be adjusted to 8.0 before every experiment.**
(100 mL)	*m*-Nitrophenol solution: dissolve 500 mg *m*-nitrophenol in 5 mL absolute ethanol. Bring the solution to 100 mL by adding 0.15 M NaCl solution; adjust to pH 8.0.
(10 mL)	Acetylcholine esterase (EC.3.1.1.7) from electric eel (Sigma C-3389): dissolve 3 mg, approximately 300 unit/mg, in 10 mL 0.15 M NaCl solution; adjust to pH 8.0.
(10 mL)	Acetylcholine substrate solution: take 1.817 g acetylcholine and dissolve completely in enough 0.15 M NaCl solution to make 10 mL. Adjust to pH 8.0.
(40 mL)	*o*-Nitrophenyl acetate solution: dissolve 181.1 mg *o*-nitrophenyl acetate in 5 mL dimethylsulfoxide. Just before the start of the experiment, add 35 mL of 0.15 M NaCl solution; adjust to pH 8.0. **Mix the solution thoroughly.**
(10 mL)	Carboxyl esterase (EC 3.1.1.1) from porcine liver, available as a 0.3 mL suspension (conc. is 15 mg protein/mL, 250 units/mg protein): add ammonium sulfate buffer to get 10 mL of solution; adjust to pH 8.0.

43 EXPERIMENT 43

Isolation and Identification of DNA from Onion

SPECIAL EQUIPMENT

(25)	Mortars
(25)	Pestles
(25)	Cheese cloth (6 × 6 in.)

CHEMICALS

(12)	Yellow onions
(500 g)	Acid-washed sand
(1 bottle)	Ivory concentrated liquid dishwashing detergent, 12.6 oz.
(250 g)	Sodium chloride, NaCl
(100 mL)	Citrate buffer: dissolve 0.88 g NaCl and 0.39 g sodium citrate in enough water to make 100 mL
(2 L)	Absolute ethanol

(200 mL) 95% ethanol

(50 mL) 1% glucose solution: dissolve 0.5 g D-glucose in enough water to make 50 mL

(50 mL) 1% ribose solution: dissolve 0.5 g D-ribose in enough water to make 50 mL

(50 mL) 1% deoxyribose solution: dissolve 0.5 g 2-deoxy-D-ribose in enough water to make 50 mL

(500 mL) Diphenylamine reagent: dissolve 7.5 g reagent grade diphenylamine (Sigma D-3409) in 50 mL glacial acetic acid. Add 7.5 mL concentrated sulfuric acid, H_2SO_4, slowly, with stirring. Prior to use, add 2.5 mL 1.6% acetaldehyde (made by dissolving 0.16 g acetaldehyde in 10 mL water); then add enough water to make 500 mL. **This reagent must be prepared shortly before use in the laboratory. Wear a face shield, rubber gloves, and a rubber apron when preparing the reagent. Do in the hood.**

As an alternative:

(200 mL) Diphenylamine solution without acetaldehyde: dissolve 2.0 g reagent-grade diphenylamine (Sigma D-3409) in 200 mL concentrated sulfuric acid, H_2SO_4. Swirl slowly with a glass rod until completely dissolved. **Prepare this reagent fresh each week. Wear a face shield, rubber gloves, and a rubber apron when preparing the reagent. Do in the hood.**

44 EXPERIMENT 44

Viscosity and Secondary Structure of DNA

EQUIPMENT

(10) Ostwald (or Cannon-Ubbelhode) capillary viscometers; 3-mL capacity, approximate capillary diameter 0.2 mm; **efflux time of water = 40–50 sec.**

(5) Stopwatches (Wristwatches can also time the efflux with sufficient precision.)

(5) Stands with utility clamps

(25) Pasteur pipets

(10) Spectroline pipet fillers

CHEMICALS

(500 mL) Buffer solution: dissolve 4.4 g sodium chloride, NaCl, and 2.2 g sodium citrate, $Na_3C_6H_5O_7 \cdot 2H_2O$, in 450 mL distilled water. Adjust the pH with either 0.1 M HCl or 0.1 M NaOH to pH 7.0. Add enough water to bring to 500 mL.

(200 mL) DNA solution: dissolve 20 mg of calf thymus Type I highly polymerized DNA (obtainable from Sigma-Aldrich Co. as well

as from other companies) in 200 mL buffer solution at pH 7.0. The purchased DNA powder should be kept in the freezer. The DNA solutions should be prepared fresh or maximum 2–3 hr. in advance of the experiment. The solution should be kept at 4°C; 1–2 hr. before the experiment, the solution should be allowed to come to room temperature. Label the solution as 0.01 g/dL concentration.

(100 mL) 1 M hydrochloric acid, 1 M HCl: add 8.3 mL concentrated HCl (12 M HCl) to 50 mL ice-cold water; add enough water to bring to 100 mL. **Wear a face shield, rubber gloves, and a rubber apron during preparation. Do in the hood.**

(100 mL) 0.1 M hydrochloric acid: add 10.0 mL 1 M HCl to 50 mL water; add enough water to bring to 100 mL. **Wear a face shield, rubber gloves, and a rubber apron during preparation. Do in the hood.**

(100 mL) 1 M sodium hydroxide: dissolve 4.00 g NaOH in 50 mL water; add enough water to bring to 100 mL. **Wear a face shield, rubber gloves, and a rubber apron during preparation. Do in the hood.**

(100 mL) 0.1 M sodium hydroxide: dissolve 0.40 g NaOH in 50 mL water; add enough water to bring to 100 mL. **Wear a face shield, rubber gloves, and a rubber apron during preparation. Do in the hood.**

45 EXPERIMENT 45

Kinetics of Urease-Catalyzed Decomposition of Urea

SPECIAL EQUIPMENT

(25) 5-mL pipets
(25) 10-mL graduated pipets
(25) 10-mL volumetric pipets
(25) 50-mL burets
(25) Buret holders
(25) Spectroline pipet fillers

CHEMICALS

(3.5 L) 0.05 M Tris buffer: dissolve 21.05 g Tris buffer in water (3 L). Adjust the pH to 7.2 with 1 M HCl solution; add sufficient water to make 3.5 L. Portions of buffer solution will be used to make urea and enzyme solutions.

(2.5 L) 0.3 M urea solution: dissolve 45 g urea in enough Tris buffer to make 2.5 L

(50 mL) 1×10^{-3} M phenylmercuric acetate: dissolve 16.5 mg phenylmercuric acetate in 40 mL water; add enough water to bring the volume to 50 mL.

CAUTION! Phenylmercuric acetate is a poison. Do not touch the chemical with your hands. Do not swallow the solution. Wear rubber gloves in the preparation. Do in the hood.

(50 mL) 1% $HgCl_2$ solution: dissolve 0.5 g $HgCl_2$ in enough water to make 50 mL solution

(100 mL) 0.04% methyl red indicator: dissolve 40 mg methyl red in enough water to make 100 mL

(500 mL) Urease solution: prepare the enzyme solution on the week of the experiment and store at 4°C. Take 1.0 g urease, dissolve in enough Tris buffer to make 500 mL. (One can buy urease with 5 to 6 units activity, for example, from Nutritional Biochemicals, Cleveland, Ohio.) The activity of the enzyme printed on the label should be checked by the stockroom personnel or instructor.

(1.0 L) 0.05 M HCl: add 4.2 mL concentrated HCl (12 M HCl) to 100 mL ice-cold water; add enough water to bring to 1.0 L volume. **Wear a face shield, rubber gloves, and a rubber apron during the preparation. Do in the hood.**

46 **EXPERIMENT 46**

Isocitrate Dehydrogenase: An Enzyme of the Citric Acid Cycle

SPECIAL EQUIPMENT

(15) Spectrophotometers with 5 cuvettes each

(25) 1-mL graduated pipets

CHEMICALS

(40 mL) Phosphate buffer at pH 7.0: mix together 25 mL 0.1 M KH_2PO_4 and 15 mL 0.1 M NaOH. To prepare 0.1 M NaOH, add 0.20 g NaOH to 20 mL water in a 50-mL volumetric flask; stir to dissolve; add enough water to bring to 50 mL. To prepare 0.1 M KH_2PO_4, add 0.68 g potassium dihydrogen phosphate to 40 mL water in a 50-mL volumetric flask; stir to dissolve; add enough water to bring to 50 mL.

(20 mL) 0.1 M $MgCl_2$: add 0.19 g magnesium chloride to enough water to make 20 mL; stir to dissolve.

(50 mL) Isocitrate dehydrogenase: commercial preparations from porcine heart are obtainable from companies such as Sigma-Aldrich Co., etc. (EC 1.1.1.42) (activity about 8 units per mg of solid). Dissolve 10 mg of the enzyme in enough water to make 50 mL. This solution should be made fresh before the lab period and kept in a refrigerator until used.

(20 mL)	6.0 mM β-nicotinamide adenine dinucleotide, β-$NADP^+$, solution: dissolve 92 mg $NADP^+$ in enough water to make 20 mL
(50 mL)	15 mM sodium isocitrate solution: dissolve 160 mg sodium isocitrate in enough water to make 50 mL

47 EXPERIMENT 47

Quantitative Analysis of Vitamin C Contained in Foods

SPECIAL EQUIPMENT

(25)	50-mL burets
(25)	Buret clamps
(25)	Ring stands
(25)	Spectroline pipet fillers
(25)	10-mL volumetric pipets
(1 box)	Cotton

CHEMICALS

(500 g)	Celite, filter aid
(1 can)	Hi-C orange drink
(1 can)	Hi-C grapefruit drink
(1 can)	Hi-C apple drink
(2 L)	0.01 M iodine solution: add 32 g KI to 800 mL water; stir to dissolve. Add 5 g I_2; stir to dissolve. Add enough water to bring to 2 L. Store in dark bottle. **Caution! Iodine is poisonous if taken internally.**
(100 mL)	3 M HCl: add 25 mL concentrated HCl (12 M HCl) to 50 mL ice-cold water; add enough water to bring to 100 mL. **Wear a face shield, rubber gloves, and a rubber apron during the preparation. Do in the hood.**
(100 mL)	2% starch solution: place 2 g soluble starch in a 50-mL beaker. Add 10 mL water. Stir vigorously to form a paste. Boil 90 mL water in a second beaker. Add the starch paste to the boiling water. Stir until the solution becomes translucent. Cool to room temperature.

48 EXPERIMENT 48

Analysis of Vitamin A in Margarine

SPECIAL EQUIPMENT

(1)	UV spectrophotometer with suitable UV light source. Preferably it should be able to read down to 200 nm.

(1 pair)	Matched quartz cells, with 1-cm internal path length
(2)	Long-wavelength UV lamp. The lamp should provide radiation in the 300-nm range (for example, #UVSL-55; LW 240 from Ultraviolet Products, Inc.)
(12)	500-mL separatory funnels
(12)	25-mL (or 50-mL) burets
(12)	Hot plates, each with a water bath
(12)	Beaker tongs
(1 jar)	Boiling stones

CHEMICALS

(0.5 lb)	Margarine
(400 mL)	50% KOH: weigh 200 g KOH and add 200 mL water, with constant stirring; add enough water to make 400 mL. **Wear a face shield, rubber gloves, and a rubber apron during preparation. Do in the hood.**
(1 L)	95% ethanol
(150 mL)	Absolute ethanol
(2 L)	Petroleum ether, b.p. 30–60°C
(3 L)	Diethyl ether
(350 g)	Alkaline aluminum oxide (alumina)
(250 g)	Sodium sulfate, anhydrous, Na_2SO_4